T0305936

DESIGN OF WATER SUPPLY PIPE NETWORKS

DESIGN OF WATER SUPPLY PIPE NETWORKS

Prabhata K. Swamee
Ashok K. Sharma

A JOHN WILEY & SONS, INC., PUBLICATION

Published by John Wiley & Sons, Inc., Hoboken, New Jersey
Published simultaneously in Canada

For general information on our other products and services or for technical support, please contact our Customer Care Department within the United States at (800) 762-2974, outside the United States at (317) 572-3993 or fax (317) 572-4002.

Wiley also publishes it books in a variety of electronic formats. Some content that appears in print may not be available in electronic formats. For more information about Wiley products, visit our web site at www.wiley.com.

Library of Congress Cataloging-in-publication Data:

Swamee, Prabhata K. (Prabhata Kumar), 1940-
   Design of water supply pipe networks / Prabhata K. Swamee, Ashok K. Sharma.
      p. cm.
   Includes bibliographical references and index.
   ISBN 978-0-470-17852-2 (cloth)
   1. Water-pipes. 2. Water—Distribution. 3. Water-supply—Management. I.   Sharma, Ashok K.
(Ashok Kumar), 1956-   II. Title.
   TD491.S93 2007
   62B.1'5—dc22

                                                                                2007023225
Printed in the United States of America

10   9   8   7   6   5   4   3   2   1

CONTENTS

PREFACE

A large amount of money is invested around the world to provide or upgrade piped water supply facilities. Even then, a vast population of the world is without safe piped water facilities. Nearly 80% to 85% of the cost of a total water supply system is contributed toward water transmission and the water distribution network. Water distribution system design has attracted many researchers due to the enormous cost.

The aim of this book is to provide the reader with an understanding of the analysis and design aspects of water distribution system. The book covers the topics related to the analysis and design of water supply systems with application to sediment-transporting pipelines. It includes the pipe flow principles and their application in analysis of water supply systems. The general principles of water distribution system design have been covered to highlight the cost aspects and the parameters required for design of a water distribution system. The other topics covered in the book relate to optimal sizing of water-supply gravity and pumping systems, reorganization and decomposition of water supply systems, and transportation of solids as sediments through pipelines. Computer programs with development details and line by line explanations have been included to help readers to develop skills in writing programs for water distribution network analysis. The application of linear and geometric programming techniques in water distribution network optimization have also been described.

Most of the designs are provided in a closed form that can be directly adopted by design engineers. A large part of the book covers numerical examples. In these examples, computations are laborious and time consuming. Experience has shown that the complete mastery of the project cannot be attained without familiarizing oneself thoroughly with numerical procedures. For this reason, it is better not to consider numerical examples as mere illustration but rather as an integral part of the general presentation.

The book is structured in such a way to enable an engineer to design functionally efficient and least-cost systems. It is also intended to aid students, professional engineers, and researchers. Any suggestions for improvement of the book will be gratefully received.

PRABHATA K. SWAMEE
ASHOK K. SHARMA

NOTATIONS

The following notations and symbols are used in this book.

A	annual recurring cost, annuity
A_e	annual cost of electricity
A_r	annual installment
a	capsule length factor
B	width of a strip zone
C	cost coefficient
C_0	initial cost of components
C_A	capitalized cost
C_c	overall or total capitalized cost
C_D	drag coefficient of particles
C_e	capitalized cost of energy
C_m	cost of pipe
C_{ma}	capitalized maintenance cost
C_N	net cost
C_P	cost of pump
C_R	cost of service reservoir, replacement cost
C_T	cost of pumps and pumping
C_v	volumetric concentration of particles
c_i	cost per meter of pipe i
D	pipe link diameter
D_e	equivalent pipe link diameter
D_{min}	minimum pipe diameter
D_n	new pipe link diameter
D_o	existing pipe link diameter
D_s	diameter of service connection pipe
D^*	optimal pipe diameter
d	confusor outlet diameter, spherical particle diameter, polynomial dual

d^*	optimal polynomial dual
E	establishment cost
F	cost function
F_A	annual averaging factor
F_D	daily averaging factor
F_g	cost of gravity main
F_P	cost of pumping main
F_s	cost of service connections
F_P^*	optimal cost of pumping main
F^*	optimal cost
f	coefficient of surface resistance
f_b	friction factor for intercapsule distance
f_c	friction factor for capsule
f_e	effective friction factor for capsule transportation
f_p	friction factor for pipe annulus
g	gravitational acceleration
H	minimum prescribed terminal head
h	pressure head
h_a	allowable pressure head in pipes
h_b	length parameter for pipe cost
h_c	extra pumping head to account for establishment cost
h_f	head loss due to surface resistance
h_j	nodal head
h_L	total head loss
h_m	minor head losses due to form resistance
h_{mi}	minor head losses due to form resistance in pipe i
h_{min}	minimum nodal pressure head in network
h_0	pumping head; height of water column in reservoir
h_0^*	optimal pumping head
h_s	staging height of service reservoir
I_k	pipe links in a loop
I_n	input source supplying to a demand node
I_p	pipe links meeting at a node
I_R	compound interest, pipes in a route connecting two input sources
I_t	flow path pipe
I_s	input source number for a pipe
i	pipe index
i_L	total number of pipe links

J_1, J_2	pipe link node
J_s	input source node of a flow path for pipe i
J_t	originating node of a flow path for pipe i
j	node index
j_L	total number of pipe nodes
k	cost coefficient, loop pipe index, capsule diameter factor
K_1, K_2	loops of pipe
k_f	form-loss coefficient for pipe fittings
k_{fp}	form-loss coefficient for fittings in pth pipe
k_L	total number of loops
k_m	pipe cost coefficient
k_n	modified pipe cost coefficient
k_p	pump cost coefficient
k_R	reservoir cost coefficient
k_s	service pipe cost coefficient
k_T	pump and pumping cost coefficient
kW	power in kilowatts
k'	capitalized cost coefficient
L	pipe link length
ℓ	index
M_1	first input point of route r
M_2	second input point of route r
MC	cut-sets in a pipe network system
m	pipe cost exponent
m_P	pump cost exponent
N_R	total pipes in route r
N_n	number of input sources supplying to a demand node
N_p	number of pipe links meeting at a node
N_t	number of pipe links in flow path of pipe i
n	input point index, number of pumping stages
n^*	optimal number of pumping stages
n_L	total number of input points
n_s	number of connections per unit length of main
P	power; population
P_i	probability of failure of pipe i
P_{NC}	net present capital cost
P_{NS}	net present salvage cost
P_{NA}	net present annual operation and maintenance cost

P_N net present value
P_s probability of failure of the system
p number of pipe breaks/m/yr
Q discharge
Q_c critical discharge
Q_e effective fluid discharge
Q_i pipe link discharge
Q_s sediment discharge, cargo transport rate
Q_T total discharge at source (s)
Q_{Tn} discharge at nth source
q nodal withdrawal
q_s service connection discharge
$\mathbf{R}$ Reynolds number
$\mathbf{R_s}$ Reynolds number for sediment particles, system reliability
R pipe bends radius
R_E cost of electricity per kilowatt hour
r rate of interest; discount rate
s ratio of mass densities of solid particles and fluid
s_b standby fraction
s_s ratio of mass densities of cargo and fluid
T fluid temperature, design period of water supply main
T_u life of component
t_c characteristic time
V velocity of flow
V_a average fluid velocity in annular space
V_b average fluid velocity between two solid transporting capsules
V_c average capsule velocity
$V_{\max}$ maximum flow velocity
V_R service reservoir volume
V_s volume of material contained in capsule
w sediment particles fall velocity, weights in geometric programming
w^* optimal weights in geometric programming
x_{i1}, x_{i2} sectional pipe link lengths
z nodal elevation
z_o nodal elevation at input point
z_L nodal elevation at supply point
z_n nodal elevation at nth node
z_x nodal elevation at point x

α	valve closer angle, pipe bend angle, salvage factor of goods
β	annual maintenance factor; distance factor between two capsules
β_i	expected number of failure per year for pipe i
λ	Lagrange multiplier, ratio of friction factors between pipe annulus and capsule
ν	kinematic viscosity of fluid
ε	roughness height of pipe wall
ρ	mass density of water
σ	peak water demand per unit area
ξ	length ratio
η	efficiency
θ	capsule wall thickness factor
θ_p	peak discharge factor
ω	rate of water supply
ΔQ_k	discharge correction in loop k

Superscript

$*$	optimal

Subscripts

e	effective, spindle depth obstructing flow in pipe
i	pipe index
$i1$	first section of pipe link
$i2$	second section of pipe link
L	terminating point or starting point
o	entry point
p	pipe
s	starting node
t	track

1

INTRODUCTION

1.1. BACKGROUND

Water and air are essential elements for human life. Even then, a large population of the world does not have access to a reliable, uncontaminated, piped water supply. Drinking water has been described as a physical, cultural, social, political, and economic resource (Salzman, 2006). The history of transporting water through pipes for human consumption begins around 3500 years ago, when for the first time pipes were used on the island of Crete. A historical perspective by James on the development of urban water systems reaches back four millennia when bathrooms and drains were common in the Indus Valley (James, 2006). Jesperson (2001) has provided a brief history of public water systems tracking back to 700 BC when sloped hillside tunnels (*qantas*) were built to transport water to Persia. Walski et al. (2001) also have published a brief history of water distribution technology beginning in 1500 BC. Ramalingam et al. (2002) refer to the early pipes made by drilling stones, wood, clay, and lead. Cast iron pipes replaced the early pipes in the 18th century, and significant developments in making pipe joints were witnessed in the 19th century. Use of different materials for pipe manufacturing increased in the 20th century.

Fluid flow through pipelines has a variety of applications. These include transport of water over long distances for urban water supply, water distribution system for a group of rural towns, water distribution network of a city, and so forth. Solids are also transported through pipelines; for example, coal and metallic ores carried in water suspension and pneumatic conveyance of grains and solid wastes. Pipeline transport of solids containerized in capsules is ideally suited for transport of seeds, chemicals that react with a carrier fluid, and toxic or hazardous substances. Compared with slurry transport, the cargo is not wetted or contaminated by the carrier fluid; no mechanism is required to separate the transported material from the fluid; and foremost it requires less power for maintaining the flow. For bulk carriage, pipeline transport can be economic in comparison with rail and road transport. Pipeline transport is free from traffic holdups and road accidents, is aesthetic because pipelines are usually buried underground, and is also free from chemical, biochemical, thermal, and noise pollution.

A safe supply of potable water is the basic necessity of mankind in the industrialized society, therefore water supply systems are the most important public utility. A colossal amount of money is spent every year around the world for providing or upgrading drinking water facilities. The major share of capital investment in a water supply system goes to the water conveyance and water distribution network. Nearly 80% to 85% of the cost of a water supply project is used in the distribution system; therefore, using rational methods for designing a water distribution system will result in considerable savings.

The water supply infrastructure varies in its complexity from a simple, rural town gravity system to a computerized, remote-controlled, multisource system of a large city; however, the aim and objective of all the water systems are to supply safe water for the cheapest cost. These systems are designed based on least-cost and enhanced reliability considerations.

1.2. SYSTEM CONFIGURATION

In general, water distribution systems can be divided into four main components: (1) water sources and intake works, (2) treatment works and storage, (3) transmission mains, and (4) distribution network. The common sources for the untreated or raw water are surface water sources such as rivers, lakes, springs, and man-made reservoirs and groundwater sources such as bores and wells. The intake structures and pumping stations are constructed to extract water from these sources. The raw water is transported to the treatment plants for processing through transmission mains and is stored in clear water reservoirs after treatment. The degree of treatment depends upon the raw water quality and finished water quality requirements. Sometimes, groundwater quality is so good that only disinfection is required before supplying to consumers. The clear water reservoir provides a buffer for water demand variation as treatment plants are generally designed for average daily demand.

Water is carried over long distances through transmission mains. If the flow of water in a transmission main is maintained by creating a pressure head by pumping, it is called a pumping main. On the other hand, if the flow in a transmission main is maintained by gravitational potential available on account of elevation difference, it is called a gravity main. There are no intermediate withdrawals in a water transmission main. Similar to transmission mains, the flow in water distribution networks is maintained either by pumping or by gravitational potential. Generally, in a flat terrain, the water pressure in a large water distribution network is maintained by pumping; however, in steep terrain, gravitational potential maintains a pressure head in the water distribution system.

A distribution network delivers water to consumers through service connections. Such a distribution network may have different configurations depending upon the layout of the area. Generally, water distribution networks have a looped and branched configuration of pipelines, but sometimes either looped or branched configurations are also provided depending upon the general layout plan of the city roads and streets. Urban water networks have mostly looped configurations, whereas rural water networks have branched configurations. On account of the high-reliability requirement of water services, looped configurations are preferred over branched configurations.

The cost of a water distribution network depends upon proper selection of the geometry of the network. The selection of street layout adopted in the planning of a city is important to provide a minimum-cost water supply system. The two most common water supply configurations of looped water supply systems are the gridiron pattern and the ring and radial pattern; however, it is not possible to find an optimal geometric pattern that minimizes the cost.

1.3. FLOW HYDRAULICS AND NETWORK ANALYSIS

The flow hydraulics covers the basic principles of flow such as continuity equation, equations of motion, and Bernoulli's equation for close conduit. Another important area of pipe flows is to understand and calculate resistance losses and form losses due to pipe fittings (i.e., bends, elbows, valves, enlargers and reducers), which are the essential parts of a pipe network. Suitable equations for form-losses calculations are required for total head-loss computation as fittings can contribute significant head loss to the system. This area of flow hydraulics is covered in Chapter 2.

The flow hydraulics of fluid transporting sediments in suspension and of capsule transport through a pipeline is complex in nature and needs specific consideration in head-loss computation. Such an area of fluid flow is of special interest to industrial engineers/designers engaged in such fluid transportation projects. Chapter 2 also covers the basics of sediment and capsule transport through pipes.

Analysis of a pipe network is essential to understand or evaluate a physical system, thus making it an integral part of the synthesis process of a network. In case of a single-input system, the input discharge is equal to the sum of withdrawals. The known parameters in a system are the pipe sizes and the nodal withdrawals. The system has to be analyzed to obtain input point discharges, pipe discharges, and nodal pressure heads. In case of a branched system, starting from a dead-end node and successively applying the node flow continuity relationship, all pipe discharges can be easily estimated. Once the pipe discharges are known, the nodal pressure heads can be calculated by applying the pipe head-loss relationship starting from an input source node with known input head. In a looped network, the pipe discharges are derived using loop head-loss relationship for known pipe sizes and nodal continuity equations for known nodal withdrawals.

Ramalingam et al. (2002) published a brief history of water distribution network analysis over 100 years and also included the chronology of pipe network analysis methods. A number of methods have been used to compute the flow in pipe networks ranging from graphical methods to the use of physical analogies and finally the use of mathematical/numerical methods.

Darcy–Weisbach and Hazen–Williams provided the equations for the head-loss computation through pipes. Liou (1998) pointed out the limitations of the Hazen–Williams equation, and in conclusion he strongly discouraged the use of the Hazen–Williams equation. He also recommended the use of the Darcy–Weisbach equation with the Colebrook–White equation. Swamee (2000) also indicated that the Hazen–Williams equation was not only inaccurate but also was conceptually incorrect. Brown (2002) examined the historical development of the Darcy–Weisbach equation for pipe flow resistance and stated that the most notable advance in the application of this equation was the publication of an explicit equation for friction factor by Swamee and Jain (1976). He concluded that due to the general accuracy and complete range of application, the Darcy–Weisbach equation should be considered the standard and the others should be left for the historians. Considering the above investigations, only the Darcy–Weisbach equation for pipe flow has been covered in this book for pipe network analysis.

Based on the application of an analysis method for water distribution system analysis, the information about pipes forming primary loops can be an essential part of the data. The loop data do not constitute information independent of the link-node information, and theoretically it is possible to generate loop data from this information. The information about the loop-forming pipes can be developed by combining flow paths. These pipe flow paths, which are the set of pipes connecting a demand (withdrawals) node to the supply (input) node, can be identified by moving opposite to the direction of flow in pipes (Sharma and Swamee, 2005). Unlike branched systems, the flow directions in looped networks are not unique and depend upon a number of factors, mainly topography, nodal demand, layout, and location and number of input (supply) points. The pipe flow patterns will vary based on these factors. Hence, combining flow paths, the flow pattern map of a water distribution network can also be generated, which is important information for an operator/manager of a water system for its efficient operation and maintenance.

The analysis of a network is also important to make decisions about the network augmentation requirements due to increase in water demand or expansion of a water servicing area. The understanding of pipe network flows and pressures is important for making such decisions for a water supply system.

Generally, the water service connections (withdrawals) are made at an arbitrary spacing from a pipeline of a water supply network. Such a network is difficult to analyze until simplified assumptions are made regarding the withdrawal spacing. The current practice is to lump the withdrawals at the nodal points; however, a distributed approach for withdrawals can also be considered. A methodology is required to calculate flow and head losses in the pipeline due to lumped and distributed withdrawals. These pipe network analysis methods are covered in Chapter 3.

1.4. COST CONSIDERATIONS

To carry out the synthesis of a water supply system, one cannot overlook cost considerations that are absent during the analysis of an existing system. Sizing of the water distribution network to satisfy the functional requirements is not enough as the solution should also be based on the least-cost considerations. Pumping systems have a large number of feasible solutions due to the trade-off between pumping head and pipe sizes. Thus, it is important to consider the cost parameters in order to synthesize a pumping system. In a water distribution system, the components sharing capital costs are pumps and pumping stations; pipes of various commercially available sizes and materials; storage reservoir; residential connections and recurring costs such as energy usage; and operation and maintenance of the system components. The development of cost functions of various components of water distribution systems is described in Chapter 4.

As the capital and recurring costs cannot be simply added to find the overall cost (life-cycle cost) of the system over its life span, a number of methods are available to combine these two costs. The capitalized cost, net present value, and annuity methods for life-cycle cost estimation are also covered in Chapter 4. Fixed costs associated with source development and treatment works for water demand are not included in the optimal design of the water supply system.

1.5. DESIGN CONSIDERATIONS

The design considerations involve topographic features of terrain, economic parameters, and fluid properties. The essential parameters for network sizing are the projection of residential, commercial, and industrial water demand; per capita water consumption; peak flow factors; minimum and maximum pipe sizes; pipe material; and reliability considerations.

Another important design parameter is the selection of an optimal design period of a water distribution system. The water systems are designed for a predecided time horizon generally called design period. For a static population, the system can be designed either for a design period equal to the life of the pipes sharing the maximum cost of the system or for the perpetual existence of the water supply system. On the other hand, for a growing population or water demand, it is always economic to design the system in stages and restrengthen the system after the end of every staging period. The design period should be based on the useful life of the component sharing maximum cost, pattern of the population growth or increase in water demand, and discount rate. The reliability considerations are also important for the design of a water distribution system as there is a trade-off between cost of the system and system reliability. The essential parameters for network design are covered in Chapter 5.

1.6. CHOICE BETWEEN PUMPING AND GRAVITY SYSTEMS

The choice between a pumping or a gravity system on a topography having mild to medium slope is difficult without an analytical methodology. The pumping system can be designed for any topographic configuration. On the other hand, a gravity system is feasible if the input point is at a higher elevation than all the withdrawal points. Large pipe diameters will be required if the elevation difference between input point and withdrawals is very small, and the design may not be economic in comparison with a pumping system. Thus, it is essential to calculate the critical elevation difference at which both pumping and gravity systems will have the same cost. The method for the selection of a gravity or pumping system for a given terrain and economic conditions are described in Chapter 6.

1.7. NETWORK SYNTHESIS

With the advent of fast digital computers, conventional methods of water distribution network design have been discarded. The conventional design practice in vogue is to analyze the water distribution system assuming the pipe diameters and the input heads and obtain the nodal pressure heads and the pipe link discharges and average velocities. The nodal pressure heads are checked against the maximum and minimum allowable pressure heads. The average pipe link velocities are checked against maximum allowable average velocity. The pipe diameters and the input heads are revised several times to ensure that the nodal pressure heads and the average pipe velocities do not cross the allowable limits. Such a design is a feasible design satisfying the functional and safety requirements. Providing a solution merely satisfying the functional and safety requirements is not enough. The cost has to be reduced to a minimum consistent with functional and safety requirements and also reliability considerations.

The main objective of the synthesis of a pipe network is to estimate design variables like pipe diameters and pumping heads by minimizing total system cost subject to a number of constraints. These constraints can be divided into *safety* and *system* constraints. The safety constraints include criteria about minimum pipe size, minimum and maximum terminal pressure heads, and maximum allowable velocity. The system constraints include criteria for nodal discharge summation and loop headloss summation in the entire distribution system. The formulation of safety and system constraints is covered in Chapter 5.

In a water distribution network synthesis problem, the cost function is the objective function of the system. The objective function and the constraints constitute a nonlinear programming problem. Such a problem can only be solved numerically and not mathematically. A number of numerical methods are available to solve such problems. Successive application of liner programming (LP) and geometric programming (GP) methods for network synthesis are covered in this book.

Broadly speaking, following are the aspects of the design of pipe network systems.

1.7.1. Designing a Piecemeal Subsystem

A subsystem can be designed piecemeal if it has a weak interaction with the remaining system. Being simplest, there is alertness in this aspect. Choosing an economic type (material) of pipes, adopting an economic size of gravity or pumping mains, adopting a minimum storage capacity of service reservoirs, and adopting the least-cost alternative of various available sources of supply are some examples that can be quoted to highlight this aspect. The design of water transmission mains and water distribution mains can be covered in this category. The water transmission main transports water from one location to another without any intermediate withdrawals. On the other hand, water distribution mains have a supply (input) point at one end and withdrawals at intermediate and end points. Chapters 6 and 7 describe the design of these systems.

1.7.2. Designing the System as a Whole

Most of the research work has been aimed at the optimization of a water supply system as a whole. The majority of the components of a water supply system have strong interaction. It is therefore not possible to consider them piecemeal. The design problem of looped network is one of the difficult problems of optimization, and a satisfactory solution methodology is in an evolving phase. The design of single-supply (input) source, branched system is covered in Chapter 8 and multi-input source, branched system in Chapter 9. Similarly, the designs of single-input source, looped system and multi-input source, looped system are discussed in Chapters 10 and 11, respectively.

1.7.3. Dividing the Area into a Number of Optimal Zones for Design

For this aspect, convenience alone has been the criterion to decompose a large network into subsystems. Of the practical considerations, certain guidelines exist to divide the network into a number of subnetworks. These guidelines are not based on any comprehensive analysis. The current practice of designing such systems is by decomposing or splitting a system into a number of subsystems. Each subsystem is separately designed and finally interconnected at the ends for reliability considerations. The decision regarding the area to be covered by each such system depends upon the designer's intuition. On the other hand, to design a large water distribution system as a single entity may have computational difficulty in terms of computer time and storage. Such a system can also be efficiently designed if it is optimally split into small subsystems (Swamee and Sharma, 1990a). The decomposition of a large water distribution system into subsubsystems and then the design of each subsystem is described in Chapter 12.

1.8. REORGANIZATION OR RESTRENGTHENING OF EXISTING WATER SUPPLY SYSTEMS

Another important aspect of water distribution system design is strengthening or reorganization of existing systems once the water demand exceeds the design capacity. Water distribution systems are designed initially for a predecided design period, and at the end of the design period, the water demand exceeds the design capacity of the existing system on account of increase in population density or extension of services to new growth areas. To handle the increase in demand, it is required either to design an entirely new system or to reorganize the existing system. As it is expensive to replace the existing system with a new system after its design life is over, the attempt should be made to improve the carrying capacity of the existing system. Moreover, if the increase in demand is marginal, then merely increasing the pumping capacity and pumping head may suffice. The method for the reorganization of existing systems (Swamee and Sharma, 1990b) is covered in Chapter 13.

1.9. TRANSPORTATION OF SOLIDS THROUGH PIPELINES

The transportation of solids apart from roads and railways is also carried out through pipelines. It is difficult to transport solids through pipelines as solids. Thus, the solids are either suspended in a carrier fluid or containerized in capsules. If suspended in a carrier fluid, the solids are separated at destination. These systems can either be gravity-sustained systems or pumping systems based on the local conditions. The design of such systems includes the estimation of carrier fluid flow, pipe size, and power requirement in case of pumping system for a given sediment flow rate. The design of such a pipe system is highlighted in Chapter 14.

1.10. SCOPE OF THE BOOK

The book is structured in such a way that it not only enables engineers to fully understand water supply systems but also enables them to design functionally efficient and least-cost systems. It is intended that students, professional engineers, and researchers will benefit from the pipe network analysis and design topics covered in this book. Hopefully, it will turn out to be a reference book to water supply engineers as some of the fine aspects of pipe network optimization are covered herein.

REFERENCES

Brown, G.O. (2002) The history of the Darcy-Weisbach equation for pipe flow resistance. In: *Environmental and Water Resources History*, edited by J.R. Rogers and A.J. Fredrich. Proceedings and Invited Papers for the ASCE 150th Anniversary, November 3–7 presented at the ASCE Civil Engineering Conference and Exposition, held in Washington, DC.

James, W. (2006). A historical perspective on the development of urban water systems. Available at http://www.soe.uoguelph.ca/webfiles/wjames/homepage/Teaching/437/wj437hi.htm.

Jesperson, K. (2001). A brief history of drinking water distribution. *On Tap* 18–19. Available at http://www.nesc.wvu.edu/ndwc/articles/OT/SP01/History_Distribution.html.

Liou, C.P. (1998). Limitations and proper use of the Hazen Williams equation. *J. Hydraul. Eng.* 124(9), 951–954.

Ramalingam, D., Lingireddy, S., and Ormsbee, L.E. (2002). History of water distribution network analysis: Over 100 years of progress. In: *Environmental and Water Resources History*, edited by J.R. Rogers and A.J. Fredrich. Proceedings and Invited Papers for the ASCE 150th Anniversary, November 3–7 presented at the ASCE Civil Engineering Conference and Exposition, held in Washington, DC.

Salzman, J. (2006). Thirst: A short history of drinking water. Duke Law School Legal Studies Paper No. 92, *Yale Journal of Law and the Humanities* 17(3). Available at http://eprints.law.duke.edu/archive/00001261/01/17_Yale_J.L._&_Human.(2006).pdf.

Sharma, A.K., and Swamee, P.K. (2005). Application of flow path algorithm in flow pattern mapping and loop data generation for a water distribution system. *J. Water Supply: Research and Technology–AQUA, 54*, IWA Publishing, London, 411–422.

Swamee, P.K. (2000). Discussion of 'Limitations and proper use of the Hazen Williams equation,' by Chyr Pyng Liou. *J. Hydraul. Eng.* 125(2), 169–170.

Swamee, P.K., and Jain, A.K. (1976). Explicit equations for pipe flow problems. *J. Hydraul. Eng.* 102(5), 657–664.

Swamee, P.K., and Sharma, A.K. (1990a). Decomposition of a large water distribution system. *J. Envir. Eng.* 116(2), 296–283.

Swamee, P.K., and Sharma, A.K. (1990b). Reorganization of a water distribution system. *J. Envir. Eng.* 116(3), 588–599.

Walski, T., Chase, D., and Savic, D. (2001). *Water Distribution Modelling*, Haestad Press, Waterbury, CT.

2

BASIC PRINCIPLES OF PIPE FLOW

Design of Water Supply Pipe Networks. By Prabhata K. Swamee and Ashok K. Sharma
Copyright © 2008 John Wiley & Sons, Inc.

Pipe flow is the most commonly used mode of carrying fluids for small to moderately large discharges. In a pipe flow, fluid fills the entire cross section, and no free surface is formed. The fluid pressure is generally greater than the atmospheric pressure but in certain reaches it may be less than the atmospheric pressure, allowing free flow to continue through siphon action. However, if the pressure is much less than the atmospheric pressure, the dissolved gases in the fluid will come out and the continuity of the fluid in the pipeline will be hampered and flow will stop.

The pipe flow is analyzed by using the continuity equation and the equation of motion. The continuity equation for steady flow in a circular pipe of diameter D is

$$Q = \frac{\pi}{4} D^2 V, \qquad (2.1)$$

where V = average velocity of flow, and Q = volumetric rate of flow, called discharge. The equation of motion for steady flow is

$$z_1 + h_1 + \frac{V_1^2}{2g} = z_2 + h_2 + \frac{V_2^2}{2g} + h_L, \qquad (2.2a)$$

where z_1 and z_2 = elevations of the centerline of the pipe (from arbitrary datum), h_1 and h_2 = pressure heads, V_1 and V_2 = average velocities at sections 1 and 2, respectively (Fig. 2.1), g = gravitational acceleration, and h_L = head loss between sections 1 and 2. The head loss h_L is composed of two parts: h_f =head loss on account of surface resistance (also called friction loss), and h_m =head loss due to form resistance, which is the head loss on account of change in shape of the pipeline (also called minor loss). Thus,

$$h_L = h_f + h_m. \qquad (2.2b)$$

The minor loss h_m is zero in Fig. 2.1, and Section 2.2 covers form (minor) losses in detail.

The term $z + h$ is called the piezometric head; and the line connecting the piezometric heads along the pipeline is called the hydraulic gradient line.

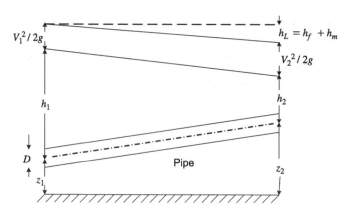

Figure 2.1. Definition sketch.

Knowing the condition at the section 1, and using Eq. (2.2a), the pressure head at section 2 can be written as

$$h_2 = h_1 + z_1 - z_2 + \frac{V_1^2 - V_2^2}{2g} - h_L. \tag{2.2c}$$

For a pipeline of constant cross section, Eq. (2.2c) is reduced to

$$h_2 = h_1 + z_1 - z_2 - h_L. \tag{2.2d}$$

Thus, h_2 can be obtained if h_L is known.

2.1. SURFACE RESISTANCE

The head loss on account of surface resistance is given by the Darcy–Weisbach equation

$$h_f = \frac{fLV^2}{2gD}, \tag{2.3a}$$

where $L =$ the pipe length, and $f =$ coefficient of surface resistance, traditionally known as *friction factor*. Eliminating V between (2.1) and (2.3a), the following equation is obtained:

$$h_f = \frac{8fLQ^2}{\pi^2 g D^5}. \tag{2.3b}$$

The coefficient of surface resistance for turbulent flow depends on the average height of roughness projection, ε, of the pipe wall. The average roughness of pipe wall for commercial pipes is listed in Table 2.1. Readers are advised to check these values with their local pipe manufacturers.

TABLE 2.1. Average Roughness Heights

Pipe Material	Roughness Height (mm)
1. Wrought iron	0.04
2. Asbestos cement	0.05
3. Poly(vinyl chloride)	0.05
4. Steel	0.05
5. Asphalted cast iron	0.13
6. Galvanized iron	0.15
7. Cast/ductile iron	0.25
8. Concrete	0.3 to 3.0
9. Riveted steel	0.9 to 9.0

The coefficient of surface resistance also depends on the Reynolds number $\mathbf{R}$ of the flow, defined as

$$\mathbf{R} = \frac{VD}{\nu}, \tag{2.4a}$$

where $\nu =$ kinematic viscosity of fluid that can be obtained using the equation given by Swamee (2004)

$$\nu = 1.792 \times 10^{-6}\left[1 + \left(\frac{T}{25}\right)^{1.165}\right]^{-1}, \tag{2.4b}$$

where T is the water temperature in °C. Eliminating V between Eqs. (2.1) and (2.4a), the following equation is obtained:

$$\mathbf{R} = \frac{4Q}{\pi\nu D}. \tag{2.4c}$$

For turbulent flow ($\mathbf{R} \geq 4000$), Colebrook (1938) found the following implicit equation for f:

$$f = 1.325\left[\ln\left(\frac{\varepsilon}{3.7D} + \frac{2.51}{\mathbf{R}\sqrt{f}}\right)\right]^{-2}. \tag{2.5a}$$

Using Eq. (2.5a), Moody (1944) constructed a family of curves between f and $\mathbf{R}$ for various values of relative roughness ε/D.

For laminar flow ($\mathbf{R} \leq 2000$), f depends on $\mathbf{R}$ only and is given by the Hagen–Poiseuille equation

$$f = \frac{64}{\mathbf{R}}. \tag{2.5b}$$

For $\mathbf{R}$ lying in the range between 2000 and 4000 (called transition range), no information is available about estimating f. Swamee (1993) gave the following equation for f valid in the laminar flow, turbulent flow, and the transition in between them:

$$f = \left\{\left(\frac{64}{\mathbf{R}}\right)^{8} + 9.5\left[\ln\left(\frac{\varepsilon}{3.7D} + \frac{5.74}{\mathbf{R}^{0.9}}\right) - \left(\frac{2500}{\mathbf{R}}\right)^{6}\right]^{-16}\right\}^{0.125}. \tag{2.6a}$$

Equation (2.6a) predicts f within 1% of the value obtained by Eqs. (2.5a). For turbulent flow, Eq. (2.6a) simplifies to

$$f = 1.325 \left[\ln \left(\frac{\varepsilon}{3.7D} + \frac{5.74}{\mathbf{R}^{0.9}} \right) \right]^{-2}. \tag{2.6b}$$

Combing with Eq. (2.4c), Eq. (2.6b) can be rewritten as:

$$f = 1.325 \left\{ \ln \left[\frac{\varepsilon}{3.7D} + 4.618 \left(\frac{vD}{Q} \right)^{0.9} \right] \right\}^{-2}. \tag{2.6c}$$

Example 2.1. Calculate friction loss in a cast iron (CI) pipe of diameter 300 mm carrying a discharge of 200 L per second to a distance of 1000 m as shown in Fig. 2.2.

Solution. Using Eq. (2.4c), the Reynolds number $\mathbf{R}$ is

$$\mathbf{R} = \frac{4Q}{\pi v D}.$$

Considering water at 20°C and using Eq. (2.4b), the kinematic viscosity of water is

$$v = 1.792 \times 10^{-6} \left[1 + \left(\frac{20}{25} \right)^{1.165} \right]^{-1} = 1.012 \times 10^{-6} \quad \mathrm{m^2/s}.$$

Substituting $Q = 0.2 \, \mathrm{m^3/s}$, $v = 1.012 \times 10^{-6} \, \mathrm{m^2/s}$, and $D = 0.3 \, \mathrm{m}$,

$$\mathbf{R} = \frac{4 \times 0.2}{3.14159 \times 1.012 \times 10^{-6} \times 0.3} = 838{,}918.$$

As the $\mathbf{R}$ is greater than 4000, the flow is turbulent. Using Table 2.1, the roughness height for CI pipes is $\varepsilon = 0.25 \, \mathrm{mm}$ (2.5×10^{-4} m). Substituting values of $\mathbf{R}$ and ε in

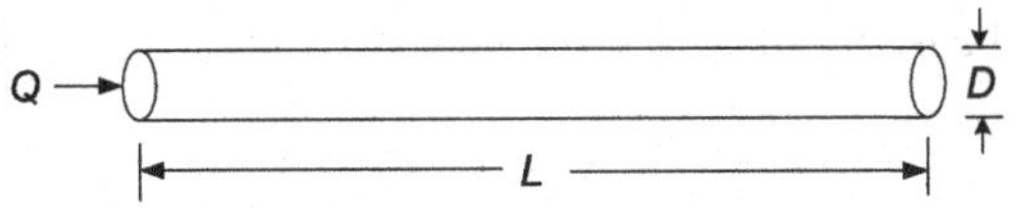

$Q = 0.2 m^3/s$, $L = 1000m$ and $D = 0.3m$

Figure 2.2. A conduit.

Eq. (2.6b) the friction factor is

$$f = 1.325 \left[\ln \left(\frac{2.5 \times 10^{-4}}{3.7 \times 0.3} + \frac{5.74}{(8.389 \times 10^5)^{0.9}} \right) \right]^{-2} 0.0193.$$

Using Eq. (2.3b), the head loss is

$$h_f = \frac{8 \times 0.0193 \times 1000 \times 0.2^2}{3.14159^2 \times 9.81 \times 0.3^5} = 26.248 \text{ m}.$$

2.2. FORM RESISTANCE

The form-resistance losses are due to bends, elbows, valves, enlargers, reducers, and so forth. Unevenness of inside pipe surface on account of imperfect workmanship also causes form loss. A form loss develops at a pipe junction where many pipelines meet. Similarly, form loss is also created at the junction of pipeline and service connection. All these losses, when added together, may form a sizable part of overall head loss. Thus, the name "minor loss" for form loss is a misnomer when applied to a pipe network. In a water supply network, form losses play a significant role. However, form losses are unimportant in water transmission lines like gravity mains or pumping mains that are long pipelines having no off-takes. Form loss is expressed in the following form:

$$h_m = k_f \frac{V^2}{2g} \tag{2.7a}$$

or its equivalent form

$$h_m = k_f \frac{8Q^2}{\pi^2 g D^4}, \tag{2.7b}$$

where k_f = form-loss coefficient. For a service connection, k_f may be taken as 1.8.

2.2.1. Pipe Bend

In the case of pipe bend, k_f depends on bend angle α and bend radius R (Fig. 2.3). Expressing α in radians, Swamee (1990) gave the following equation for the form-loss coefficient:

$$k_f = \left[0.0733 + 0.923 \left(\frac{D}{R} \right)^{3.5} \right] \alpha^{0.5}. \tag{2.8}$$

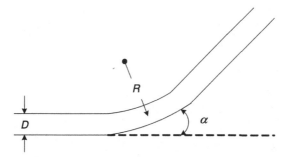

Figure 2.3. A pipe bend.

It should be noticed that Eq. (2.8) does not hold good for near zero bend radius. In such a case, Eq. (2.9) should be used for loss coefficient for elbows.

2.2.2. Elbows

Elbows are used for providing sharp turns in pipelines (Fig. 2.4). The loss coefficient for an elbow is given by

$$k_f = 0.442\alpha^{2.17}, \tag{2.9}$$

where $\alpha =$ elbow angle in radians.

2.2.3. Valves

Valves are used for regulating the discharge by varying the head loss accrued by it. For a 20% open sluice valve, loss coefficient is as high as 31. Even for a fully open valve, there is a substantial head loss. Table 2.2 gives k_f for fully open valves. The most commonly used valves in the water supply systems are the sluice valve and the rotary valve as shown in Fig. 2.5 and Fig. 2.6, respectively.

For partly closed valves, Swamee (1990) gave the following loss coefficients:

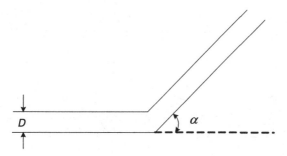

Figure 2.4. An elbow.

TABLE 2.2. Form-Loss Coefficients for Valves

Valve Type	Form-Loss Coefficient k_f
Sluice valve	0.15
Switch valve	2.4
Angle valve	5.0
Globe valve	10.0

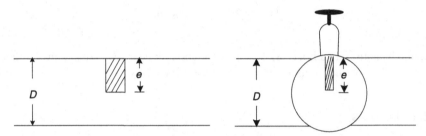

Figure 2.5. A sluice valve.

2.2.3.1. Sluice Valve. A partly closed sluice valve is shown in Fig. 2.5. Swamee (1990) developed the following relationship for loss coefficients:

$$k_f = 0.15 + 1.91\left(\frac{e}{D - e}\right)^2, \tag{2.10}$$

where e is the spindle depth obstructing flow in pipe.

2.2.3.2. Rotary Valve. A partly closed rotary valve is shown in Fig. 2.6. The loss coefficients can be estimated using the following equation (Swamee, 1990):

$$k_f = 133\left(\frac{\alpha}{\pi - 2\alpha}\right)^{2.3}, \tag{2.11}$$

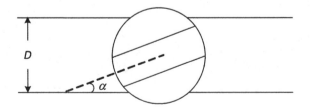

Figure 2.6. A rotary valve.

where α = valve closure angle in radians. Partly or fully closed valves are not considered at the design stage, as these situations develop during the operation and maintenance of the water supply systems.

2.2.4. Transitions

Transition is a gradual expansion (called enlarger) or gradual contraction (called reducer). In the case of transition, the head loss is given by

$$h_m = k_f \frac{(V_1 - V_2)^2}{2g} \tag{2.12a}$$

or its equivalent form

$$h_m = k_f \frac{8(D_2^2 - D_1^2)^2 Q^2}{\pi^2 g D_1^4 D_2^4}, \tag{2.12b}$$

where the suffixes 1 and 2 refer to the beginning and end of the transition, respectively. The loss coefficient depends on how gradual or abrupt the transition is. For straight gradual transitions, Swamee (1990) gave the following equations for k_f:

2.2.4.1. Gradual Contraction.
A gradual pipe contraction is shown in Fig. 2.7. The loss coefficient can be obtained using the following equation:

$$k_f = 0.315 \, \alpha_c^{1/3}. \tag{2.13a}$$

The contraction angle α_c (in radians) is given by

$$\alpha_c = 2 \tan^{-1} \left(\frac{D_1 - D_2}{2L} \right), \tag{2.13b}$$

where L = transition length.

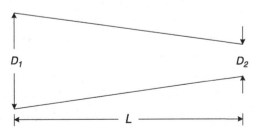

Figure 2.7. A gradual contraction transition.

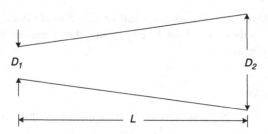

Figure 2.8. A gradual expansion transition.

2.2.4.2. Gradual Expansion. A gradual expansion is depicted in Fig. 2.8. The following relationship can be used for the estimation of loss coefficient:

$$k_f = \left\{ \frac{0.25}{\alpha_e^3} \left[1 + \frac{0.6}{r^{1.67}} \left(\frac{\pi - \alpha_e}{\alpha_e} \right) \right]^{0.533r - 2.6} \right\}^{-0.5}, \tag{2.13c}$$

where r = expansion ratio D_2/D_1, and α_e = expansion angle (in radians) given by

$$\alpha_e = 2 \tan^{-1} \left(\frac{D_2 - D_1}{2L} \right). \tag{2.13d}$$

2.2.4.3. Optimal Expansions Transition. Based on minimizing the energy loss, Swamee et al. (2005) gave the following equation for optimal expansion transition in pipes and power tunnels as shown in Fig. 2.9:

$$D = D_1 + (D_2 - D_1) \left[\left(\frac{L}{x} - 1 \right)^{1.786} + 1 \right]^{-1}, \tag{2.13e}$$

where x = distance from the transition inlet.

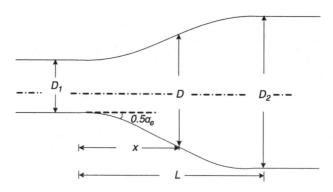

Figure 2.9. Optimal transition profile.

Figure 2.10. An abrupt expansion transition.

2.2.4.4. Abrupt Expansion. The loss coefficient for abrupt expansion as shown in Fig. 2.10 is

$$k_f = 1. \tag{2.14a}$$

2.2.4.5. Abrupt Contraction. Swamee (1990) developed the following expression for the loss coefficient of an abrupt pipe contraction as shown in Fig. 2.11:

$$k_f = 0.5\left[1 - \left(\frac{D_2}{D_1}\right)^{2.35}\right]. \tag{2.14b}$$

2.2.5. Pipe Junction

Little information is available regarding the form loss at a pipe junction where many pipelines meet. The form loss at a pipe junction may be taken as

$$h_m = k_f \frac{V_{\max}^2}{2g}, \tag{2.15}$$

where $V_{\max}$ = maximum velocity in a pipe branch meeting at the junction. In the absence of any information, k_f may be assumed as 0.5.

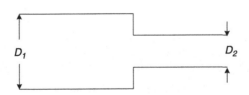

Figure 2.11. An abrupt contraction transition.

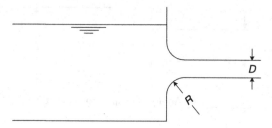

Figure 2.12. Entrance transition.

2.2.6. Pipe Entrance

There is a form loss at the pipe entrance (Fig. 2.12). Swamee (1990) obtained the following equation for the form-loss coefficient at the pipe entrance:

$$k_f = 0.5\left[1 + 36\left(\frac{R}{D}\right)^{1.2}\right]^{-1}, \tag{2.16}$$

where R = radius of entrance transition. It should be noticed that for a sharp entrance, $k_f = 0.5$.

2.2.7. Pipe Outlet

A form loss also generates at an outlet. For a confusor outlet (Fig. 2.13), Swamee (1990) found the following equation for the head-loss coefficient:

$$k_f = 4.5\frac{D}{d} - 3.5, \tag{2.17}$$

where d = outlet diameter. Putting $D/d = 1$ in Eq. (2.17), for a pipe outlet, $k_f = 1$.

Figure 2.13. A confusor outlet.

2.2.8. Overall Form Loss

Knowing the various loss coefficients k_{f1}, k_{f2}, k_{f3}, ..., k_{fn} in a pipeline, overall form-loss coefficient k_f can be obtained by summing them, that is,

$$k_f = k_{f1} + k_{f2} + k_{f3} + \cdots + k_{fn}. \tag{2.18}$$

Knowing the surface resistance loss h_f and the form loss h_m, the net loss h_L can be obtained by Eq. (2.2b). Using Eqs. (2.3a) and (2.18), Eq. (2.2b) reduces to

$$h_L = \left(k_f + \frac{fL}{D} \right) \frac{V^2}{2g} \tag{2.19a}$$

or its counterpart

$$h_L = \left(k_f + \frac{fL}{D} \right) \frac{8Q^2}{\pi^2 g D^4}. \tag{2.19b}$$

2.2.9. Pipe Flow Under Siphon Action

A pipeline that rises above its hydraulic gradient line is termed a *siphon*. Such a situation can arise when water is carried from one reservoir to another through a pipeline that crosses a ridge. As shown in Fig. 2.14, the pipeline between the points b and c crosses a ridge at point e. If the pipe is long, head loss due to friction is large and the form losses can be neglected. Thus, the hydraulic gradient line is a straight line that joins the water surfaces at points A and B.

The pressure head at any section of the pipe is represented by the vertical distance between the hydraulic gradient line and the centerline of the pipe. If the hydraulic gradient line is above the centerline of pipe, the water pressure in the pipeline is

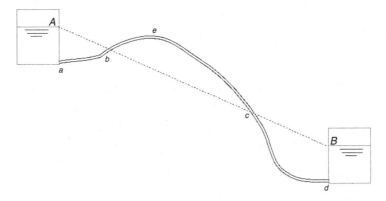

Figure 2.14. Pipe flow under siphon action.

above atmospheric. On the other hand if it is below the centerline of the pipe, the pressure is below atmospheric. Thus, it can be seen from Fig. 2.14 that at points b and c, the water pressure is atmospheric, whereas between b and c it is less than atmospheric. At the highest point e, the water pressure is the lowest. If the pressure head at point e is less than -2.5 m, the water starts vaporizing and causes the flow to stop. Thus, no part of the pipeline should be more than 2.5 m above the hydraulic gradient line.

Example 2.2A. A pumping system with different pipe fittings is shown in Fig. 2.15. Calculate residual pressure head at the end of the pipe outlet if the pump is generating an input head of 50 m at 0.1 m^3/s discharge. The CI pipe diameter D is 0.3 m. The contraction size at point 3 is 0.15 m; pipe size between points 6 and 7 is 0.15 m; and confusor outlet size $d = 0.15$ m. The rotary valve at point 5 is fully open. Consider the following pipe lengths between points:

Points 1 and 2 $=100$ m, points 2 and 3 $= 0.5$ m; and points 3 and 4 $= 0.5$ m
Points 4 and 6 $= 400$ m, points 6 and 7 $= 20$ m; and points 7 and 8 $= 100$ m

Solution

1. *Head loss between points 1 and 2.*
   Pipe length 100 m, flow 0.1 m^3/s, and pipe diameter 0.3 m.
   Using Eq. (2.4b), v for 20°C is 1.012×10^{-6} m^2/s, similarly using Eq. (2.4c), Reynolds number $\mathbf{R} = 419{,}459$. Using Table 2.1 for CI pipes, ε is 0.25 mm. The friction factor f is calculated using Eq. (2.6b) $= 0.0197$. Using Eq. 2.3b the head loss h_{f12} in pipe (1–2) is

$$h_{f12} = \frac{8fLQ^2}{\pi^2 g D^5} = \frac{8 \times 0.0197 \times 100 \times 0.1^2}{3.14159^2 \times 9.81 \times 0.3^5} = 0.670 \text{ m.}$$

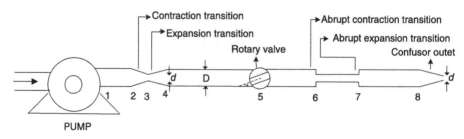

Figure 2.15. A pumping system with different pipe fittings.

2. *Head loss between points 2 and 3 (a contraction transition).*
   For $D = 0.3$, $d = 0.15$, and transition length $= 0.5$ m, the contraction angle α_c can be calculated using Eq. (2.13b):

$$\alpha_c = 2\tan^{-1}\left(\frac{D_1 - D_2}{2L}\right) = 2\tan^{-1}\left(\frac{0.3 - 0.15}{2 \times 0.5}\right) = 0.298 \text{ radians.}$$

Using Eq. (2.13a), the form-loss coefficient is

$$k_f = 0.315\,\alpha_c^{1/3} = 0.315 \times 0.298^{1/3} = 0.210$$

Using Eq. (2.12b), the head loss $h_{m23} = 0.193$ m.

3. *Head loss between points 3 and 4 (an expansion transition).*
   For $d = 0.15$, $D = 0.3$, the expansion ratio $r = 2$, and transition length $= 0.5$ m. Using Eq. (2.13d), the expansion angle $\alpha_e = 0.298$ radians. Using Eq. (2.13c), the form-loss coefficient $= 0.716$. Using Eq. (2.12b), the head loss $h_{m34} = 0.657$ m.

4. *Headloss between points 4 and 6.*
   Using Eq. (2.4c), with $\nu = 1.012 \times 10^{-6}$ m^2/s, diameter 0.3, and discharge 0.1 m^3/s, the Reynolds number $= 419{,}459$. With $\varepsilon = 0.25$ mm using Eq. (2.6b), $f = 0.0197$. Thus, for pipe length 400 m, using Eq. (2.3b), head loss $h_f = 2.681$ m.

5. *Head loss at point 5 due to rotary valve (fully open).*
   For fully open valve $\alpha = 0$. Using Eq. (2.11), form-loss coefficient $k_f = 0$ and using Eq. (2.7b), the form loss $h_m = 0.0$ m.

6. *Head loss at point 6 due to abrupt contraction.*
   For $D = 0.3$ m and $d = 0.15$ m, using Eq. (2.14b), the form-loss coefficient

$$k_f = 0.5\left[1 - \left(\frac{0.15}{0.3}\right)^{2.35}\right] = 0.402.$$

Using Eq. (2.12b), the form loss $h_m = 0.369$ m.

7. *Head loss in pipe between points 6 and 7.*
   Pipe length $= 20$ m, pipe diameter $= 0.15$ m, and roughness height $= 0.25$ mm.

   Reynolds number $= 838{,}914$ and pipe friction factor $f = 0.0227$, head loss $h_{f67} = 4.930$ m.

8. *Head loss at point 7 (an abrupt expansion).*
   An abrupt expansion from 0.15 m pipe size to 0.30 m.
   Using Eq. (2.14a), $k_f = 1$ and using Eq. (2.12b), $h_m = 0.918$ m.

TABLE 2.3. Pipe Transition Computations x versus D

x	D (optimal)	D (linear)
0.0	1.000	1.000
0.2	1.019	1.100
0.4	1.078	1.200
0.6	1.180	1.300
0.8	1.326	1.400
1.0	1.500	1.500
1.2	1.674	1.600
1.4	1.820	1.700
1.6	1.922	1.800
1.8	1.981	1.900
2.0	2.000	2.000

9. *Head loss in pipe between points 7 and 8.*
   Pipe length = 100 m, pipe diameter = 0.30 m, and roughness height = 0.25 mm.
   Reynolds number = 423,144 and pipe friction factor $f = 0.0197$.
   Head loss $h_{f78} = 0.670$ m.
10. *Head loss at outlet point 8 (confusor outlet).*
    Using Eq. (2.17), the form-loss coefficient

$$k_f = 4.5\frac{D}{d} - 3.5 = 4.5 \times \frac{0.30}{0.15} - 3.5 = 5.5. \text{ Using Eq. (2.12b), } h_m$$
$$= 0.560 \text{ m.}$$

Total head loss $h_L = 0.670 + 0.193 + 0.657 + 2.681 + 0.369 + 0 +$
$4.930 + 0.918 + 0.670 + 0.560 = 11.648$ m.
Thus, the residual pressure at the end of the pipe outlet $= 50 - 11.648 = 38.352$ m.

Example 2.2B. Design an expansion for the pipe diameters 1.0 m and 2.0 m over a distance of 2 m for Fig. 2.9.

Solution. Equation (2.13e) is used for the calculation of optimal transition profile. The geometry profile is $D_1 = 1.0$ m, $D_2 = 2.0$ m, and $L = 2.0$ m.
   Substituting various values of x, the corresponding values of D using Eq. (2.13e) and with linear expansion were computed and are tabulated in Table 2.3.

2.3. PIPE FLOW PROBLEMS

In pipe flow, there are three types of problems pertaining to determination of (a) the nodal head; (b) the discharge through a pipe link; and (c) the pipe diameter. Problems

(a) and (b) belong to analysis, whereas problem (c) falls in the category of synthesis/design.

2.3.1. Nodal Head Problem

In the nodal head problem, the known quantities are L, D, h_L, Q, ε, ν, and k_f. Using Eqs. (2.2b) and (2.7b), the nodal head h_2 (as shown in Fig. 2.1) is obtained as

$$h_2 = h_1 + z_1 - z_2 - \left(k_f + \frac{fL}{D}\right)\frac{8Q^2}{\pi^2 gD^4}. \tag{2.20}$$

2.3.2. Discharge Problem

For a long pipeline, form losses can be neglected. Thus, in this case the known quantities are L, D, h_f, ε, and ν. Swamee and Jain (1976) gave the following solution for turbulent flow through such a pipeline:

$$Q = -0.965 D^2 \sqrt{gD\,h_f/L}\,\ln\left(\frac{\varepsilon}{3.7D} + \frac{1.78\nu}{D\sqrt{gD\,h_f/L}}\right) \tag{2.21a}$$

Equation (2.21a) is exact. For laminar flow, the Hagen–Poiseuille equation gives the discharge as

$$Q = \frac{\pi g D^4 h_f}{128\nu L}. \tag{2.21b}$$

Swamee and Swamee (2008) gave the following equation for pipe discharge that is valid under laminar, transition, and turbulent flow conditions:

$$
Q = D^2\sqrt{gDh_f/L}\left\{\left(\frac{128\nu}{\pi D\sqrt{gDh_f/L}}\right)^4\right.
$$
$$
\left. + 1.153\left[\left(\frac{415\nu}{D\sqrt{gDh_f/L}}\right)^8 - \ln\left(\frac{\varepsilon}{3.7D} + \frac{1.775\nu}{D\sqrt{gDh_f/L}}\right)^{-4}\right]^{-0.25}\right\} \tag{2.21c}
$$

Equation (2.21c) is almost exact as the maximum error in the equation is 0.1%.

2.3.3. Diameter Problem

In this problem, the known quantities are L, h_f, ε, Q, and ν. For a pumping main, head loss is not known, and one has to select the optimal value of head loss by minimizing the cost. This has been dealt with in Chapter 6. However, for turbulent flow in a long gravity main, Swamee and Jain (1976) obtained the following solution for the pipe diameter:

$$D = 0.66 \left[\varepsilon^{1.25} \left(\frac{LQ^2}{gh_f} \right)^{4.75} + vQ^{9.4} \left(\frac{L}{gh_f} \right)^{5.2} \right]^{0.04}. \tag{2.22a}$$

In general, the errors involved in Eq. (2.22a) are less than 1.5%. However, the maximum error occurring near transition range is about 3%. For laminar flow, the Hagen–Poiseuille equation gives the diameter as

$$D = \left(\frac{128vQL}{\pi g h_f} \right)^{0.25}. \tag{2.22b}$$

Swamee and Swamee (2008) gave the following equation for pipe diameter that is valid under laminar, transition, and turbulent flow conditions

$$D = 0.66 \left[\left(214.75 \frac{vLQ}{gh_f} \right)^{6.25} + \varepsilon^{1.25} \left(\frac{LQ^2}{gh_f} \right)^{4.75} + vQ^{9.4} \left(\frac{L}{gh_f} \right)^{5.2} \right]^{0.04}. \tag{2.22c}$$

Equation (2.22c) yields D within 2.75%. However, close to transition range, the error is around 4%.

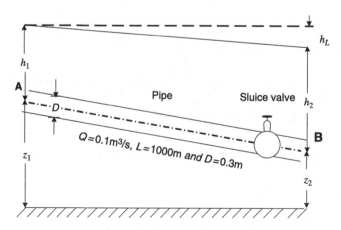

Figure 2.16. A gravity main.

Example 2.3. As shown in Fig. 2.16, a discharge of 0.1 m³/s flows through a CI pipe main of 1000 m in length having a pipe diameter 0.3 m. A sluice valve of 0.3 m size is placed close to point B. The elevations of points A and B are 10 m and 5 m, respectively. Assume water temperature as 20°C. Calculate:

(A) Terminal pressure h_2 at point B and head loss in the pipe if terminal pressure h_1 at point A is 25 m.
(B) The discharge in the pipe if the head loss is 10 m.
(C) The CI gravity main diameter if the head loss in the pipe is 10 m and a discharge of 0.1 m³/s flows in the pipe.

Solution

(A) The terminal pressure h_2 at point B can be calculated using Eq. (2.20). The friction factor f can be calculated applying Eq. (2.6a) and the roughness height of CI pipe $= 0.25$ mm is obtained from Table 2.1. The form-loss coefficient for sluice valve from Table 2.2 is 0.15. The viscosity of water at 20°C can be calculated using Eq. (2.4b). The coefficient of surface resistance depends on the Reynolds number **R** of the flow:

$$\mathbf{R} = \frac{4Q}{\pi \nu D} = 419,459.$$

Thus, substituting values in Eq. (2.6a), the friction factor

$$f = \left\{ \left(\frac{64}{\mathbf{R}}\right)^8 + 9.5\left[\ln\left(\frac{\varepsilon}{3.7D} + \frac{5.74}{\mathbf{R}^{0.9}}\right) - \left(\frac{2500}{\mathbf{R}}\right)^6\right]^{-16} \right\}^{0.125} = 0.0197.$$

Using Eq. (2.20), the terminal head h_2 at point B is

$$h_2 = h_1 + z_1 - z_2 - \left(k_f + \frac{fL}{D}\right)\frac{8Q^2}{\pi^2 gD^4}$$

$$= 25 + 10 - 5 - \left(0.15 + \frac{0.0197 \times 1000}{0.3}\right)\frac{8 \times 0.1^2}{3.14159^2 \times 9.81 \times 0.3^5}$$

$$= 30 - (0.015 + 6.704) = 23.281 \text{ m}.$$

(B) If the total head loss in the pipe is predecided equal to 10 m, the discharge in CI pipe of size 0.3 m can be calculated using Eq. (2.21a):

$$Q = -0.965 D^2 \sqrt{gDh_f/L} \ln\left(\frac{\varepsilon}{3.7D} + \frac{1.78v}{D\sqrt{gDh_f/L}}\right)$$

$$= -0.965 \times 0.3^2 \sqrt{9.81 \times (10/1000)} \ln\left(\frac{0.25 \times 10^{-3}}{3.7 \times 0.3}\right.$$

$$+ \left.\frac{1.78 \times 1.012 \times 10^{-6}}{0.3\sqrt{9.81 \times 0.3 \times (10/1000)}}\right)$$

$$= 0.123 \text{ m}^3/\text{s}.$$

(C) Using Eq. (2.22a), the gravity main diameter for preselected head loss of 10 m and known pipe discharge 0.1 m^3/s is

$$D = 0.66 \left[\varepsilon^{1.25}\left(\frac{LQ^2}{gh_f}\right)^{4.75} + vQ^{9.4}\left(\frac{L}{gh_f}\right)^{5.2}\right]^{0.04}$$

$$= 0.66 \left[0.00025^{1.25}\left(\frac{1000 \times 0.1^2}{9.81 \times 10}\right)^{4.75} + 1.012 \times 10^{-6}\right.$$

$$\left. \times 0.1^{9.4}\left(\frac{1000}{9.81 \times 10}\right)^{5.2}\right]^{0.04}$$

$$= 0.284 \text{ m}.$$

Also, if head loss is considered = 6.72 m, the pipe diameter is 0.306 m and flow is 0.1 m^3/s.

2.4. EQUIVALENT PIPE

In the water supply networks, the pipe link between two nodes may consist of a single uniform pipe size (diameter) or a combination of pipes in series or in parallel. As shown in Fig. 2.17a, the discharge Q flows from node A to B through a pipe of uniform diameter D and length L. The head loss in the pipe can simply be calculated using Darcy–Weisbach equation (2.3b) rewritten considering $h_L = h_f$ as:

$$h_L = \frac{8fLQ^2}{\pi^2 gD^5}. \tag{2.23}$$

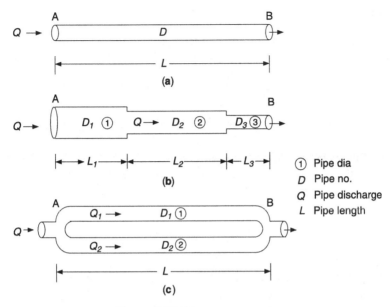

Figure 2.17. Pipe arrangements.

Figure 2.17b depicts that the same discharge Q flows from node A to node B through a series of pipes of lengths L_1, L_2, and L_3 having pipe diameters D_1, D_2, and D_3, respectively. It can be seen that the uniform discharge flows through the various pipes but the head loss across each pipe will be different. The total head loss across node A and node B will be the sum of the head losses in the three individual pipes as

$$h_L = h_{L1} + h_{L2} + h_{L3}.$$

Similarly Fig. 2.17c shows that the total discharge Q flows between parallel pipes of length L and diameters D_1 and D_2 as

$$Q = Q_1 + Q_2.$$

As the pressure head at node A and node B will be constant, hence the head loss between both the pipes will be the same.

The set of pipes arranged in parallel and series can be replaced with a single pipe having the same head loss across points A and B and also the same total discharge Q. Such a pipe is defined as an equivalent pipe.

2.4.1. Pipes in Series

In case of a pipeline made up of different lengths of different diameters as shown in Fig. 2.17b, the following head loss and flow conditions should be satisfied:

$$h_L = h_{L1} + h_{L2} + h_{L3} + \cdots$$
$$Q = Q_1 = Q_2 = Q_3 = \cdots.$$

Using the Darcy–Weisbach equation with constant friction factor f, and neglecting minor losses, the head loss in N pipes in series can be calculated as:

$$h_L = \sum_{i=1}^{N} \frac{8fL_iQ^2}{\pi^2 gD_i^5}. \tag{2.24}$$

Denoting equivalent pipe diameter as D_e, the head loss can be rewritten as:

$$h_L = \frac{8fQ^2}{\pi^2 gD_e^5} \sum_{i=1}^{N} L_i. \tag{2.25}$$

Equating these two equations of head loss, one gets

$$D_e = \left(\frac{\sum_{i=1}^{N} L_i}{\sum_{i=1}^{N} \frac{L_i}{D_i^5}} \right)^{0.2}. \tag{2.26}$$

Example 2.4. An arrangement of three pipes in series between tank A and B is shown in Fig. 2.18. Calculate equivalent pipe diameter and the corresponding flow. Assume Darcy–Weisbach's friction factor $f = 0.02$ and neglect entry and exit (minor) losses.

Solution. The equivalent pipe D_e can be calculated using Eq. (2.26):

$$D_e = \left(\frac{\sum_{i=1}^{N} L_i}{\sum_{i=1}^{N} \frac{L_i}{D_i^5}} \right)^{0.2}.$$

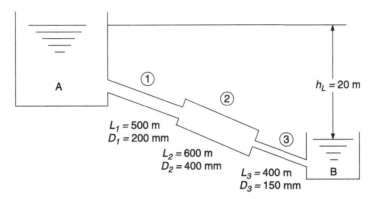

Figure 2.18. Pipes in series.

Substituting values,

$$D_e = \left(\frac{500 + 600 + 400}{\dfrac{500}{0.2^5} + \dfrac{600}{0.4^5} + \dfrac{400}{0.15^5}} \right)^{0.2} = 0.185 \, \text{m}$$

and

$$K_e = \left[\frac{8fL_e}{\pi^2 g D_e^5} \right] = \frac{8 \times 0.02 \times 1500}{3.14^2 \times 9.81 \times 0.185^5} = 11{,}450.49 \, \text{s}^2/\text{m}^5.$$

where $L_e = \Sigma L_i$ and K_e a pipe constant.

The discharge in pipe can be calculated:

$$Q = \left[\frac{h_L}{K_e} \right]^{0.2} = \left[\frac{20}{11{,}385.64} \right]^{0.2} = 0.042 \, \text{m}^3/\text{s}.$$

The calculated equivalent pipe size 0.185 m is not a commercially available pipe diameter and thus has to be manufactured specially. If this pipe is replaced by a commercially available nearest pipe size of 0.2 m, the pipe discharge should be recalculated for revised diameter.

2.4.2. Pipes in Parallel

If the pipes are arranged in parallel as shown in Fig. 2.17c, the following head loss and flow conditions should be satisfied:

$$h_L = h_{L1} = h_{L2} = h_{L3} = \cdots\cdots$$
$$Q = Q_1 + Q_2 + Q_3 + \cdots\cdots.$$

The pressure head at nodes A and B remains constant, meaning thereby that head loss in all the parallel pipes will be the same.

Using the Darcy–Weisbach equation and neglecting minor losses, the discharge Q_i in pipe i can be calculated as

$$Q_i = \pi D_i^2 \left(\frac{gD_i h_L}{8fL_i} \right)^{0.5}.$$ (2.27)

Thus for N pipes in parallel,

$$Q = \pi \sum_{i=1}^{N} D_i^2 \left(\frac{gD_i h_L}{8fL_i} \right)^{0.5}.$$ (2.28)

The discharge Q flowing in the equivalent pipe is

$$Q = \pi D_e^2 \left(\frac{gD_e h_L}{8fL} \right)^{0.5},$$ (2.29)

where L is the length of the equivalent pipe. This length may be different than any of the pipe lengths L_1, L_2, L_3, and so forth. Equating these two equations of discharge

$$D_e = \left[\sum_{i=1}^{N} \left(\frac{L}{L_i} \right)^{0.5} D_i^{2.5} \right]^{0.4}.$$ (2.30)

Example 2.5. For a given parallel pipe arrangement in Fig. 2.19, calculate equivalent pipe diameter and corresponding flow. Assume Darcy–Weisbach's friction factor $f = 0.02$ and neglect entry and exit (minor) losses. Length of equivalent pipe can be assumed as 500 m.

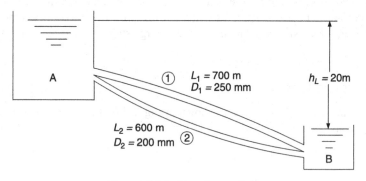

Figure 2.19. Pipes in parallel.

Solution. The equivalent pipe D_e can be calculated using Eq. (2.30):

$$D_e = \left[\sum_{i=1}^{N} \left(\frac{L}{L_i} \right)^{0.5} D_i^{2.5} \right]^{0.4}.$$

Substituting values in the above equation:

$$D_e = \left[\left(\frac{500}{700} \right)^{0.5} 0.25^{2.5} + \left(\frac{500}{600} \right)^{0.5} 0.20^{2.5} \right]^{0.4} = 0.283 \approx 0.28 \text{ m}.$$

Similarly, the discharge Q flowing in the equivalent pipe is

$$Q = \pi D_e^2 \left(\frac{g D_e h_L}{8fL} \right)^{0.5}.$$

Substituting values in the above equation

$$Q = 3.14 \times 0.28 \times 0.28 \left(\frac{9.81 \times 0.28 \times 20}{8 \times 0.02 \times 500} \right)^{0.5} = 0.204 \text{ m}^3/\text{s}.$$

The calculated equivalent pipe size 0.28 m is not a commercially available pipe diameter and thus has to be manufactured specially. If this pipe is replaced by a commercially available nearest pipe size of 0.3 m, the pipe discharge should be recalculated for revised diameter.

2.5. RESISTANCE EQUATION FOR SLURRY FLOW

The resistance equation (2.3a) is not applicable to the fluids carrying sediment in suspension. Durand (Stepanoff, 1969) gave the following equation for head loss for flow of fluid in a pipe with heterogeneous suspension of sediment particles:

$$h_f = \frac{fLV^2}{2gD} + \frac{81(s-1)C_v fL\sqrt{(s-1)gD}}{2C_D^{0.75}V} \tag{2.31}$$

where s = ratio of mass densities of particle and fluid, C_v = volumetric concentration, C_D = drag coefficient of particle, and f = friction factor of sediment fluid, which can be determined by Eq. (2.6a). For spherical particle of diameter d, Swamee and Ojha

(1991) gave the following equation for C_D:

$$C_D = 0.5\left\{16\left[\left(\frac{24}{\mathbf{R_s}}\right)^{1.6}+\left(\frac{130}{\mathbf{R_s}}\right)^{0.72}\right]^{2.5}+\left[\left(\frac{40{,}000}{\mathbf{R_s}}\right)^{2}+1\right]^{-0.25}\right\}^{0.25}, \qquad (2.32)$$

where $\mathbf{R_s}$ = sediment particle Reynolds number given by

$$\mathbf{R_s} = \frac{wd}{v}, \qquad (2.33)$$

where w = fall velocity of sediment particle, and d = sediment particle diameter. Equation (2.33) is valid for $\mathbf{R_s} \le 1.5 \times 10^5$. Denoting $v* = v/[d\sqrt{(s-1)gd}]$, the fall velocity can be obtained applying the following equation (Swamee and Ojha, 1991):

$$w = \sqrt{(s-1)gd}\left\{\left[(18v*)^2+(72v*)^{0.54}\right]^5+\left[(10^8v*)^{1.7}+1.43\times10^6\right]^{-0.346}\right\}^{-0.1}. \qquad (2.34)$$

A typical slurry transporting system is shown in Fig. 2.20.

Example 2.6. A CI pumping main of 0.3 m size and length 1000 m carries a slurry of average sediment particle size of 0.1 mm with mass densities of particle and fluid ratio as 2.5. If the volumetric concentration of particles is 20% and average temperature of water 20°C, calculate total head loss in the pipe.

Solution. The head loss for flow of fluid in a pipe with heterogeneous suspension of sediment particles can be calculated using Eq. (2.31).

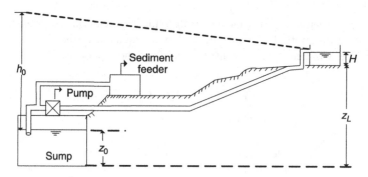

Figure 2.20. A typical slurry transporting system.

The fall velocity of sediment particles w can be obtained using Eq. (2.34) as

$$w = \sqrt{(s-1)gd}\left\{\left[(18v*)^2+(72v*)^{0.54}\right]^5+\left[(10^8 v*)^{1.7}+1.43\times10^6\right]^{-0.346}\right\}^{-0.1},$$

where $v* = v/[d\sqrt{(s-1)gd}]$. Substituting $s = 2.5$, $d = 0.0001$ m, $g = 9.81$ m/s^2 the v^* is 0.2637 and sediment particle fall velocity w = 0.00723 m/s. The sediment particle Reynolds number is given by

$$R_s = \frac{wd}{v} = \frac{0.00732\times1\times10^{-4}}{1.012\times10^{-6}} = 0.723.$$

The drag coefficient C_D for 0.1-mm-diameter spherical particle can be calculated using Eq. (2.32) for $R_s = 0.723$:

$$C_D = 0.5\left\{16\left[\left(\frac{24}{R_s}\right)^{1.6}+\left(\frac{130}{R_s}\right)^{0.72}\right]^{2.5}+\left[\left(\frac{40,000}{R_s}\right)^2+1\right]^{-0.25}\right\}^{0.25}$$

$$= 36.28.$$

The head loss in pipe is calculated using Eq. (2.31)

$$h_f = \frac{fLV^2}{2gD}+\frac{81(s-1)C_v fL\sqrt{(s-1)gD}}{2C_D^{0.75}V}$$

For flow velocity in pipe $V = \dfrac{0.1}{\pi\times0.3^2/4} = 1.414$ m/s, the head loss

$$h_f = \frac{0.0197\times1000\times1.414^2}{2\times9.81\times0.3}$$

$$+\frac{81\times(2.5-1)\times0.2\times0.0197\times1000\sqrt{(2.5-1)\times9.81\times0.3}}{2\times36.28^{0.75}\times1.414}$$

$$= 6.719\,\text{m}+24.062\,\text{m} = 30.781\,\text{m}.$$

2.6. RESISTANCE EQUATION FOR CAPSULE TRANSPORT

Figure 2.21 depicts the pipeline carrying cylindrical capsules. The capsule has diameter kD, length aD, and wall thickness θD. The distance between two consecutive capsules is βaD. Capsule transport is most economic when capsules are made neutrally buoyant or nearly so, to avoid contact with pipe wall. In such a case, the capsule mass density is equal to the carrier fluid mass density ρ. With this condition, the volume V_s of the

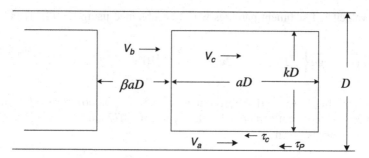

Figure 2.21. Capsule and its surroundings.

material contained in the capsule is obtained as

$$V_s = \frac{\pi}{4s_s}D^3\{k^2a - 2s_c\theta[k(k + 2a) - 2\theta(2k + a - 2\theta)]\}, \qquad (2.35)$$

where s_s is the ratio of mass densities of cargo and fluid.

The flow pattern in capsule transport repeats after the distance $(1 + \beta)aD$ called *characteristic length*. Considering the capsule velocity as V_c, the capsule covers the characteristic length in the characteristic time t_c given by

$$t_c = \frac{(1 + \beta)aD}{V_c}. \qquad (2.36)$$

The volumetric cargo transport rate Q_s is the volume of cargo passing in time t_c. Thus, using Eq. (2.35), the characteristic time t_c is obtained as

$$t_c = \frac{\pi}{4s_sQ_s}D^3\{k^2a - 2s_c\theta[k(k + 2a) - 2\theta(2k + a - 2\theta)]\}, \qquad (2.37)$$

where $s_s\rho$ = mass density of cargo. Equating Eqs. (2.36) and (2.37), the capsule velocity is obtained as

$$V_c = \frac{4a(1 + \beta)s_sQ_s}{\pi D^2\{k^2a - 2s_c\theta[k(k + 2a) - 2\theta(2k + a - 2\theta)]\}}. \qquad (2.38)$$

Swamee (1998) gave the resistance equation for pipe flow carrying neutrally buoyant capsules as

$$h_f = \frac{8f_eLQ_s^2}{\pi^2 gD^5}, \qquad (2.39)$$

where f_e = effective friction factor given by

$$f_e = \frac{a(1+\beta)s_s^2\left[f_p a + f_b \beta a\left(1 + k^2\sqrt{k\lambda}\right)^2 + k^5\lambda\right]}{(1+\sqrt{k\lambda})^2\{k^2 a - 2s_c\theta[k(k+2a) - 2\theta(2k+a-2\theta)]\}^2}, \qquad (2.40)$$

where s_c = ratio of mass densities of capsule material and fluid, and f_b, f_c, and f_p = the friction factors for intercapsule distance, capsule, and pipe annulus, respectively. These friction factors can be obtained by Eq. (2.6a) using $\mathbf{R} = V_b D/\nu$, $(1 - k)(V_c - V_a)D/\nu$ and $(1 - k)V_a D/\nu$, respectively. Further, $\lambda = f_p/f_c$, and V_a = average fluid velocity in annular space between capsule and pipe wall given by

$$V_a = \frac{V_c}{1 + \sqrt{k\lambda}}, \qquad (2.41)$$

and V_b = average fluid velocity between two capsules, given by

$$V_b = \frac{1 + k^2\sqrt{k\lambda}}{1 + \sqrt{k\lambda}} V_c. \qquad (2.42)$$

The power consumed in overcoming the surface resistance is $\rho g Q_e h_f$, where Q_e is the effective fluid discharge given by

$$Q_e = \frac{a(1+\beta)s_s Q_s}{k^2 a - 2s_c\theta[k(k+2a) - 2\theta(2k+a-2\theta)]} \qquad (2.43)$$

The effective fluid discharge includes the carrier fluid volume and the capsule fluid volume in one characteristic length divided by characteristic time t_c.

It has been found that at an optimal $k = k^*$, the power loss is minimum. Depending upon the other parameters, k^* varied in the range $0.984 \le k^* \le 0.998$. Such a high value of k cannot be provided as it requires perfect straight alignment. Subject to topographic constraints, maximum k should be provided. Thus, k can be selected in the range $0.85 \le k \le 0.95$.

Example 2.7. Calculate the energy required to transport cargo at a rate of 0.01 m^3/s through an 0.5-m poly(vinyl chloride) pipeline of length 4000 m. The elevation difference between two reservoirs Z_{EL} is 15 m and the terminal head $H = 5$ m. The gravitational acceleration is 9.81 m/s^2, ratio of mass densities of cargo and fluid $s_s = 1.75$, ratio of mass densities of capsule walls and fluid $s_c = 2.7$ and the fluid density is 1000 kg/m^3. The nondimensional capsule length $\alpha = 1.5$, nondimensional distance between capsules $\beta = 15$, nondimensional capsule diameter $k = 0.9$, and capsule wall thickness is 10 mm. The schematic representation of the system is shown in Fig. 2.22.

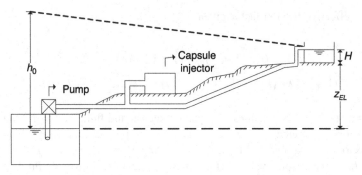

Figure 2.22. A capsule transporting system.

Solution. Considering water at 20°C and using Eq. (2.4b), the kinematic viscosity of water is $\nu = 1.012 \times 10^{-6}\, \mathrm{m^2/s}$. Using Eq. (2.38), the capsule velocity is obtained as

$$V_c = \frac{4a(1 + \beta)s_s Q_s}{\pi D^2 \{k^2 a - 2 s_c \theta[k(k + 2a) - 2\theta(2k + a - 2\theta)]\}}$$

$$V_c = \frac{4 \times 1.5(1 + 1.5) \times 1.75 \times 0.01}{3.14 \times 0.5^2 \begin{Bmatrix} 0.9^2 \times 1.5 - 2 \times 2.7 \times 0.022[0.9(0.9 + 2 \times 1.5) \\ -2 \times 0.022(2 \times 0.9 + 1.5 - 2 \times 0.022)] \end{Bmatrix}}$$

$$= 0.393\, \mathrm{m/s},$$

and $t_c = \dfrac{(1 + \beta)aD}{V_c} = \dfrac{((1 + 1.5) \times 1.5 \times 0.5)}{0.393} = 4.76\,\mathrm{s}.$

The friction factors f_b, f_c, and f_p are obtained by Eq. (2.6a) using $\mathbf{R} = V_b D/\nu$, $(1 - k)$ $(V_c - V_a)D/\nu$ and $(1 - k)V_a D/\nu$, respectively. The $\lambda = f_p/f_c$ is obtained iteratively as 0.983 with starting value as 1.

Using Eq. (2.41), $V_a = 0.203\,\mathrm{m^3/s}$, and Eq. (2.42), $V_b = 0.357\,\mathrm{m^3/s}$ can be calculated.

Thus, for calculated $\mathbf{R}$ values, $f_b = 0.0167$, $f_c = 0.0.0316$, and $f_p = 0.0311$ are calculated.

Using Eq. (2.40), the effective friction factor f_e is obtained as 3.14 and the head loss in pipe:

$$h_f = \frac{8 f_e L Q_s^2}{\pi^2 g D^5} = \frac{8 \times 3.14 \times 4000 \times 0.01^2}{3.1415^2 \times 9.81 \times 0.5^5} = 3.32\,\mathrm{m}.$$

Using Eq. (2.43), the effective fluid discharge Q_e is calculated as $0.077\,\mathrm{m^3/s}$. Considering pump efficiency η as 75%, the power consumed in kwh $= \rho g Q_e h_f/ (1000\eta) = 1000 \times 9.81 \times 0.077 \times (3.32 + 20)/(1000 \times 0.75) = 23.55$ kwh.

EXERCISES

2.1. Calculate head loss in a 500-m-long CI pipe of diameter 0.4 m carrying a discharge of 0.2 m^3/s. Assume water temperature equal to 20°C.

2.2. Calculate form-resistance coefficient and form loss in the following pipe specials if the pipe discharge is 0.15 m^3/s:

(a) Pipe bend of 0.3-m diameter, bend radius of 1.0 m, and bend angle as 0.3 radians.

(b) A 2/3 open sluice valve of diameter 0.4 m.

(c) A gradual expansion fitting (enlarger) of end diameters of 0.2 m and 0.3 m with transition length of 0.5 m.

(d) An abrupt contraction transition from 0.4-m diameter to 0.2-m diameter.

2.3. The pump of a 500-m-long rising main develops a pressure head of 30 m. The main size is 0.3 m and carries a discharge of 0.15 m^3/s. A sluice valve is fitted in the main, and the main has a confusor outlet of size 0.2 m. Calculate terminal head.

2.4. Water is transported from a reservoir at higher elevation to another reservoir through a series of three pipes. The first pipe of 0.4-m diameter is 500 m long, the second pipe 600 m long, size 0.3 m, and the last pipe is 500 m long of diameter 0.2 m. If the elevation difference between two reservoirs is 30 m, calculate equivalent pipe size and flow in the pipe.

2.5. Water between two reservoirs is transmitted through two parallel pipes of length 800 m and 700 m having diameters of 0.3 m and 0.25 m, respectively. It the elevation difference between two reservoirs is 35 m, calculate the equivalent pipe diameter and the flow in the pipe. Neglect minor losses and water columns in reservoirs. The equivalent length of pipe can be assumed as 600 m.

2.6. A CI pumping main of 0.4 m in size and length 1500 m carries slurry of average sediment particle size of 0.2 mm with mass densities of particle and fluid ratio as 2.5. If the volumetric concentration of particles is 30% and average temperature of water 20°C, calculate total head loss in the pipe.

2.7. Solve Example 2.7 for cargo transport rate of 0.0150 through a 0.65 m poly(vinyl chloride) pipeline of length 5000 m. Consider any other data similar to Example 2.7.

REFERENCES

Colebrook, C.F. (1938–1939). Turbulent flow in pipes with particular reference to the transition region between smooth and rough pipe laws. *J. Inst. Civ. Engrs. London* 11, 133–156.

Moody, L.F. (1944). Friction factors for pipe flow. *Trans. ASME* 66, 671–678.

Stepanoff, A.J. (1969). *Gravity Flow of Solids and Transportation of Solids in Suspension.* John Wiley & Sons, New York.

Swamee, P.K. (1990). Form resistance equations for pipe flow. *Proc. National Symp. on Water Resource Conservation, Recycling and Reuse.* Indian Water Works Association, Nagpur, India, Feb. 1990, pp. 163–164.

Swamee, P.K. (1993). Design of a submarine oil pipeline. *J. Transp. Eng.* 119(1), 159–170.

Swamee, P.K. (1998). Design of pipelines to transport neutrally buoyant capsules. *J. Hydraul. Eng.* 124(11), 1155–1160.

Swamee, P.K. (2004). Improving design guidelines for class-I circular sedimentation tanks. *Urban Water Journal* 1(4), 309–314.

Swamee, P.K., and Jain, A.K. (1976). Explicit equations for pipe flow problems. *J. Hydraul. Eng. Div.* 102(5), 657–664.

Swamee, P.K., and Ojha, C.S.P. (1991). Drag coefficient and fall velocity of nonspherical particles. *J. Hydraul. Eng.* 117(5), 660–667.

Swamee, P.K., Garg, A., and Saha, S. (2005). Design of straight expansive pipe transitions. *J. Hydraul. Eng.* 131(4), 305–311.

Swamee, P.K., and Swamee, N. (2008). Full range pipe-flow equations. *J. Hydraul. Res.*, IAHR, 46(1) (in press).

3

PIPE NETWORK ANALYSIS

Design of Water Supply Pipe Networks. By Prabhata K. Swamee and Ashok K. Sharma
Copyright © 2008 John Wiley & Sons, Inc.

A pipe network is analyzed for the determination of the nodal pressure heads and the link discharges. As the discharges withdrawn from the network vary with time, it results in a continuous change in the nodal pressure heads and the link discharges. The network is analyzed for the worst combination of discharge withdrawals that may result in low-pressure heads in some areas. The network analysis is also carried out to find deficiencies of a network for remedial measures. It is also required to identify pipe links that would be closed in an emergency to meet firefighting demand in some localities due to limited capacity of the network. The effect of closure of pipelines on account of repair work is also studied by analyzing a network. Thus, network analysis is critical for proper operation and maintenance of a water supply system.

3.1. WATER DEMAND PATTERN

Houses are connected through service connections to water distribution network pipelines for water supply. From these connections, water is drawn as any of the water taps in a house opens, and the withdrawal stops as the tap closes. Generally, there are many taps in a house, thus the withdrawal rate varies in an arbitrary manner. The maximum withdrawal rates occur in morning and evening hours. The maximum discharge (withdrawal rate) in a pipe is a function of the number of houses (persons) served by the service connections. In the analysis and design of a pipe network, this maximum withdrawal rate is considered.

The service connections are taken at arbitrary spacing from a pipeline of a water supply network (Fig. 3.1a). It is not easy to analyze such a network unless simplifying assumptions are made regarding the withdrawal spacing. A conservative assumption is to consider the withdrawals to be lumped at the two end points of the pipe link. With this assumption, half of the withdrawal from the link is lumped at each node (Fig. 3.1b). A more realistic assumption is to consider the withdrawals to be distributed along the link (Fig. 3.1c). The current practice is to lump the discharges at the nodal points.

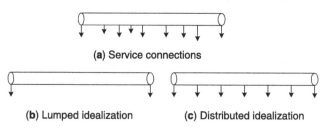

(a) Service connections

(b) Lumped idealization   (c) Distributed idealization

Figure 3.1. Withdrawal patterns.

3.2. HEAD LOSS IN A PIPE LINK

3.2.1. Head Loss in a Lumped Equivalent

Considering q to be the withdrawal rate per unit length of a link, the total withdrawal rate from the pipe of length L is qL. Lumping the discharges at the two pipe link ends, the head loss on account of surface resistance is given by

$$h_f = \frac{8fLQ^2}{\pi^2 gD^5}\left(1 - \frac{qL}{2Q}\right)^2,$$ (3.1a)

where $Q =$ discharge entering the link. Equation (2.6b) can be used for calculation of f, where Reynolds number $\mathbf{R}$ is to be taken as

$$\mathbf{R} = \frac{4(Q - 0.5qL)}{\pi v D}.$$ (3.1b)

3.2.2. Head Loss in a Distributed Equivalent

The discharge at a distance x from the pipe link entrance end is $Q - qx$, and the corresponding head loss in a distance dx is given by

$$dh_f = \frac{8f(Q - qx)^2 dx}{\pi^2 gD^5}.$$ (3.2)

Integrating Eq. (3.2) between the limits $x = 0$ and L, the following equation is obtained:

$$h_f = \frac{8fLQ^2}{\pi^2 gD^5}\left[1 - \frac{qL}{Q} + \frac{1}{3}\left(\frac{qL}{Q}\right)^2\right].$$ (3.3)

For the calculation of f, $\mathbf{R}$ can be obtained by Eq. (3.1b).

Example 3.1. Calculate head loss in a CI pipe of length $L = 500$ m, discharge Q at entry node $= 0.1\,\mathrm{m}^3/\mathrm{s}$, pipe diameter $D = 0.25$ m if the withdrawal (Fig. 3.1) is at a rate of $0.0001\,\mathrm{m}^3/\mathrm{s}$ per meter length. Assume (a) lumped idealized withdrawal and (b) distributed idealized withdrawal patterns.

Solution. Using Table 2.1 and Eq. (2.4b), roughness height ε of CI pipe $= 0.25$ mm and kinematic viscosity v of water at $20°C = 1.0118 \times 10^{-6}\,\mathrm{m}^2/\mathrm{s}$.

(a) Lumped idealized withdrawal (Fig. 3.1b): Applying Eq. (3.2),

$$\mathbf{R} = \frac{4(Q - 0.5qL)}{\pi v D} = \frac{4(0.1 - 0.5 \times 0.0001 \times 500)}{3.1415 \times 1.01182 \times 10^{-6} \times 0.25} = 377{,}513.$$

Using Eq. (2.6a) for $\mathbf{R} = 377{,}513$, the friction factor $f = 0.0205$.
Using Eq. (3.1a), the head loss

$$h_f = \frac{8fLQ^2}{\pi^2 g D^5}\left(1 - \frac{qL}{2Q}\right)^2 = \frac{8 \times 0.0205 \times 500 \times 0.1^2}{3.1415^2 \times 9.81 \times 0.25^5}\left(1 - \frac{0.0001 \times 500}{2 \times 0.1}\right)^2$$

$$= 4.889\,\text{m}.$$

(b) Distributed idealized withdrawal (Fig. 3.1c): As obtained by Eq. (3.1b), $\mathbf{R} = 377{,}513$, and $f = 0.0205$. Using Eq. (3.3), the head loss is

$$h_f = \frac{8fLQ^2}{\pi^2 g D^5}\left[1 - \frac{qL}{Q} + \frac{1}{3}\left(\frac{qL}{Q}\right)^2\right] = \frac{8 \times 0.0205 \times 500 \times 0.1^2}{3.1415^2 \times 9.81 \times 0.25^5}$$

$$\times \left[1 - \frac{0.0001 \times 500}{0.1} + \frac{1}{3}\left(\frac{0.0001 \times 500}{0.1}\right)^2\right]$$

$$= 5.069\,\text{m}.$$

3.3. ANALYSIS OF WATER TRANSMISSION LINES

Water transmission lines are long pipelines having no withdrawals. If water is carried by gravity, it is called a gravity main (see Fig. 3.2). In the analysis of a gravity main, it is

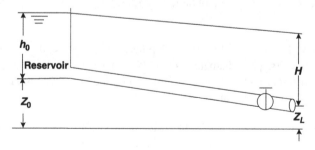

Figure 3.2. A gravity main.

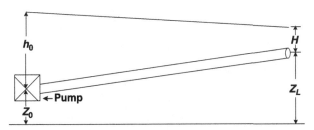

Figure 3.3. A pumping main.

required to find the discharge carried by the pipeline. The available head in the gravity main is $h_0 + z_0 - z_L$, and almost the entire head is lost in surface resistance. Thus,

$$h_0 + z_0 - z_L = \frac{8fLQ^2}{\pi^2 gD^5},\qquad (3.4)$$

where f is given by Eq. (2.6b). It is difficult to solve Eq. (2.6a) and Eq. (3.4), however, using Eq. (2.21a) the discharge is obtained as:

$$Q = -0.965D^2 \left[\frac{gD(h_0 + z_0 - z_L)}{L}\right]^{0.5} \ln\left\{\frac{\varepsilon}{3.7D} + \frac{1.78v}{D}\left[\frac{L}{gD(h_0 + z_0 - z_L)}\right]^{0.5}\right\}.$$
$$(3.5)$$

If water is pumped from an elevation z_0 to z_L, the pipeline is called a pumping main (Fig. 3.3). In the analysis of a pumping main, one is required to find the pumping head h_0 for a given discharge Q. This can be done by a combination of Eqs. (2.2b), (2.2d), and (2.19b). That is,

$$h_0 = H + z_L - z_0 + \left(k_f + \frac{fL}{D}\right)\frac{8Q^2}{\pi^2 gD^4},\qquad (3.6)$$

where H = the terminal head (i.e., the head at $x = L$.). Neglecting the form loss for a long pumping main, Eq. (3.6) reduces to

$$h_0 = H + z_L - z_0 + \frac{8fLQ^2}{\pi^2 gD^5}.\qquad (3.7)$$

Example 3.2. For a polyvinyl chloride (PVC) gravity main (Fig. 3.2), calculate flow in a pipe of length 600 m and size 0.3 m. The elevations of reservoir and outlet are 20 m and 10 m, respectively. The water column in reservoir is 5 m, and a terminal head of 5 m is required at outlet.

Solution. At 20°C, $v = 1.012 \times 10^{-6}$; and from Table 2.1, $\varepsilon = 0.05$ mm. With $L = 600$ m, $h_0 = 5$ m, $z_o = 20$ m, $z_L = 10$ m, and $D = 0.3$ m Eq. (3.5) gives

$$Q = -0.965D^2 \left[\frac{gD(h_0 + z_0 - z_L)}{L}\right]^{0.5} \ln\left\{\frac{\varepsilon}{3.7D} + \frac{1.78v}{D}\left[\frac{L}{gD(h_0 + z_0 - z_L)}\right]^{0.5}\right\}$$

$Q = 0.227 \text{ m}^3/\text{s}.$

3.4. ANALYSIS OF DISTRIBUTION MAINS

A pipeline in which there are multiple withdrawals is called a distribution main. In a distribution main, water may flow on account of gravity (Fig. 3.4) or by pumping (Fig. 3.5) with withdrawals $q_1, q_2, q_3, \ldots, q_n$ at the nodal points $1, 2, 3, \ldots, n$. In the analysis of a distribution main, one is required to find the nodal heads $h_1, h_2, h_3, \ldots, H$. The discharge flowing in the jth pipe link Q_j is given by

$$Q_j = \sum_{p=0}^{j} q_{n-p} \tag{3.8}$$

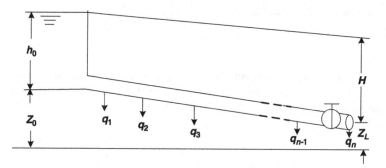

Figure 3.4. A gravity sustained distribution main.

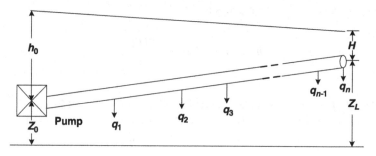

Figure 3.5. A pumping distribution main.

and the nodal head h_j is given by

$$h_j = h_0 + z_0 - z_i - \frac{8}{\pi^2 g}\sum_{p=1}^{j}\left(\frac{f_p L_p}{D_p} + k_{fp}\right)\frac{Q_p^2}{D_p^4},$$ (3.9)

where the suffix p stands for pth pipe link. For a gravity main, $h_0 =$ head in the intake chamber and for a pumping main it is the pumping head. The value of f for pth pipe link is given by

$$f_p = 1.325\left\{\ln\left[\frac{\varepsilon}{3.7D} + 4.618\left(+\frac{\nu D_p}{Q_p}\right)^{0.9}\right]\right\}^{-2}.$$ (3.10)

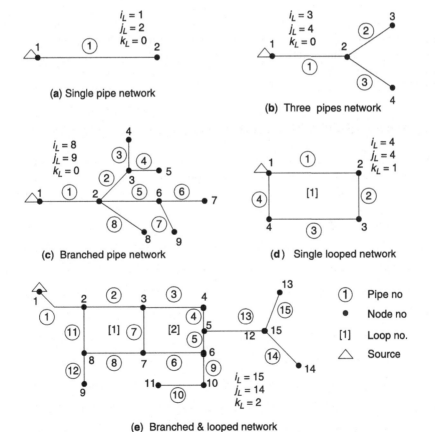

(a) Single pipe network

(b) Three pipes network

(c) Branched pipe network

(d) Single looped network

(e) Branched & looped network

Figure 3.6. Pipe node connectivity.

3.5. PIPE NETWORK GEOMETRY

The water distribution networks have mainly the following three types of configurations:

- Branched or tree-like configuration
- Looped configuration
- Branched and looped configuration

Figure 3.6a–c depicts some typical branched networks. Figure 3.6d is single looped network and Figure 3.6e represents a branched and looped configuration. It can be seen from the figures that the geometry of the networks has a relationship between total number of pipes (i_L), total number of nodes (j_L), and total number of *primary* loops (k_L). Figure 3.6a represents a system having a single pipeline and two nodes. Figure 3.6b has three pipes and four nodes, and Fig. 3.6c has eight pipes and nine nodes. Similarly, Fig. 3.6d has four pipes, four nodes, and one closed loop. Figure 3.6e has 15 pipes, 14 nodes, and 2 primary loops. The primary loop is the smallest closed loop while higher-order loop or secondary loop consists of more than one primary loop. For example, in Fig. 3.6e, pipes 2, 7, 8, and 11 form a primary loop and on the other hand pipes 2, 3, 4, 5, 6, 8, and 11 form a secondary loop. All the networks satisfy a geometry relationship that the total number of pipes are equal to total number of nodes + total number of loops − 1. Thus, in a network, $i_L = j_L + k_L - 1$.

3.6. ANALYSIS OF BRANCHED NETWORKS

A branched network, or a tree network, is a distribution system having no loops. Such a network is commonly used for rural water supply. The simplest branched network is a radial network consisting of several distribution mains emerging out from a common input point (see Fig. 3.7). The pipe discharges can be determined for each radial branch using Eq. (3.8), rewritten as:

$$Q_{ij} = \sum_{p=o}^{j} q_{i,n-p}. \qquad (3.11)$$

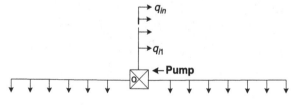

Figure 3.7. A radial network.

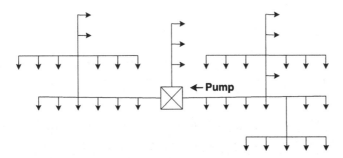

Figure 3.8. A branched network.

The power consumption will depend on the total discharge pumped Q_I given by

$$Q_T = \sum_{i=1}^{i_L} Q_{oi}. \tag{3.12}$$

In a typical branched network (Fig. 3.8), the pipe discharges can be obtained by adding the nodal discharges and tracing the path from tail end to the input point until all the tail ends are covered. The nodal heads can be found by proceeding from the input point and adding the head losses (friction loss and form loss) in each link until a tail end is reached. The process has to be repeated until all tail ends are covered. Adding the terminal head to the maximum head loss determines the pumping head.

3.7. ANALYSIS OF LOOPED NETWORKS

A pipe network in which there are one or more closed loops is called a looped network. A typical looped network is shown in Fig. 3.9. Looped networks are preferred from the reliability point of view. If one or more pipelines are closed for repair, water can still reach the consumer by a circuitous route incurring more head loss. This feature is

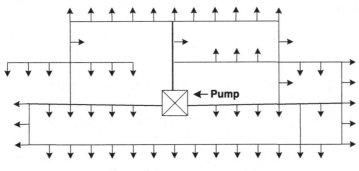

Figure 3.9. Looped network.

absent in a branched network. With the changing demand pattern, not only the magnitudes of the discharge but also the flow directions change in many links. Thus, the flow directions go on changing in a large looped network.

Analysis of a looped network consists of the determination of pipe discharges and the nodal heads. The following laws, given by Kirchhoff, generate the governing equations:

- The algebraic sum of inflow and outflow discharges at a node is zero; and
- The algebraic sum of the head loss around a loop is zero.

On account of nonlinearity of the resistance equation, it is not possible to solve network analysis problems analytically. Computer programs have been written to analyze looped networks of large size involving many input points like pumping stations and elevated reservoirs.

The most commonly used looped network analysis methods are described in detail in the following sections.

3.7.1. Hardy Cross Method

Analysis of a pipe network is essential to understand or evaluate a pipe network system. In a branched pipe network, the pipe discharges are unique and can be obtained simply by applying discharge continuity equations at all the nodes. However, in case of a looped pipe network, the number of pipes is too large to find the pipe discharges by merely applying discharge continuity equations at nodes. The analysis of looped network is carried out by using additional equations found from the fact that while traversing along a loop, as one reaches at the starting node, the net head loss is zero. The analysis of looped network is involved, as the loop equations are nonlinear in discharge.

Hardy Cross (1885–1951), who was professor of civil engineering at the University of Illinois, Urbana-Champaign, presented in 1936 a method for the analysis of looped pipe network with specified inflow and outflows (Fair et al., 1981). The method is based on the following basic equations of continuity of flow and head loss that should be satisfied:

1. The sum of inflow and outflow at a node should be equal:

$$\sum Q_i = q_j \qquad \text{for all nodes } j = 1, 2, 3, \ldots, j_L, \tag{3.13}$$

where Q_i is the discharge in pipe i meeting at node (junction) j, and q_j is nodal withdrawal at node j.

2. The algebraic sum of the head loss in a loop must be equal to zero:

$$\sum_{\text{loop } k} K_i Q_i |Q_i| = 0 \qquad \text{for all loops } k = 1, 2, 3, \ldots, k_L, \tag{3.14}$$

where

$$K_i = \frac{8 f_i L_i}{\pi^2 g D_i^5},\tag{3.15}$$

where i = pipe link number to be summed up in the loop k.

In general, it is not possible to satisfy Eq. (3.14) with the initially assumed pipe discharges satisfying nodal continuity equation. The discharges are modified so that Eq. (3.14) becomes closer to zero in comparison with initially assumed discharges. The modified pipe discharges are determined by applying a correction ΔQ_k to the initially assumed pipe flows. Thus,

$$\sum_{\text{loop } k} K_i (Q_i + \Delta Q_k)|(Q_i + \Delta Q_k)| = 0.\tag{3.16}$$

Expanding Eq. (3.16) and neglecting second power of ΔQ_k and simplifying Eq. (3.16), the following equation is obtained:

$$\Delta Q_k = -\frac{\displaystyle\sum_{\text{loop } k} K_i Q_i |Q_i|}{2 \displaystyle\sum_{\text{loop } k} K_i |Q_i|}.\tag{3.17}$$

Knowing ΔQ_k, the corrections are applied as

$$Q_{i\,new} = Q_{i\,old} + \Delta Q_k \qquad \text{for all } k.\tag{3.18}$$

The overall procedure for the looped network analysis can be summarized in the following steps:

1. Number all the nodes and pipe links. Also number the loops. For clarity, pipe numbers are circled and the loop numbers are put in square brackets.
2. Adopt a sign convention that a pipe discharge is positive if it flows from a lower node number to a higher node number, otherwise negative.
3. Apply nodal continuity equation at all the nodes to obtain pipe discharges. Starting from nodes having two pipes with unknown discharges, assume an arbitrary discharge (say 0.1 m³/s) in one of the pipes and apply continuity equation (3.13) to obtain discharge in the other pipe. Repeat the procedure until all the pipe flows are known. If there exist more than two pipes having unknown discharges, assume arbitrary discharges in all the pipes except one and apply continuity equation to get discharge in the other pipe. The total number of pipes having arbitrary discharges should be equal to the total number of primary loops in the network.

4. Assume friction factors $f_i = 0.02$ in all pipe links and compute corresponding K_i using Eq. (3.15). However, f_i can be calculated iteratively using Eq. (2.6a).

5. Assume loop pipe flow sign convention to apply loop discharge corrections; generally, clockwise flows positive and counterclockwise flows negative are considered.

6. Calculate ΔQ_k for the existing pipe flows and apply pipe corrections algebraically.

7. Apply the similar procedure in all the loops of a pipe network.

Repeat steps 6 and 7 until the discharge corrections in all the loops are relatively very small.

Example 3.3. A single looped network as shown in Fig. 3.10 has to be analyzed by the Hardy Cross method for given inflow and outflow discharges. The pipe diameters D and lengths L are shown in the figure. Use Darcy–Weisbach head loss–discharge relationship assuming a constant friction factor $f = 0.02$.

Solution

    Step 1: The pipes, nodes, and loop are numbered as shown in Fig. 3.10.

    Step 2: Adopt the following sign conventions:

    A positive pipe discharge flows from a lower node to a higher node.

    Inflow into a node is positive withdrawal negative.

    In the summation process of Eq. (3.13), a positive sign is used if the discharge in the pipe is out of the node under consideration. Otherwise, a negative sign will be

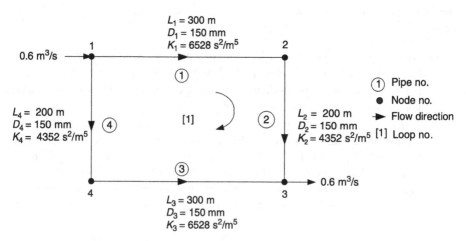

Figure 3.10. Single looped network.

attached to the discharge. For example in Fig. 3.10 at node 2, the flow in pipe 1 is toward node 2, thus the Q_1 at node 2 will be negative while applying Eq. (3.13).

Step 3: Apply continuity equation to obtain pipe discharges. Scanning the figure for node 1, the discharges in pipes 1 and 4 are unknown. The nodal inflow q_1 is $0.6\,\mathrm{m^3/s}$ and nodal outflow $q_3 = -0.6\,\mathrm{m^3/s}$. The q_2 and q_3 are zero. Assume an arbitrary flow of $0.1\,\mathrm{m^3/s}$ in pipe 1 ($Q_1 = 0.1\,\mathrm{m^3/s}$), meaning thereby that the flow in pipe 1 is from node 1 to node 2. The discharge in pipe Q_4 can be calculated by applying continuity equation at node 1 as

$$Q_1 + Q_4 = q_1 \text{ or } Q_4 = q_1 - Q_1, \text{ hence } Q_4 = 0.6 - 0.1 = 0.5\,\mathrm{m^3/s}.$$

The discharge in pipe 4 is positive meaning thereby that the flow will be from node 1 to node 4 (toward higher numbering node).

Also applying continuity equation at node 2:

$$-Q_1 + Q_2 = q_2 \text{ or } Q_2 = q_2 + Q_1, \text{ hence } Q_2 = 0 + 0.1 = 0.1\,\mathrm{m^3/s}.$$

Similarly applying continuity equation at node 3, flows in pipe $Q_3 = -0.5\,\mathrm{m^3/s}$ can be calculated. The pipe flow directions for the initial flows are shown in the figure.

Step 4: For assumed pipe friction factors $f_i = 0.02$, the calculated K values as $K = 8fL/\pi^2 gD^5$ for all the pipes are given in the Fig. 3.10.

Step 5: Adopted clockwise flows in pipes positive and counterclockwise flows negative.

Step 6: The discharge correction for the initially assumed pipe discharges can be calculated as follows:

Iteration 1

Pipe	Flow in Pipe Q $(\mathrm{m^3/s})$	K $(\mathrm{s^2/m^5})$	$KQ\lvert Q\rvert$ (m)	$2K\lvert Q\rvert$ $(\mathrm{s/m^2})$	Corrected Flow $Q = Q + \Delta Q$ $(\mathrm{m^3/s})$
1	0.10	6528.93	65.29	1305.79	0.30
2	0.10	4352.62	43.53	870.52	0.30
3	−0.50	6528.93	−1632.23	6528.93	−0.30
4	−0.50	4352.62	−1088.15	4352.62	−0.30
Total			−2611.57	13,057.85	
ΔQ			$-(-2611.57/13{,}057) = 0.20\,\mathrm{m^3/s}$		

Repeat the process again for the revised pipe discharges as the discharge correction is quite large in comparison to pipe flows:

Iteration 2

Pipe	Flow in Pipe Q (m³/s)	K (s²/m⁵)	$KQ\|Q\|$ (m)	$2K\|Q\|$ (s/m²)	Corrected Flow $Q = Q + \Delta Q$ (m³/s)
1	0.30	6528.93	587.60	3917.36	0.30
2	0.30	4352.62	391.74	2611.57	0.30
3	−0.30	6528.93	−587.60	3917.36	−0.30
4	−0.30	4352.62	−391.74	2611.57	−0.30
Total			0.00	13,057.85	
ΔQ			$= -(0/13{,}057) = 0.00 \text{ m}^3/\text{s}$		

As the discharge correction $\Delta Q = 0$, the final discharges are

$Q_1 = 0.3 \text{ m}^3/\text{s}$
$Q_2 = 0.3 \text{ m}^3/\text{s}$
$Q_3 = 0.3 \text{ m}^3/\text{s}$
$Q_4 = 0.3 \text{ m}^3/\text{s}.$

Example 3.4. The pipe network of two loops as shown in Fig. 3.11 has to be analyzed by the Hardy Cross method for pipe flows for given pipe lengths L and pipe diameters D. The nodal inflow at node 1 and nodal outflow at node 3 are shown in the figure. Assume a constant friction factor $f = 0.02$.

Solution. Applying steps 1–7, the looped network analysis can be conducted as illustrated in this example. The K values for Darcy–Weisbach head loss–discharge relationship are also given in Fig. 3.11.

To obtain initial pipe discharges applying nodal continuity equation, the arbitrary pipe discharges equal to the total number of loops are assumed. The total number of

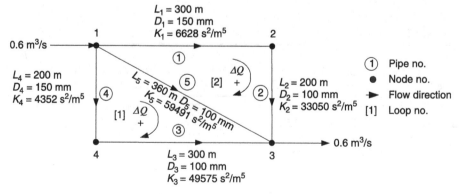

Figure 3.11. Looped network.

loops in a network can be obtained from the following geometric relationship:

Total number of loops = Total number of pipes − Total number of nodes + 1

Moreover, in this example there are five pipes and four nodes. One can apply nodal continuity equation at three nodes (total number of nodes − 1) only as, on the outcome of the other nodal continuity equations, the nodal continuity at the fourth node (last node) automatically gets satisfied. In this example there are five unknown pipe discharges, and to obtain pipe discharges there are three known nodal continuity equations and two loop head-loss equations.

To apply continuity equation for initial pipe discharges, the discharges in pipes 1 and 5 equal to $0.1 \text{ m}^3/\text{s}$ are assumed. The obtained discharges are

$Q_1 = 0.1 \text{ m}^3/\text{s}$ (flow from node 1 to node 2)
$Q_2 = 0.1 \text{ m}^3/\text{s}$ (flow from node 2 to node 3)
$Q_3 = 0.4 \text{ m}^3/\text{s}$ (flow from node 4 to node 3)
$Q_4 = 0.4 \text{ m}^3/\text{s}$ (flow from node 1 to node 4)
$Q_5 = 0.1 \text{ m}^3/\text{s}$ (flow from node 1 to node 3)

The discharge correction ΔQ is applied in one loop at a time until the ΔQ is very small in all the loops. ΔQ in Loop 1 (loop pipes 3, 4, and 5) and corrected pipe discharges are given in the following table:

Loop 1: Iteration 1

Pipe	Flow in Pipe Q (m^3/s)	K (s^2/m^5)	$KQ\|Q\|$ (m)	$2K\|Q\|$ (s/m^2)	Corrected Flow $Q = Q + \Delta Q$ (m^3/s)
3	−0.40	49,576.12	−7932.18	39,660.89	−0.25
4	−0.40	4352.36	−696.38	3481.89	−0.25
5	0.10	59,491.34	594.91	11,898.27	0.25
Total			−8033.64	55,041.05	
ΔQ			$0.15 \text{ m}^3/\text{s}$		

Thus the discharge correction ΔQ in loop 1 is $0.15 \text{ m}^3/\text{s}$. The discharges in loop pipes are corrected as shown in the above table. Applying the same methodology for calculating ΔQ for Loop 2:

Loop 2: Iteration 1

Pipe	Flow in Pipe Q (m^3/s)	K (s^2/m^5)	$KQ\|Q\|$ (m)	$2K\|Q\|$ (s/m^2)	Corrected Flow $Q = Q + \Delta Q$ (m^3/s)
1	0.10	6528.54	65.29	1305.71	0.19
2	0.10	33,050.74	330.51	6610.15	0.19
5	−0.25	59,491.34	−3598.93	29,264.66	−0.16
Total			−3203.14	37,180.52	
ΔQ			$0.09 \text{ m}^3/\text{s}$		

The process of discharge correction is in repeated until the ΔQ value is very small as shown in the following tables:

Loop 1: Iteration 2

Pipe	Flow in Pipe Q (m^3/s)	K (s^2/m^5)	$KQ\lvert Q\rvert$ (m)	$2K\lvert Q\rvert$ (s/m^2)	Corrected Flow $Q = Q + \Delta Q$ (m^3/s)
3	−0.25	49,576.12	−3098.51	24,788.06	−0.21
4	−0.25	4352.36	−272.02	2176.18	−0.21
5	0.16	59,491.34	1522.98	19,037.23	0.20
Total			−1847.55	46,001.47	
ΔQ			0.04 m^3/s		

Loop 2: Iteration 2

Pipe	Flow in Pipe Q (m^3/s)	K (s^2/m^5)	$KQ\lvert Q\rvert$ (m)	$2K\lvert Q\rvert$ (s/m^2)	Corrected Flow $Q = Q + \Delta Q$ (m^3/s)
1	0.19	6528.54	226.23	2430.59	0.21
2	0.19	33,050.74	1145.28	12,304.85	0.21
5	−0.20	59,491.34	−2383.53	23,815.92	−0.17
Total			−1012.02	38,551.36	
ΔQ			0.03 m^3/s		

Loop 1: Iteration 3

Pipe	Flow in Pipe Q (m^3/s)	K (s^2/m^5)	$KQ\lvert Q\rvert$ (m)	$2K\lvert Q\rvert$ (s/m^2)	Corrected Flow $Q = Q + \Delta Q$ (m^3/s)
3	−0.210	49,576.12	−2182.92	20,805.82	−0.197
4	−0.210	4352.36	−191.64	1826.57	−0.197
5	0.174	59,491.34	1799.33	20,692.47	0.187
Total			−575.23	43,324.86	
ΔQ			0.01 m^3/s		

Loop 2: Iteration 3

Pipe	Flow in Pipe Q (m^3/s)	K (s^2/m^5)	$KQ\lvert Q\rvert$ (m)	$2K\lvert Q\rvert$ (s/m^2)	Corrected Flow $Q = Q + \Delta Q$ (m^3/s)
1	0.212	6528.54	294.53	2773.35	0.220
2	0.212	33,050.74	1491.07	14,040.10	0.220
5	−0.187	59,491.34	−2084.55	22,272.21	−0.180
Total			−298.95	39,085.67	
ΔQ			0.008 m^3/s		

Loop 1: Iteration 4

Pipe	Flow in Pipe Q (m^3/s)	K (s^2/m^5)	$KQ\|Q\|$ (m)	$2K\|Q\|$ (s/m^2)	Corrected Flow $Q = Q + \Delta Q$ (m^3/s)
3	−0.197	49,576.12	−1915.41	19,489.36	−0.193
4	−0.197	4352.36	−168.16	1711.00	−0.193
5	0.180	59,491.34	1917.68	21,362.18	0.183
Total			−165.89	42,562.54	
ΔQ			0.004 m^3/s		

Loop 2: Iteration 4

Pipe	Flow in Pipe Q (m^3/s)	K (s^2/m^5)	$KQ\|Q\|$ (m)	$2K\|Q\|$ (s/m^2)	Corrected Flow $Q = Q + \Delta Q$ (m^3/s)
1	0.220	6528.54	316.13	2873.22	0.222
2	0.220	33,050.74	1600.39	14,545.68	0.222
5	−0.183	59,491.34	−2001.85	21,825.91	−0.181
Total			−85.33	39,244.81	
ΔQ			0.002 m^3/s		

Loop 1: Iteration 5

Pipe	Flow in Pipe Q (m^3/s)	K (s^2/m^5)	$KQ\|Q\|$ (m)	$2K\|Q\|$ (s/m^2)	Corrected Flow $Q = Q + \Delta Q$ (m^3/s)
3	−0.193	49,576.12	−1840.21	19,102.92	−0.192
4	−0.193	4352.36	−161.55	1677.07	−0.192
5	0.181	59,491.34	1954.67	21,567.21	0.182
Total			−47.09	42,347.21	
ΔQ			0.001 m^3/s		

Loop 2: Iteration 5

Pipe	Flow in Pipe Q (m^3/s)	K (s^2/m^5)	$KQ\|Q\|$ (m)	$2K\|Q\|$ (s/m^2)	Corrected Flow $Q = Q + \Delta Q$ (m^3/s)
1	0.222	6528.54	322.40	2901.61	0.223
2	0.222	33,050.74	1632.17	14,689.40	0.223
5	-0.182	59,491.34	−1978.73	21,699.52	−0.182
Total			−24.15	39,290.53	
ΔQ			0.001 m^3/s		

The discharge corrections in the loops are very small after five iterations, thus the final pipe discharges in the looped pipe network in Fig. 3.11 are

$$Q_1 = 0.223 \text{ m}^3/\text{s}$$
$$Q_2 = 0.223 \text{ m}^3/\text{s}$$
$$Q_3 = 0.192 \text{ m}^3/\text{s}$$
$$Q_4 = 0.192 \text{ m}^3/\text{s}$$
$$Q_5 = 0.182 \text{ m}^3/\text{s}$$

3.7.2. Newton–Raphson Method

The pipe network can also be analyzed using the Newton–Raphson method, where unlike the Hardy Cross method, the entire network is analyzed altogether. The Newton–Raphson method is a powerful numerical method for solving systems of nonlinear equations. Suppose that there are three nonlinear equations $F_1(Q_1, Q_2, Q_3) = 0$, $F_2(Q_1, Q_2, Q_3) = 0$, and $F_3(Q_1, Q_2, Q_3) = 0$ to be solved for Q_1, Q_2, and Q_3. Adopt a starting solution (Q_1, Q_2, Q_3). Also consider that $(Q_1 + \Delta Q_1, Q_2 + \Delta Q_2, Q_3 + \Delta Q_3)$ is the solution of the set of equations. That is,

$$F_1(Q_1 + \Delta Q_1, Q_2 + \Delta Q_2, Q_3 + \Delta Q_3) = 0$$
$$F_2(Q_1 + \Delta Q_1, Q_2 + \Delta Q_2, Q_3 + \Delta Q_3) = 0 \qquad (3.19a)$$
$$F_3(Q_1 + \Delta Q_1, Q_2 + \Delta Q_2, Q_3 + \Delta Q_3) = 0.$$

Expanding the above equations as Taylor's series,

$$F_1 + [\partial F_1/\partial Q_1]\Delta Q_1 + [\partial F_1/\partial Q_2]\Delta Q_2 + [\partial F_1/\partial Q_3]\Delta Q_3 = 0$$
$$F_2 + [\partial F_2/\partial Q_1]\Delta Q_1 + [\partial F_2/\partial Q_2]\Delta Q_2 + [\partial F_2/\partial Q_3]\Delta Q_3 = 0 \qquad (3.19b)$$
$$F_3 + [\partial F_3/\partial Q_1]\Delta Q_1 + [\partial F_3/\partial Q_2]\Delta Q_2 + [\partial F_3/\partial Q_3]\Delta Q_3 = 0.$$

Arranging the above set of equations in matrix form,

$$\begin{bmatrix} \partial F_1/\partial Q_1 & \partial F_1/\partial Q_2 & \partial F_1/\partial Q_3 \\ \partial F_2/\partial Q_1 & \partial F_2/\partial Q_2 & \partial F_2/\partial Q_3 \\ \partial F_3/\partial Q_1 & \partial F_3/\partial Q_2 & \partial F_3/\partial Q_3 \end{bmatrix} \begin{bmatrix} \Delta Q_1 \\ \Delta Q_2 \\ \Delta Q_3 \end{bmatrix} = - \begin{bmatrix} F_1 \\ F_2 \\ F_3 \end{bmatrix}. \qquad (3.19c)$$

Solving Eq. (3.19c),

$$\begin{bmatrix} \Delta Q_1 \\ \Delta Q_2 \\ \Delta Q_3 \end{bmatrix} = - \begin{bmatrix} \partial F_1/\partial Q_1 & \partial F_1/\partial Q_2 & \partial F_1/\partial Q_3 \\ \partial F_2/\partial Q_1 & \partial F_2/\partial Q_2 & \partial F_2/\partial Q_3 \\ \partial F_3/\partial Q_1 & \partial F_3/\partial Q_2 & \partial F_3/\partial Q_3 \end{bmatrix}^{-1} \begin{bmatrix} F_1 \\ F_2 \\ F_3 \end{bmatrix}. \qquad (3.20)$$

Knowing the corrections, the discharges are improved as

$$
\begin{bmatrix} Q_1 \\ Q_2 \\ Q_3 \end{bmatrix} = \begin{bmatrix} Q_1 \\ Q_2 \\ Q_3 \end{bmatrix} + \begin{bmatrix} \Delta Q_1 \\ \Delta Q_2 \\ \Delta Q_3 \end{bmatrix}.
\tag{3.21}
$$

It can be seen that for a large network, it is time consuming to invert the matrix again and again. Thus, the inverted matrix is preserved and used for at least three times to obtain the corrections.

The overall procedure for looped network analysis by the Newton–Raphson method can be summarized in the following steps:

Step 1: Number all the nodes, pipe links, and loops.

Step 2: Write nodal discharge equations as

$$
F_j = \sum_{n=1}^{j_n} Q_{jn} - q_j = 0 \quad \text{for all nodes} - 1,
\tag{3.22}
$$

where Q_{jn} is the discharge in nth pipe at node j, q_j is nodal withdrawal, and j_n is the total number of pipes at node j.

Step 3: Write loop head-loss equations as

$$
F_k = \sum_{n=1}^{k_n} K_n Q_{kn} |Q_{kn}| = 0 \quad \text{for all the loops } (n = 1, k_n).
\tag{3.23}
$$

where K_n is total pipes in kth loop.

Step 4: Assume initial pipe discharges $Q_1, Q_2, Q_3, \ldots$ satisfying continuity equations.

Step 5: Assume friction factors $f_i = 0.02$ in all pipe links and compute corresponding K_i using Eq. (3.15).

Step 6: Find values of partial derivatives $\partial F_n / \partial Q_i$ and functions F_n, using the initial pipe discharges Q_i and K_i.

Step 7: Find ΔQ_i. The equations generated are of the form $Ax = b$, which can be solved for ΔQ_i.

Step 8: Using the obtained ΔQ_i values, the pipe discharges are modified and the process is repeated again until the calculated ΔQ_i values are very small.

Example 3.5. The configuration of Example 3.3 is considered in this example for illustrating the use of the Newton–Raphson method. For convenience, Fig. 3.10 is repeated as Fig. 3.12.

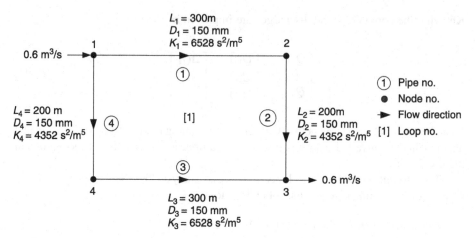

Figure 3.12. Single looped network.

Solution. The nodal discharge functions F are

$$F_1 = Q_1 + Q_4 - 0.6 = 0$$
$$F_2 = -Q_1 + Q_2 = 0$$
$$F_3 = Q_2 + Q_3 - 0.6 = 0,$$

and loop head-loss function

$$F_4 = 6528|Q_1|Q_1 + 4352|Q_2|Q_2 - 6528|Q_3|Q_3 - 4352|Q_2|Q_2 = 0.$$

The derivatives are

$$\partial F_1/\partial Q_1 = 1 \qquad \partial F_1/\partial Q_2 = 0 \qquad \partial F_1/\partial Q_3 = 0 \qquad \partial F_1/\partial Q_4 = 1$$
$$\partial F_2/\partial Q_1 = -1 \qquad \partial F_2/\partial Q_2 = 1 \qquad \partial F_2/\partial Q_3 = 0 \qquad \partial F_2/\partial Q_4 = 0$$
$$\partial F_3/\partial Q_1 = 0 \qquad \partial F_3/\partial Q_2 = 1 \qquad \partial F_3/\partial Q_3 = 1 \qquad \partial F_3/\partial Q_4 = 0$$
$$\partial F_4/\partial Q_1 = 6528Q_1 \quad \partial F_4/\partial Q_2 = 4352Q_2 \quad \partial F_4/\partial Q_3 = -6528Q_3 \quad \partial F_4/\partial Q_4 = -4352Q_4$$

The generated equations are assembled in the following matrix form:

$$
\begin{bmatrix} \Delta Q_1 \\ \Delta Q_2 \\ \Delta Q_3 \\ \Delta Q_4 \end{bmatrix} = -
\begin{bmatrix}
\partial F_1/\partial Q_1 & \partial F_1/\partial Q_2 & \partial F_1/\partial Q_3 & \partial F_1/\partial Q_4 \\
\partial F_2/\partial Q_1 & \partial F_2/\partial Q_2 & \partial F_2/\partial Q_3 & \partial F_2/\partial Q_4 \\
\partial F_3/\partial Q_1 & \partial F_3/\partial Q_2 & \partial F_3/\partial Q_3 & \partial F_3/\partial Q_4 \\
\partial F_4/\partial Q_1 & \partial F_4/\partial Q_2 & \partial F_4/\partial Q_3 & \partial F_4/\partial Q_4
\end{bmatrix}^{-1}
\begin{bmatrix} F_1 \\ F_2 \\ F_3 \\ F_4 \end{bmatrix}.
$$

Substituting the derivatives, the following form is obtained:

$$
\begin{bmatrix} \Delta Q_1 \\ \Delta Q_2 \\ \Delta Q_3 \\ \Delta Q_4 \end{bmatrix} = -\begin{bmatrix} 1 & 0 & 0 & 1 \\ -1 & 1 & 0 & 0 \\ 0 & 1 & 1 & 0 \\ 6528Q_1 & 4352Q_2 & -6528Q_3 & -4352Q_4 \end{bmatrix}^{-1} \begin{bmatrix} F_1 \\ F_2 \\ F_3 \\ F_4 \end{bmatrix}.
$$

Assuming initial pipe discharge in pipe 1 $Q_1 = 0.5\,\mathrm{m}^3/\mathrm{s}$, the other pipe discharges obtained by continuity equation are

$Q_2 = 0.5\,\mathrm{m}^3/\mathrm{s}$
$Q_3 = 0.1\,\mathrm{m}^3/\mathrm{s}$
$Q_4 = 0.1\,\mathrm{m}^3/\mathrm{s}$

Substituting these values in the above equation, the following form is obtained:

$$
\begin{bmatrix} \Delta Q_1 \\ \Delta Q_2 \\ \Delta Q_3 \\ \Delta Q_4 \end{bmatrix} = -\begin{bmatrix} 1 & 0 & 0 & 1 \\ -1 & 1 & 0 & 0 \\ 0 & 1 & 1 & 0 \\ 3264 & 2176 & -652.8 & -435.2 \end{bmatrix}^{-1} \begin{bmatrix} 0 \\ 0 \\ 0 \\ 2611.2 \end{bmatrix}.
$$

Using Gaussian elimination method, the solution is obtained as

$\Delta Q_1 = -0.2\,\mathrm{m}^3/\mathrm{s}$
$\Delta Q_2 = -0.2\,\mathrm{m}^3/\mathrm{s}$
$\Delta Q_3 = 0.2\,\mathrm{m}^3/\mathrm{s}$
$\Delta Q_4 = 0.2\,\mathrm{m}^3/\mathrm{s}$

Using these discharge corrections, the revised pipe discharges are

$Q_1 = Q_1 + \Delta Q_1 = 0.5 - 0.2 = 0.3\,\mathrm{m}^3/\mathrm{s}$
$Q_2 = Q_2 + \Delta Q_2 = 0.5 - 0.2 = 0.3\,\mathrm{m}^3/\mathrm{s}$
$Q_3 = Q_3 + \Delta Q_3 = 0.1 + 0.2 = 0.3\,\mathrm{m}^3/\mathrm{s}$
$Q_4 = Q_4 + \Delta Q_4 = 0.1 + 0.2 = 0.3\,\mathrm{m}^3/\mathrm{s}$

The process is repeated with the new pipe discharges. Revised values of F and derivative $\partial F/\partial Q$ values are obtained. Substituting the revised values, the following new solution is generated:

$$
\begin{bmatrix} \Delta Q_1 \\ \Delta Q_2 \\ \Delta Q_3 \\ \Delta Q_4 \end{bmatrix} = -\begin{bmatrix} 1 & 0 & 0 & 1 \\ -1 & 1 & 0 & 0 \\ 0 & 1 & 1 & 0 \\ 1958.4 & 1305.6 & -1958.4 & -1305.6 \end{bmatrix}^{-1} \begin{bmatrix} 0 \\ 0 \\ 0 \\ 0 \end{bmatrix}.
$$

As the right-hand side is operated upon null vector, all the discharge corrections $\Delta Q = 0$. Thus, the final discharges are

$Q_1 = 0.3 \ \text{m}^3/\text{s}$
$Q_2 = 0.3 \ \text{m}^3/\text{s}$
$Q_3 = 0.3 \ \text{m}^3/\text{s}$
$Q_4 = 0.3 \ \text{m}^3/\text{s}$

The solution obtained by the Newton–Raphson method is the same as that obtained by the Hardy Cross method (Example 3.3).

3.7.3. Linear Theory Method

The linear theory method is another looped network analysis method presented by Wood and Charles (1972). The entire network is analyzed altogether like the Newton–Raphson method. The nodal flow continuity equations are obviously linear but the looped head-loss equations are nonlinear. In the method, the looped energy equations are modified to be linear for previously known discharges and solved iteratively. The process is repeated until the two solutions are close to the allowable limits. The nodal discharge continuity equations are

$$F_j = \sum_{n=1}^{j_n} Q_{jn} - q_j = 0 \quad \text{for all nodes } -1. \tag{3.24}$$

Equation (3.24) can be generalized in the following form for the entire network:

$$F_j = \sum_{n=1}^{i_L} a_{jn} Q_{jn} - q_j = 0, \tag{3.25}$$

where a_{jn} is $+1$ if positive discharge flows in pipe n, -1 if negative discharge flows in pipe n, and 0 if pipe n is not connected to node j. The total pipes in the network are i_L. The loop head-loss equation are

$$F_k = \sum_{n=1}^{k_n} K_n |Q_{kn}| Q_{kn} = 0 \quad \text{for all the loops.} \tag{3.26}$$

The above equation can be linearized as

$$F_k = \sum_{n=1}^{k_n} b_{kn} Q_{kn} = 0, \tag{3.27}$$

where $b_{kn} = K_n|Q_{kn}|$ for initially known pipe discharges. Equation (3.27) can be generalized for the entire network in the following form:

$$F_k = \sum_{n=1}^{i_L} b_{kn} Q_{kn} = 0, \qquad (3.28)$$

where $b_{kn} = K_{kn}|Q_{kn}|$ if pipe n is in loop k or otherwise $b_{kn} = 0$. The coefficient b_{kn} is revised with current pipe discharges for the next iteration. This results in a set of linear equations, which are solved by using any standard method for solving linear equations. Thus, the total set of equations required for i_L unknown pipe discharges are

- Nodal continuity equations for $n_L - 1$ nodes
- Loop head-loss equations for k_L loops

The overall procedure for looped network analysis by the linear theory method can be summarized in the following steps:

Step 1: Number pipes, nodes, and loops.

Step 2: Write nodal discharge equations as

$$F_j = \sum_{n=1}^{j_n} Q_{jn} - q_j = 0 \quad \text{for all nodes} - 1,$$

where Q_{jn} is the discharge in the nth pipe at node j, q_j is nodal withdrawal, and j_n the total number of pipes at node j.

Step 3: Write loop head-loss equations as

$$F_k = \sum_{n=1}^{k_n} b_{kn} Q_{kn} = 0 \quad \text{for all the loops.}$$

Step 4: Assume initial pipe discharges $Q_1, Q_2, Q_3, \ldots$. It is not necessary to satisfy continuity equations.

Step 5: Assume friction factors $f_i = 0.02$ in all pipe links and compute corresponding K_i using Eq. (3.15).

Step 6: Generalize nodal continuity and loop equations for the entire network.

Step 7: Calculate pipe discharges. The equation generated is of the form $Ax = b$, which can be solved for Q_i.

Step 8: Recalculate coefficients b_{kn} from the obtained Q_i values.

Step 9: Repeat the process again until the calculated Q_i values in two consecutive iterations are close to predefined limits.

Example 3.6. For sake of comparison, the configuration of Example 3.3 is considered in this example. For convenience, Fig. 3.10 is repeated here as Fig. 3.13.

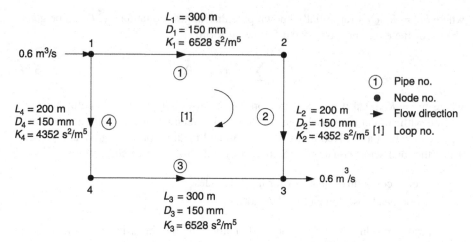

Figure 3.13. Single looped network.

Solution. The nodal discharge functions F for Fig. 3.13 can be written as

$$F_1 = Q_1 + Q_4 - 0.6 = 0$$
$$F_2 = -Q_1 + Q_2 = 0$$
$$F_3 = Q_2 + Q_3 - 0.6 = 0,$$

and loop head-loss function

$$F_4 = K_1|Q_1|Q_1 + K_2|Q_2|Q_2 - K_3|Q_3|Q_3 - K_4|Q_4|Q_4 = 0,$$

which is linearized as

$$F_4 = b_1Q_1 + b_2Q_2 - b_3Q_3 - b_4Q_2 = 0.$$

Assuming initial pipe discharges as $0.1 \text{ m}^3/\text{s}$ in al the pipes, the coefficients for head-loss function are calculated as

$$b_1 = K_1Q_1 = 6528 \times 0.1 = 652.8$$
$$b_2 = K_2Q_2 = 4352 \times 0.1 = 435.2$$
$$b_3 = K_3Q_3 = 6528 \times 0.1 = 652.8$$
$$b_4 = K_4Q_4 = 4352 \times 0.1 = 435.2.$$

Thus the matrix of the form $Ax = B$ can be written as

$$
\begin{bmatrix}
1 & 0 & 0 & 1 \\
-1 & 1 & 0 & 0 \\
0 & 1 & 1 & 0 \\
652.8 & 435.2 & -6528.8 & -435.2
\end{bmatrix}
\begin{bmatrix}
Q_1 \\
Q_2 \\
Q_3 \\
Q_4
\end{bmatrix}
=
\begin{bmatrix}
0.6 \\
0 \\
0.6 \\
0
\end{bmatrix}.
$$

Solving the above set of linear equations, the pipe discharges obtained are

$Q_1 = 0.3 \text{ m}^3/\text{s}$
$Q_2 = 0.3 \text{ m}^3/\text{s}$
$Q_3 = 0.3 \text{ m}^3/\text{s}$
$Q_4 = 0.3 \text{ m}^3/\text{s}$

Repeating the process, the revised head-loss coefficients are

$$b_1 = K_1 Q_1 = 6528 \times 0.3 = 1958.4$$
$$b_2 = K_2 Q_2 = 4352 \times 0.3 = 1305.6$$
$$b_3 = K_3 Q_3 = 6528 \times 0.3 = 1958.4$$
$$b_4 = K_4 Q_4 = 4352 \times 0.3 = 1305.6$$

Thus, the matrix of the form $Ax = B$ is written as

$$
\begin{bmatrix}
1 & 0 & 0 & 1 \\
-1 & 1 & 0 & 0 \\
0 & 1 & 1 & 0 \\
1958.4 & 1305.6 & -1958.4 & -1305.6
\end{bmatrix}
\begin{bmatrix}
Q_1 \\
Q_2 \\
Q_3 \\
Q_4
\end{bmatrix}
=
\begin{bmatrix}
0.6 \\
0 \\
0.6 \\
0
\end{bmatrix}.
$$

Solving the above set of linear equations, the pipe discharges obtained are

$Q_1 = 0.3 \text{ m}^3/\text{s}$
$Q_2 = 0.3 \text{ m}^3/\text{s}$
$Q_3 = 0.3 \text{ m}^3/\text{s}$
$Q_4 = 0.3 \text{ m}^3/\text{s}$

Thus, the above are the final pipe discharges as the two iterations provide the same solution.

3.8. MULTI-INPUT SOURCE WATER NETWORK ANALYSIS

Generally, urban water distribution systems have looped configurations and receive water from multi-input points (sources). The looped configuration of pipelines is preferred over branched configurations due to high reliability (Sarbu and Kalmar, 2002) and low risk from the loss of services. The analysis of a single-input water system is simple. On the other hand in a multi-input point water system, it is difficult to evaluate the input point discharges, based on input head, topography, and pipe layout. Such an analysis requires either search methods or formulation of additional nonlinear equations between input points.

A simple alternative method for the analysis of a multi-input water network is described in this section. In order to describe the algorithm properly, a typical water distribution network as shown in Fig. 3.14 is considered. The geometry of the network is described by the following data structure.

3.8.1. Pipe Link Data

The pipe link i has two end points with the nodes $J_1(i)$ and $J_2(i)$ and has a length L_i for $i = 1, 2, 3, \ldots, i_L$; i_L being the total number of pipe links in the network. The pipe nodes are defined such that $J_1(i)$ is a lower-magnitude node and $J_2(i)$ is a higher-magnitude node of pipe i. The total number of nodes in the network is J_L. The elevations of the end points are $z(J_{1i})$ and $z(J_{2i})$. The pipe link population load is $P(i)$, diameter of pipe i is $D(i)$, and total form-loss coefficient due to valves and fittings is $k_f(i)$. The pipe data structure is shown in Table 3.1.

3.8.2. Input Point Data

The nodal number of the input point is designated as $S(n)$ for $n = 1$ to n_L (total number of input points). The two input points at nodes 1 and 13 are shown in Fig. 3.14. The

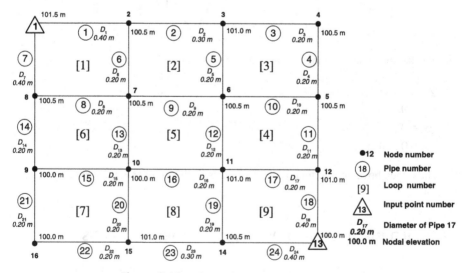

Figure 3.14. A looped water supply network.

TABLE 3.1. Network Pipe Link Data

Pipe (i)/Node (j)	Node 1 $J_1(i)$	Node 2 $J_2(i)$	Loop 1 $K_1(i)$	Loop 2 $K_2(i)$	Length $L(i)$ (m)	Form-Loss Coefficient $k_f(i)$	Population $P(i)$	Pipe Size $D(i)$ (m)	Elevation $z(j)$ (m)
1	1	2	1	0	800	0.15	400	0.40	101.5
2	2	3	2	0	800	0.15	400	0.30	100.5
3	3	4	3	0	800	0	400	0.20	101.0
4	4	5	3	0	600	0	300	0.20	100.5
5	3	6	2	3	600	0	300	0.20	100.5
6	2	7	1	2	600	0	300	0.20	100.5
7	1	8	1	0	600	0	300	0.40	100.5
8	7	8	1	6	800	0	400	0.20	100.5
9	6	7	2	5	800	0	400	0.20	100.0
10	5	6	3	4	800	0.2	400	0.20	100.0
11	5	12	4	0	600	0	300	0.20	101.0
12	6	11	4	5	600	0	300	0.20	101.0
13	7	10	5	6	600	0	300	0.20	100.0
14	8	9	6	0	600	0	300	0.20	100.5
15	9	10	6	7	800	0	400	0.20	101.0
16	10	11	5	8	800	0	400	0.20	100.0
17	11	12	4	9	800	0	400	0.20	—
18	12	13	9	0	600	0.15	300	0.20	—
19	11	14	8	9	600	0	300	0.20	—
20	10	15	7	8	600	0	300	0.20	—
21	9	16	7	0	600	0	300	0.20	—
22	15	16	7	0	800	0	400	0.20	—
23	14	15	8	0	800	0	400	0.30	—
24	13	13	9	0	800	0.15	400	0.40	—

TABLE 3.2. Input Point Nodes and Input Heads

Input Point Number $S(n)$	Input Point Node $j(S(n))$	Input Point Head (m) $h_0(n)$
1	1	19
2	13	22

corresponding input point pressure heads $h_0(S(n))$ for $n = 1$ to n_L for analysis purposes are given in Table 3.2.

3.8.3. Loop Data

The pipe link i can be the part of two loops $K_1(i)$ and $K_2(i)$. In case of a branched pipe configuration, $K_1(i)$ and $K_2(i)$ are zero. However, the description of loops is not independent information and can be generated from pipe–node connectivity data.

3.8.4. Node–Pipe Connectivity

There are $N_p(j)$ number of pipe links meeting at the node j. These pipe links are numbered as $I_p(j,\ell)$ with ℓ varying from 1 to $N_p(j)$. Scanning Table 3.1, the node pipe connectivity data can be formed. For example, pipes 6, 8, 9, and 13 are connected to node 7. Thus, $N_p(j = 7) = 4$ and pipe links are $I_p(7,1) = 6$, $I_p(7,2) = 8$, $I_p(7,3) = 9$, and $I_p(7,4) = 13$. The generated node–pipe connectivity data are given in Table 3.3.

TABLE 3.3. Node–Pipe Connectivity

j	$N_p(j)$	$I_p(j, \ell)$ $\ell = 1$ to $N_p(j)$			
		1	2	3	4
1	2	1	7		
2	3	1	2	6	
3	3	2	3	5	
4	2	3	4		
5	3	4	10	11	
6	4	5	9	10	12
7	4	6	8	9	13
8	3	7	8	14	
9	3	14	15	21	
10	4	13	15	16	20
11	4	12	16	17	19
12	3	11	17	18	
13	2	18	24		
14	3	19	23	24	
15	3	20	22	23	
16	2	21			

3.8.5. Analysis

Analysis of a pipe network is essential to understand or evaluate a physical system. In case of a single-input system, the input discharge is equal to the sum of withdrawals. The known parameters in a system are the input pressure heads and the nodal withdrawals. In the case of a multi-input network system, the system has to be analyzed to obtain input point discharges, pipe discharges, and nodal pressure heads. Walski (1995) indicated the numerous pipe sizing problems that are faced by practicing engineers. Similarly, there are many pipe network analysis problems faced by water engineers, and the analysis of a multi-input points water system is one of them. Rossman (2000) described the analysis method used in EPANet to estimate pipe flows for the given input point heads.

To analyze the network, the population served by pipe link i was distributed equally to both nodes at the ends of pipe i, $J_1(i)$, and $J_2(i)$. For pipes having one of their nodes as an input node, the complete population load of the pipe is transferred to another node. Summing up the population served by the various half-pipes connected at a particular node, the total nodal population P_j is obtained. Multiplying P_j by the per-capita water demand w and peak discharge factor θ_P, the nodal withdrawals q_j are obtained. If ω is in liters per person per day and q_j is in cubic meters per second, the results can be written as

$$q_j = \frac{\theta_p \omega P_j}{86{,}400{,}000}. \tag{3.29}$$

The nodal water demand due to industrial and firefighting usage if any can be added to nodal demand. The nodal withdrawals are assumed to be positive and input discharges as negative. The total water demand of the system Q_T is

$$Q_T = \sum_{j=1}^{j_L - n_L} q(j). \tag{3.30}$$

The most important aspect of multiple-input-points water distribution system analysis is to distribute Q_T among all the input nodes $S(n)$ such that the computed head $h(S(n))$ at input node is almost equal to given head $h_0(S(n))$.

For starting the algorithm, initially total water demand is divided equally on all the input nodes as

$$Q_{Tn} = \frac{Q_T}{n_L}. \tag{3.31}$$

In a looped network, the pipe discharges are derived using loop head-loss relationships for known pipe sizes and nodal linear continuity equations for known nodal withdrawals. A number of methods are available to analyze such systems as described in this chapter. Assuming an arbitrary pipe discharge in one of the pipes of all the loops and using continuity equation, the pipe discharges are calculated. The discharges in loop pipes are corrected using the Hardy Cross method, however, any other analysis method can also be used. To apply nodal continuity equation, a sign convention for pipe flows is assumed that a positive discharge in a pipe flows from a lower-magnitude node to a higher-magnitude node.

The head loss in the pipes is calculated using Eqs. (2.3b) and (2.7b):

$$h_{fi} = \frac{8 f_i L_i Q_i^2}{\pi^2 g D_i^5} + k_{fi} \frac{8 Q_i^2}{\pi^2 g D_i^4}, \tag{3.32}$$

where h_{fi} is the head loss in the ith link in which discharge Q_i flows, g is gravitational acceleration, k_{fi} is form-loss coefficient for valves and fittings, and f_i is a coefficient of surface resistance. The friction factor f_i can be calculated using Eq. (2.6a).

Thus, the computed pressure heads of all the nodes can be calculated with reference to an input node of maximum piezometric head (input point at node 13 in this case). The calculated pressure head at other input point nodes will depend upon the correct division of input point discharges. The input point discharges are modified until the computed pressure heads at input points other than the reference input point are equal to the given input point heads $h_0(n)$.

A discharge correction ΔQ, which is initially taken equal to $0.05 Q_T/n_L$, is applied at all the point nodes discharges, other than that of highest piezometric head input node. The correction is subtractive if $h(S(n)) > h_0(n)$ and it is additive otherwise. The input discharge of highest piezometric head input node is obtained by continuity considerations. The process of discharge correction and network analysis is repeated until the

$$\text{error} = \frac{|h_0(n) - h(S(n))|}{h_0(n)} \le 0.01 \quad \text{for all values of } n \text{ (input points).} \tag{3.33}$$

The designer can select any other suitable value of minimum error for input head correction. The next ΔQ is modified as half of the previous iteration to safeguard against any repetition of input point discharge values in alternative iterations. If such a repetition is not prevented, Eq. (3.33) will never be satisfied and the algorithm will never terminate.

The water distribution network as shown in Fig. 3.14 was analyzed using the described algorithm. The rate of water supply 300 liters per person per day and a peak factor of 2.0 were adopted for the analysis. The final input point discharges obtained are given in Table 3.4.

TABLE 3.4. Input Point Discharges

Input Point	Input Point Node	Input Point Head (m)	Input Point Discharge m³/s
1	1	19	0.0204
2	13	22	0.0526

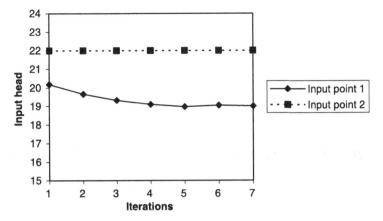

Figure 3.15. Variation of computed input heads with analysis iterations.

The variation of computed input point head with analysis iterations is shown in Fig. 3.15. The constant head line for input point 2 indicates the reference point head. Similarly, the variation of input point discharges with analysis iterations is shown in Fig. 3.16. It can be seen that input discharge at input point 2 (node 13) is higher due to higher piezometric head meaning thereby that it will supply flows to a larger population than the input node of lower piezometric head (input point 1).

The computed pipe discharges are given in Table 3.5. The sum of discharges in pipes 1 and 7 is equal to discharge of source node 1, and similarly the sum of discharges in pipes 18 and 24 is equal to the discharge of source node 2. The negative discharge in pipes indicates that the flow is from a higher-magnitude node to a lower-magnitude node of the pipe. For example, discharge in pipe 4 is -0.003 meaning thereby that the flow in the pipe is from pipe node number 5 to node number 4.

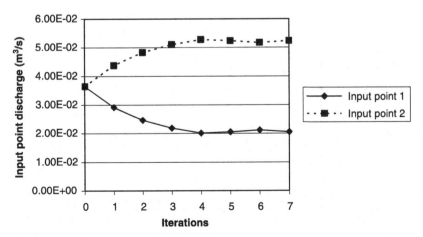

Figure 3.16. Variation of computed input discharges with analysis iterations.

TABLE 3.5. Pipe Discharges

Pipe (i)	1	2	3	4	5	6	7	8
$Q(i)\,\mathrm{m^3/s}$	0.0105	0.0025	0.0001	−0.003	−0.0024	0.0015	0.0099	−0.0023
Pipe (i)	9	10	11	12	13	14	15	16
$Q(i)\,\mathrm{m^3/s}$	0.00001	0.0014	−0.0087	−0.007	−0.0024	0.002	−0.0014	−0.0056
Pipe (i)	17	18	19	20	21	22	23	24
$Q(i)\,\mathrm{m^3/s}$	−0.0045	−0.0189	−0.0142	−0.0043	−0.0009	0.0039	0.013	0.0337

3.9. FLOW PATH DESCRIPTION

Sharma and Swamee (2005) developed a method for flow path identification. A node (nodal point) receives water through various paths. These paths are called *flow paths*. Knowing the discharges flowing in the pipe links, the flow paths can be obtained by starting from a node and proceeding in a direction opposite to the flow. The advantages of these flow paths are described in this section.

Unlike branched systems, the flow directions in looped networks are not unique and depend upon a number of factors, mainly topography nodal demand and location and number of input (supply) points.

The flow path is a set of pipes through which a pipe is connected to an input point. Generally, there are several paths through which a node *j* receives the discharge from an input point, and similarly there can be several paths through which a pipe is connected to an input point for receiving discharge. Such flow paths can be obtained by proceeding in a direction opposite to the flow until an input source is encountered. To demonstrate the flow path algorithm, the pipe number, node numbers, and the discharges in pipes as listed in Table 3.5 are shown in Fig. 3.17.

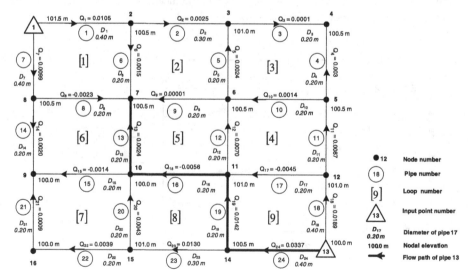

Figure 3.17. Flow paths in a water supply system.

Considering pipe $i = 13$ at node $j = 7$, it is required to find a set of pipes through which pipe 13 is connected to the input point. As listed in Table 3.1, the other node of pipe 13 is 10. Following Table 3.5, the discharge in the pipe is negative meaning thereby that the water flows from node 10 to 7. Thus, if one travels from node 7 to node 10, it will be in a direction opposite to flow. In this manner, one reaches at node 10.

Scanning Table 3.3 for node 10, one finds that it connects four pipes, namely 13, 15, 16, and 20. One has already traveled along pipe 13, therefore, consider pipes 15, 16, and 20 only. One finds from Table 3.5 that the discharge in pipe 15 is negative and from Table 3.1 that the other node of this pipe 15 is 9, thus a negative discharge flows from node 10 to node 9. Also by similar argument, one may discover that the discharge in pipe 16 flows from node 11 to 10 and the discharge in pipe 20 flows from node 15 to 10. Thus, for moving against the flow from node 10, one may select one of the pipes, namely 16 and 20, except pipe 15 in which the movement will be in the direction of flow. Selecting a pipe with higher magnitude of flow, one moves along the pipe 16 and reaches the node 11. Repeating this procedure, one moves along the pipes 19 and 24 and reaches node 13 (input point). The flow path for pipe 13 thus obtained is shown in Fig. 3.17.

TABLE 3.6. List of Flow Paths of Pipes

	$I_t(i,\ell)$ $\ell = 1, N_t(i)$						
i	1	2	3	4	$N_t(i)$	$J_t(i)$	$J_s(i)$
1	1				1	2	1
2	2	1			2	3	1
3	3	2	1		3	4	1
4	4	11	18		3	5	13
5	5	12	19	24	4	3	13
6	6	1			2	7	1
7	7				1	8	1
8	8	7			2	7	1
9	9	12	19	24	4	7	13
10	10	11	18		3	6	13
11	11	18			2	5	13
12	12	19	24		3	6	13
13	13	16	19	24	4	7	13
14	14	7			2	9	1
15	15	16	19	24	4	9	13
16	16	19	24		3	10	13
17	17	18			2	11	13
18	18				1	12	13
19	19	24			2	11	13
20	20	23	24		3	10	13
21	21	22	23	24	4	9	13
22	22	23	24		3	16	13
23	23	24			2	15	13
24	24				1	14	13

Thus starting from pipe $i = 13$, one encounters four pipes before reaching the input point. The total number of pipes in the track N_t is a function of pipe i, in this case pipe 13, the total number of pipes in path $N_t(13) = 4$, and the flow path is originating from the node $J_t(i = 13) = 7$. The flow path terminates at node 13, which is one of the input sources. Thus, the source node $J_s(i = 13)$ is 13. The pipes encountered on the way are designated $I_t(i,\ell)$ with ℓ varying from 1 to $N_t(i)$. In this case, the following $I_t(i,\ell)$ were obtained: $I_t(13,1) = 13$, $I_t(13,2) = 16$, $I_t(13,3) = 19$, and $I_t(13,4) = 24$.

The flow paths of pipes in the water supply system in Fig. 3.17 and their corresponding originating nodes and source nodes are given in Table 3.6.

The advantages of flow path generation are

1. The flow paths of pipes generate flow pattern of water in pipes of a water distribution system. This information will work as a decision support system for operators/managers of water supply systems in efficient operation and maintenance of the system.

2. This information can be used for generating head-loss constraint equations for the design of a water distribution network having single or multi-input sources.

EXERCISES

3.1. Calculate head loss in a CI pipe of length $L = 100$ m, with discharge Q at entry node $= 0.2$ m^3/s, and pipe diameter $D = 0.3$ m, if the idealized withdrawal as shown in Fig. 3.1b is at a rate of 0.0005 m^3/s per meter length.

3.2. For a CI gravity main (Fig. 3.2), calculate flow in a pipe of length 300 m and size 0.2 m. The elevations of reservoir and outlet are 15 m and 5 m, respectively. The water column in reservoir is 5 m, and a terminal head of 6 m is required at outlet.

3.3. Analyze a single looped pipe network as shown in Fig. 3.18 for pipe discharges using Hardy Cross, Newton–Raphson, and linear theory methods. Assume a constant friction factor $f = 0.02$ for all pipes in the network.

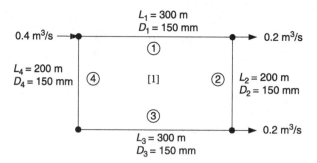

Figure 3.18. Single looped network.

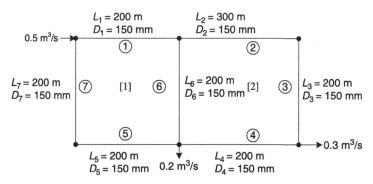

Figure 3.19. A pipe network with two loops.

3.4. Analyze a looped pipe network as shown in Fig. 3.19 for pipe discharges using Hardy Cross, Newton–Raphson, and linear theory methods. Assume a constant friction factor $f = 0.02$ for all pipes in the network.

REFERENCES

Fair, G.M., Geyer, J.C., and Okun, D.A. (1981). *Elements of Water Supply and Wastewater Disposal.* John Wiley & Sons, New York.

Rossman, L.A. (2000). *EPANET Users Manual.* EPA/600/R-00/057. U.S. EPA, Cincinnati OH.

Sarbu, I., and Kalmar, F. (2002). Optimization of looped water supply networks. *Periodica Polytechnica Ser. Mech. Eng.* 46(1), 75–90.

Sharma, A.K., and Swamee, P.K. (2005) Application of flow path algorithm in flow pattern mapping and loop data generation for a water distribution system. *J. Water Supply: Research & Technology–AQUA, IWA* 54, 411–422.

Walski, T.M. (1995). Optimization and pipe sizing decision. *J. Water Resource Planning and Mangement* 121(4), 340–343.

Wood, D.J., and Charles, C.O.A. (1972). Hydraulic network analysis using Linear Theory. *J. Hydr. Div.* 98(7), 1157–1170.

4

COST CONSIDERATIONS

In order to synthesize a pipe network system in a rational way, one cannot overlook the cost considerations that are altogether absent during the analysis of an existing system. All the pipe system designs that can transport the fluid, or fluid with solid material in suspension, or containerized in capsules in planned quantity are feasible designs. Had

Design of Water Supply Pipe Networks. By Prabhata K. Swamee and Ashok K. Sharma
Copyright © 2008 John Wiley & Sons, Inc.

79

there been only one feasible design as in the case of a gravity main, the question of selecting the best design would not have arisen. Unfortunately, there is an extremely large number of feasible designs, of which one has to select the best.

What is a best design? It is not easy to answer this question. There are many aspects of this question. For example, the system must be economic and reliable. Economy itself is no virtue; it is worthwhile to pay a little more if as a result the gain in value exceeds the extra cost. For increasing reliability, the cost naturally goes up. Thus, a trade-off between the cost and the reliability is required for arriving at the best design. In this chapter, cost structure of a pipe network system is discussed for constructing an objective function based on the cost. This function can be minimized by fulfilling the fluid transport objective at requisite pressure.

Figure 4.1 shows the various phases of cost calculations in a water supply project. For the known per capita water requirement, population density, and topography of the area, the decision is taken about the terminal pressure head, minimum pipe diameters, and the pipe materials used before costing a water supply system. The financial resources and the borrowing rates are also known initially. Based on this information, the water distribution network can be planned in various types of geometry, and the large areas can be divided into various zones. Depending on the geometric planning, one can arrive at a primitive value of the cost called *the forecast of cost.* This cost gives an idea about the magnitude of expenditure incurred without going into the design aspect. If the forecast of cost is not suitable, one may review the infrastructure planning and the requirements. The process can be repeated until the forecast of cost is suitable. The financially infeasible projects are normally dropped at this stage. Once the forecast of cost is acceptable, one may proceed for the detailed design of the water supply system and obtain the pipe diameters, power required for pumping, staging and capacity of service reservoirs and so forth. Based on detailed design, the cost of the project can be worked out in detail. This cost is called *the estimated cost.* Knowing the estimated cost, all the previous stages can be reviewed again, and the estimated cost is revised if unsuitable. The process can be repeated until the estimated cost is acceptable. At this stage, the water supply project can be constructed. One gets the *actual cost* of the project after its execution. Thus, it can be seen that the engineering decisions are based on the forecast of cost and the estimated cost.

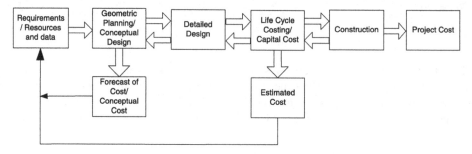

Figure 4.1. Interaction of different types of costs.

The forecast of cost is expressed in terms of the population served, the area covered and the per capita water demand, and the terminal pressure head and minimum pipe diameter provisions. Generally, the equations used for forecasting the cost are thumb-rule type and may not involve all the parameters described herein. On the other hand, the estimated cost is precise as it is based on the project design. Thus, the estimated cost may be selected as the objective of the design that has to be minimized.

It is important to understand that the accuracy of cost estimates is dependent on the amount of data and its precision including suitability of data to the specific site and the project. Hence, reliable construction cost data is important for proper planning and execution of any water supply project. The forecast of cost and estimated cost also have some degree of uncertainty, which is usually addressed through the inclusion of lump-sum allowances and contingencies. Apart from construction costs, other indirect costs like engineering fees, administrative overhead, and land costs should also be considered. The planner should take a holistic approach as the allowances for indirect costs vary with the size and complexity of the project. The various components of a water supply system are discussed in the following sections.

4.1. COST FUNCTIONS

The cost function development methodology for some of the water supply components is described in the following section (Sharma and Swamee, 2006). The reader is advised to collect current cost data for his or her geographic location to develop such cost functions because of the high spatial and temporal variation of such data.

4.1.1. Source and Its Development

The water supply source may be a river or a lake intake or a well field. The other pertinent works are the pumping plant and the pump house. The cost of the pump house is not of significance to be worked out as a separate function. The cost of pumping plant C_p, along with all accessories and erection, is proportional to its power P. That is,

$$C_p = k_p P^{m_p}, \qquad (4.1)$$

where P is expressed in kW, k_p = a coefficient, and m_p = an exponent. The power of the pump is given by

$$P = \frac{\rho g Q h_0}{1000 \eta}, \qquad (4.2a)$$

where ρ = mass density of fluid, and η = combined efficiency of pump and prime mover. For reliability, actual capacity of the pumping plant should be more than the capacity calculated by Eq. (4.2a). That is,

$$P = \frac{(1 + s_b) \rho g Q h_0}{1000 \eta}, \qquad (4.2b)$$

where s_b = standby fraction. Using Eqs. (4.1) and (4.2b), the cost of the pumping plant is obtained as

$$C_p = k_p \left[\frac{(1 + s_b)\rho g Q h_0}{1000\eta} \right]^{m_p}. \tag{4.2c}$$

The parameters k_p and m_p in Eq. (4.1) vary spatially and temporally. The k_p is influenced by inflation. On the other hand, m_p is influenced mainly by change in construction material and production technology. For a known set of pumping capacities and cost data, the k_p and m_p can be obtained by plotting a log-log curve. For illustration purposes, the procedure is depicted by using the data (Samra and Essery, 2003) as listed in Table 4.1. Readers are advised to plot a similar curve based on the current price structure at their geographic locations.

The pump and pumping station cost data is plotted in Fig. 4.2 and can be represented by the following equation:

$$C_p = 5560 P^{0.723}. \tag{4.3}$$

Thus, $k_p = 5560$ and $m_p = 0.723$. As the cost of the pumping plant is considerably less than the cost of energy, by suitably adjusting the coefficient k_p, the exponent m_p can be made as unity. This makes the cost to linearly vary with P.

4.1.2. Pipelines

Usually, pipelines are buried underground with 1 m of clear cover. The width of the trench prepared to lay the pipeline is kept to 60 cm plus the pipe diameter. This criterion may vary based on the machinery used during the laying process and the local guidelines. The cost of fixtures, specials, and appurtenances are generally found to be of the order 10% to 15% of the cost of the pipeline. The cost of completed pipeline C_m shows the following relationship with the pipe length L and pipe diameter D:

$$C_m = k_m L D^m, \tag{4.4}$$

TABLE 4.1. Pump and Pumping Station Cost

Pump Power (kW)	Pump and Pumping Station Cost (A$)
10	36,000
20	60,000
30	73,000
50	105,000
100	185,000
200	305,000
400	500,000
600	685,000
800	935,000

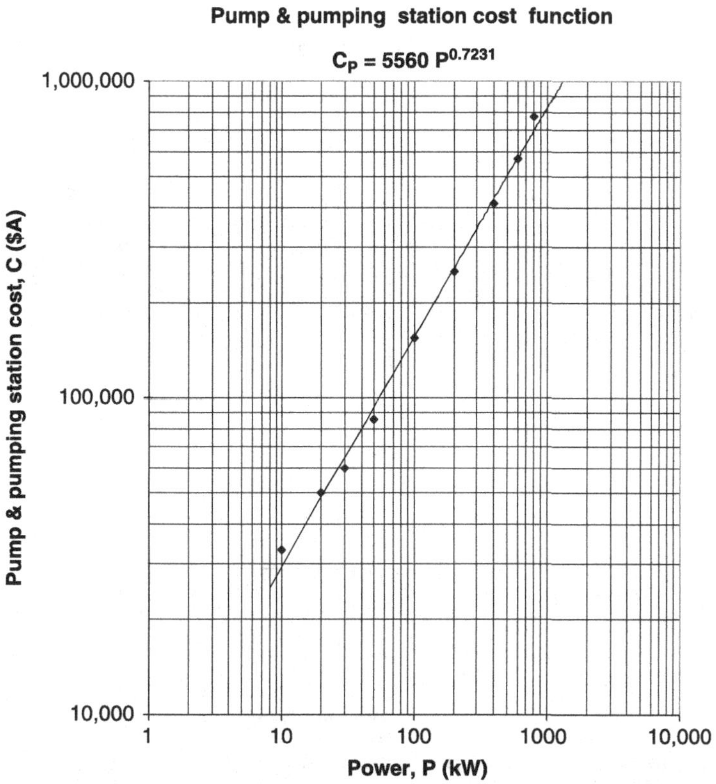

Pump & pumping station cost function

$$C_P = 5560\ P^{0.7231}$$

Figure 4.2. Variation of pump and pumping station cost with pump power.

where k_m = a coefficient, and m = an exponent. The pipe cost parameters k_m and m depend on the pipe material, the monetary unit of the cost, and the economy. To illustrate the methodology for developing pipe cost relationship, the per meter cost of various sizes of ductile iron cement lined (DICL) pipes is plotted on log-log scale as shown in Fig. 4.3 using a data set (Samra and Essery, 2003). Cast Iron (CI) pipe cost parameters $k_m = 480$ and $m = 0.935$ have been used as data in the book for various examples.

It is not always necessary to get a single cost function for the entire set of data. The cost data may generate more than one straight line while plotted on a log-log scale. Sharma (1989) plotted the local CI pipe cost data to develop the cost function. The variation is depicted in Fig. 4.4. It was found that the entire data set was represented by two cost functions.

The following function was valid for pipe diameters ranging from 0.08 m to 0.20 m,

$$C_{m1} = 1320D^{0.866}, \tag{4.5a}$$

and the per meter pipe cost of diameters 0.25 m to 0.75 m was represented by

$$C_{m2} = 4520D^{1.632}. \tag{4.5b}$$

84

COST CONSIDERATIONS

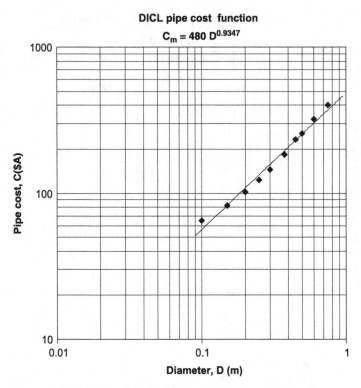

Figure 4.3. Variation of DICL pipe per meter cost with pipe diameter.

The Eqs. (4.5a), and (4.5b) were combined into a single cost function representing the entire set of data as:

$$C_m = 1320D^{0.866}\left[1 + \left(\frac{D}{0.2}\right)^{9.7}\right]^{0.08}. \tag{4.5c}$$

Also, the cost analysis of high-pressure pipes indicated that the cost function can be represented by the following equation:

$$C_m = k_m\left(1 + \frac{h_a}{h_b}\right)LD^m, \tag{4.6}$$

where h_a = allowable pressure head, and h_b = a length parameter. The length parameter h_b depends on the pipe material. For cast iron pipes, it is 55–65 m, whereas for asbestos cement pipes, it is 15–20 m. h_b can be estimated for plotting known k_m values for pipes with various allowable pressures (k_m vs. allowable pressure plot).

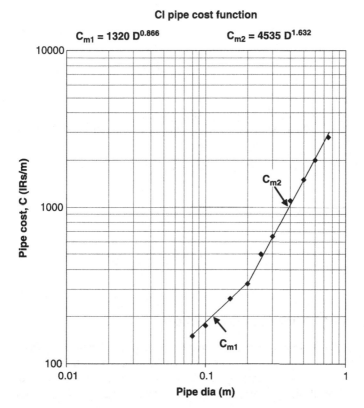

Figure 4.4. Variation of CI pipe cost per meter with diameter.

4.1.3. Service Reservoir

The cost functions for service reservoirs are developed in this section. It is not always possible to develop a cost function simply by plotting the cost data on a log-log scale, as any such function would not represent the entire data set within a reasonable error. The analytical methods are used to represent such data sets. On the basis of analysis of cost of service reservoirs of various capacities and staging heights, Sharma (1979), using the Indian data, obtained the following equation for the service reservoir cost C_R:

$$C_R = k_R V_R^{0.5} \left\{ \left(1 + \frac{V_R}{100}\right) \left[1 + \frac{h_s}{4}\right]^{3.2} \right\}^{0.2}, \tag{4.7a}$$

where V_R = reservoir capacity in m³, h_s = the staging height in m, and k_R = a coefficient, and for large capacities and higher staging, Eq. (4.7a) is converted to

$$C_R = 0.164 \, k_R V_R^{0.7} h_s^{0.64}. \tag{4.7b}$$

For a surface reservoir, Eq. (4.7a) reduces to

$$C_R = k_R V_R^{0.5} \left(1 + \frac{V_R}{100}\right)^{0.2}. \tag{4.7c}$$

The cost function for a surface concrete reservoir was developed using the Australian data (Samra and Essery, 2003) listed in Table 4.2.

Using the analytical methods, the following cost function for surface reservoir was developed:

$$C_R = \frac{290 V_R}{\left[1 + \left(\dfrac{V_R}{1100}\right)^{5.6}\right]^{0.075}}. \tag{4.7d}$$

Comparing Eqs. (4.7c) and (4.7d), it can be seen that depending on the prevailing cost data, the functional form may be different.

4.1.4. Cost of Residential Connection

The water supply system optimization should also include the cost of service connections to residential units as this component contributes a significant cost to the total cost. Swamee and Kumar (2005) gave the following cost function for the estimation of cost C_s of residential connections (ferrule) from water mains through a service main of diameter D_s:

$$C_s = k_s L D_s^{m_s}. \tag{4.7e}$$

TABLE 4.2. Service Reservoir Cost

Reservoir Capacity (m³)	Cost (A$)
100	28,000
200	55,000
400	125,000
500	160,000
1000	300,000
2000	435,000
4000	630,000
5000	750,000
8000	1,000,000
10,000	1,150,000
15,000	1,500,000
20,000	1,800,000

4.1.5. Cost of Energy

The annual recurring cost of energy consumed in maintaining the flow depends on the discharge pumped and the pumping head h_0 produced by the pump. If $Q =$ the peak discharge, the effective discharge will be $F_A F_D Q$, where $F_A =$ the annual averaging factor, and $F_D =$ the daily averaging factor for the discharge. The average power P, developed over a year, will be

$$P = \frac{\rho g Q h_0 F_A F_D}{1000 \eta}.$$ (4.8)

Multiplying the power by the number of hours in a year (8760) and the rate of electricity per kilowatt-hour, R_E, the annual cost of energy A_e consumed in maintaining the flow is worked out to be

$$A_e = \frac{8.76 \rho g Q h_0 F_A F_D R_E}{\eta}.$$ (4.9)

4.1.6. Establishment Cost

Swamee (1996) introduced the concept of establishment cost E in the formulation of cost function. The establishment cost includes the cost of the land and capitalized cost of operational staff and other facilities that are not included elsewhere in the cost function. In case of a pumping system, it can be expressed in terms of additional pumping head $h = E/\rho g k_T Q$, where k_T is relative cost factor described in Section 4.5.

4.2. LIFE-CYCLE COSTING

Life-cycle costing (LCC) is an economic analysis technique to estimate the total cost of a system over its life span or over the period a service is provided. It is a systematic approach that includes all the cost of the infrastructure facilities incurred over the analysis period. The results of a LCC analysis are used in the decision making to select an option from available alternatives to provide a specified service. Figure 4.5 depicts the conceptual variation of system costs for alternative configurations. The optimal system configuration is the one with least total cost. The LCC analysis also provides the information to the decision maker about the trade-off between high capital (construction) and lower operating and maintenance cost of alternative systems. The methodologies for combining capital and recurring costs are described under the next section.

4.3. UNIFICATION OF COSTS

The cost of pumps, buildings, service reservoirs, treatment plants, and pipelines are incurred at the time of construction of the water supply project, whereas the cost of power and the maintenance and repair costs of pipelines and pumping plants have to

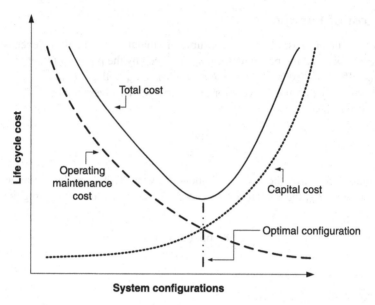

Figure 4.5. Variation of total cost with system configuration.

be incurred every year. The items involving the capital cost have a finite life: a pipeline lasts for 60–90 years, whereas a pumping plant has a life of 12–15 years. After the life of a component is over, it has to be replaced. The replacement cost has also to be considered as an additional recurring cost. Thus, there are two types of costs: (1) capital cost or the initial investment that has to be incurred for commissioning of the project, and (2) the recurring cost that has to be incurred continuously for keeping the project in operating condition.

These two types of costs cannot be simply added to find the overall cost or life-cycle cost. These costs have to be brought to the same units before they can be added. For combining these costs, the methods generally used are the capitalization method, the annuity method, and the net present value method. These methods are described in the following sections.

4.3.1. Capitalization Method

In this method, the recurring costs are converted to capital costs. This method estimates the amount of money to be kept in a bank yielding an annual interest equal to the annual recurring cost. If an amount C_A is kept in a bank with an annual interest rate of lending r per unit of money, the annual interest on the amount will be rC_A. Equating the annual interest to the annual recurring cost A, the capitalized cost C_A is obtained as

$$C_A = \frac{A}{r}.$$

(4.10a)

A component of a pipe network system has a finite life T. The replacement cost C_R has to be kept in a bank for T years so that its interest is sufficient to get the new component. If the original cost of a component is C_0, by selling the component after T years as scrap, an amount αC_0 is recovered, where $\alpha =$ salvage factor. Thus, the net liability after T years, C_N, is

$$C_N = (1 - \alpha)C_0. \tag{4.10b}$$

On the other hand, the amount C_R with interest rate r yields the compound interest I_R given by

$$I_R = \{(1+r)^T - 1\}C_R. \tag{4.10c}$$

Equating I_R and C_N, the replacement cost is obtained as

$$C_R = \frac{(1 - \alpha)C_0}{(1+r)^T - 1}. \tag{4.11}$$

Denoting the annual maintenance factor as β, the annual maintenance cost is given by βC_0. Using Eq. (4.10a), the capitalized cost of maintenance C_{ma}, works out to be

$$C_{ma} = \frac{\beta C_0}{r}. \tag{4.12}$$

Adding C_0, C_R, and C_{ma}, the overall capitalized cost C_c is obtained as

$$C_c = C_0 \left[1 + \frac{1 - \alpha}{(1+r)^T - 1} + \frac{\beta}{r} \right]. \tag{4.13}$$

Using Eqs. (4.10a) and (4.13), all types of costs can be capitalized to get the overall cost of the project.

4.3.2. Annuity Method

This method converts the capital costs into recurring costs. The capital investment is assumed to be incurred by borrowing the money that has to be repaid in equal annual installments throughout the life of the component. These installments are paid along with the other recurring costs. The annual installments (called annuity) can be combined with the recurring costs to find the overall annual investment.

If annual installments A_r for the system replacement are deposited in a bank up to T years, the first installment grows to $A_r (1+r)^{T-1}$, the second installment to

$A_r(1 + r)^{T-2}$, and so on. Thus, all the installments after T years add to C_a given by

$$C_a = A_r\left[1 + (1 + r) + (1 + r)^2 + \cdots + (1 + r)^{T-1}\right]. \qquad (4.14a)$$

Summing up the geometric series, one gets

$$C_a = A_r \frac{(1 + r)^T - 1}{r}. \qquad (4.14b)$$

Using Eqs. (4.10b) and (4.14b), A_r is obtained as

$$A_r = \frac{(1 - \alpha)r}{(1 + r)^T - 1} C_0. \qquad (4.15a)$$

The annuity A_0 for the initial capital investment is given by

$$A_0 = rC_0. \qquad (4.15b)$$

Adding up A_0, A_r, and the annual maintenance cost βC_0, the annuity A is

$$A = rC_0\left[1 + \frac{1 - \alpha}{(1 + r)^T - 1} + \frac{\beta}{r}\right]. \qquad (4.16)$$

Comparing Eqs. (4.13) and (4.16), it can be seen that the annuity is r times the capitalized cost. Thus, one can use either the annuity or the capitalization method.

4.3.3. Net Present Value or Present Value Method

The net present value analysis method is one of the most commonly used tools to determine the current value of future investments to compare alternative water system options. In this method, if the infrastructure-associated future costs are known, then using a suitable discount rate, the current worth (value) of the infrastructure can be calculated. The net present capital cost P_{NC} of a future expenditure can be derived as

$$P_{NC} = F(1 + r)^{-T}, \qquad (4.17a)$$

where F is future cost, r is discount rate, and T is the analysis period. It is assumed that the cost of component C_0 will remain the same over the analysis period, and it is customary in such analysis to assume present cost C_0 and future cost F of a component the same due to uncertainties in projecting future cost and discount rate. Thus,

Eq. (4.17a) can be written as:

$$P_{NC} = C_0(1 + r)^{-T}.$$ (4.17b)

The salvage value of a component at the end of the analysis period can be represented as αC_0°, the current salvage cost P_{NS} over the analysis period can be computed as

$$P_{NS} = \alpha C_0(1 + r)^{-T}.$$ (4.17c)

The annual recurring expenditure for operation and maintenance $A_r = \beta C_0$ over the period T, can be converted to net present value P_{NA} as:

$$P_{NA} = \beta C_0 \left[(1 + r)^{-1} + (1 + r)^{-2} + \cdots + (1 + r)^{-(T-1)} + (1 + r)^{-T}\right].$$ (4.17d)

Summing up the geometric series,

$$P_{NA} = \beta C_0 \frac{(1 + r)^T - 1}{r(1 + r)^T}.$$ (4.17e)

The net present value of the total system P_N is the sum of Eqs. (4.17c), (4.17e), and initial cost C_0 of the component as

$$P_N = C_0 \left[1 - \alpha(1 + r)^{-T} + \beta \frac{(1 + r)^T - 1}{r(1 + r)^T}\right].$$ (4.18)

4.4. COST FUNCTION PARAMETERS

The various cost coefficients like k_p, k_m, k_R, and so forth, refer to the capital cost of the components like the pump, the pipeline, the service reservoir, and so on. Using Eq. (4.13), the initial cost coefficient k can be converted to the capitalized cost coefficient k'. Thus,

$$k' = k \left[1 + \frac{1 - \alpha}{(1 + r)^T - 1} + \frac{\beta}{r}\right].$$ (4.19)

The formulation in the subsequent chapters uses capitalized coefficients in which primes have been dropped for convenience. For calculating capitalized coefficients, one requires various parameters of Eq. (4.13). These parameters are listed in Table 4.3. Additional information on life of pipes is available in Section 5.4.8. The readers are advised to modify Table 4.3 for their geographic locations.

TABLE 4.3. Cost Parameters

Component	α	β	T (years)
1. Pipes			
(a) Asbestos cement (AC)	0.0	0.005	60
(b) Cast iron (CI)	0.2	0.005	120
(c) Galvanized iron (GI)	0.2	0.005	120
(d) Mild steel (MS)	0.2	0.005	120
(e) Poly(vinyl chloride) (PVC)	0.0	0.005	60
(f) Reinforced concrete (RCC)	0.0	0.005	60–100
2. Pump house	0.0	0.015	50–60
3. Pumping plant	0.2	0.030	12–15
4. Service reservoir	0.0	0.015	100–120

4.5. RELATIVE COST FACTOR

Using Eqs. (4.9) and (4.10a), the capitalized cost of energy consumed, C_e, is obtained as

$$C_e = \frac{8.76 \rho g Q h_0 F_A F_D R_E}{\eta r}.$$

(4.20)

Combining Eqs. (4.2c) with $m_P = 1$ and (4.20), the cost of pumps and pumping, C_T is found to be

$$C_T = k_T \rho g Q h_0,$$

(4.21a)

where

$$k_T = \frac{(1 + s_b)k_p}{1000\eta} + \frac{8.76 F_A F_D R_E}{\eta r}.$$

(4.21b)

It has been observed that in the equations for optimal diameter and the pumping head, the coefficients k_m and k_T appear as k_T/k_m. Instead of their absolute magnitude, this ratio is an important parameter in a pipe network design problem.

4.6. EFFECT OF INFLATION

In the foregoing developments, the effect of inflation has not been considered. If inflation is considered in the formulation of capitalized cost, annuity, or net present value, physically unrealistic results, like salvage value greater than the initial cost, is obtained. Any economic analysis based on such results would not be acceptable for engineering systems.

The effect of inflation is to dilute the money in the form of cash. On the other hand, the value of real estate, like the water supply system, remains unchanged. Moreover, the real worth of revenue collected from a water supply project remains unaffected by inflation. Because Eqs. (4.13), (4.16), and (4.18) are based on money in cash, these equations are not valid for an inflationary economy. However, these equations are useful in evaluating the overall cost of an engineering project with the only change in the interpretation of interest rate. In Eqs. (4.13), (4.16), and (4.18), r is a hypothetical parameter called *social recovery factor* or *discount rate*, which the designer may select according to his judgment. Generally, it is taken as interest rate − inflation. Prevailing interest rate should not be taken as the interest rate; it should be equal to the interest rate at which states (government) provide money to water authorities for water systems. These interest rates are generally very low in comparison with prevailing interest rate.

Example 4.1. Find the capitalized cost of a 5000-m-long, cast iron pumping main of diameter 0.5 m. It carries a discharge of 0.12 m³/s throughout the year. The pumping head developed is 30 m; unit cost of energy = $0.0005k_m$ units; combined efficiency of pump and prime mover = 0.75; $k_p/k_m = 1.6$ units; $s_b = 0.5$; adopt $r = 0.07$ per year.

Solution. Cost of pipeline $C_m = k_m L D^m = k_m 5000 \times 0.5^{1.64} = 1604.282 k_m$.

$$\text{Installed power } P = \frac{(1 + s_b)\rho g Q h_0}{1000 \eta} = \frac{(1 + 0.5) \times 1000 \times 9.79 \times 0.12 \times 30}{1000 \times 0.75}$$

$$= 70.488 \text{ kW}.$$

Cost of pumping plant = $1.6 k_m 70.488 = 112.781 k_m$.

$$\text{Annual cost of energy} = \frac{8.76 \rho g Q h_0 R_E}{\eta}$$

$$= \frac{8.76 \times 1000 \times 9.79 \times 0.12 \times 30 \times 0.0005\, k_m}{0.75}$$

$$= 205.825 k_m.$$

From Table 4.3, the life of pipes and pumps and, the salvage and maintenance factors can be obtained.

$$\text{Capitalized cost of pipeline} = 1604.282\, k_m \left[1 + \frac{1 - 0.2}{(1 + 0.07)^{60} - 1} + \frac{0.005}{0.07} \right]$$

$$= 1741.411 k_m.$$

$$\text{Capitalized cost of pumps} = 112.781\, k_m \left[1 + \frac{1 - 0.2}{(1 + 0.07)^{15} - 1} + \frac{0.03}{0.07} \right]$$

$$= 212.408 k_m.$$

Capitalized cost of energy = $205.825 k_m / 0.07 = 2940.357 k_m$.

Therefore, capitalized cost of pumping main = $1741.411 k_m + 212.408 k_m$
$$+ 2940.357 k_m$$

$$= 4894.176k_m.$$

Example 4.2. Find net present value (NPV) of the pumping system as described in Example 4.1.

Solution.

$$\text{NPV pipeline} = 1604.282\,k_m\left[1 - 0.2(1 + 0.07)^{-60} + 0.005\frac{(1 + 0.07)^{60} - 1}{0.07(1 + 0.07)^{60}}\right]$$

$$= 1711.361k_m.$$

As per Table 4.3, the life of pumping plant is 15 years. Thus, four sets of pumping plants will be required over the 60-year, analysis period.

$$\text{NPV pumping plants} = 112.781\,k_m\left[1 + (1 + 0.07)^{-15} + (1 + 0.07)^{-30} + (1 + r)^{-45}\right]$$

$$-0.2 \times 112.781\,k_m\left[(1 + 0.07)^{-15} + (1 + 0.07)^{-30} + (1 + 0.7)^{-45}\right.$$

$$\left.+ (1 + 0.7)^{-60}\right] + \frac{0.03 \times 112.781\,k_m}{0.07}\frac{(1 + 0.07)^{60} - 1}{(1 + 0.07)^{60}}$$

$$= 173.750k_m - 12.600k_m + 47.500k_m = 208.650k_m.$$

$$\text{NPV annual energy cost} = \frac{205.82\,k_m}{0.07}\frac{(1 + 0.07)^{60} - 1}{(1 + 0.07)^{60}} = 2889.544k_m.$$

NPV of pumping system $= 1711.361k_m + 208.650k_m + 2889.544k_m = 4809.555k_m.$

Example 4.3. Find the relative cost factor k_T/k_m for a water distribution system consisting of cast iron pipes and having a pumping plant of standby 0.5. The combined efficiency of pump and prime mover $= 0.75$. The unit cost of energy $= 0.0005k_m$ units. The annual and daily averaging factors are 0.8 and 0.4, respectively; $k_p/k_m = 1.6$ units. Adopt $r = 0.05$ per year.

Solution. Dropping primes, Eq. (4.19) can be written as

$$k \Leftarrow k\left[1 + \frac{1 - \alpha}{(1 + r)^T - 1} + \frac{\beta}{r}\right]. \tag{4.22}$$

Thus, using Eq. (4.22), k_m is replaced by $1.145k_m$. Similarly, k_p is replaced by $2.341k_p = 1.6 \times 2.341k_m$. Thus, k_p is replaced by $3.746k_m$. Using (4.21b),

$$k_T = \frac{(1 + 0.5) \times 3.746\,k_m}{1000 \times 0.75} + \frac{8.76 \times 0.8 \times 0.4 \times 0.0005\,k_m}{0.75 \times 0.05}$$

$$= 0.00749k_m + 0.0374k_m = 0.0449k_m.$$

Thus, the relative cost factor $k_T/k_m = 0.0449k_m/1.145k_m = 0.0392$ units.

As seen in later chapters, k_T/k_m occurs in many optimal design formulations.

EXERCISES

4.1. Find the capitalized cost of an 8000-m-long, cast iron pumping main of diameter 0.65 m. It carries a discharge of $0.15\,\mathrm{m^3/s}$ throughout the year. The pumping head developed is 40 m; unit cost of energy = $0.0005k_m$ units; combined efficiency of pump and prime mover = 0.80; $k_p/k_m = 1.7$ units; $s_b = 0.5$; adopt $r = 0.07$ per year. Use Table 4.3 for necessary data.

4.2. Find net present value (NPV) of the pumping system having a 2000-m-long, cast-iron main of diameter 0.65 m. It carries a discharge of $0.10\,\mathrm{m^3/s}$ throughout the year. The pumping head developed is 35 m; unit cost of energy = $0.0006k_m$ units; combined efficiency of pump and prime mover = 0.80; $k_p/k_m = 1.8$ units; $s_b = 0.75$; adopt $r = 0.06$ per year. Compare the results with capitalized cost of this system and describe the reasons for the difference in the two life-cycle costs. Use Table 4.3 for necessary data.

4.3. Find the relative cost factor k_T/k_m for a water distribution system consisting of cast iron pipes and having a pumping plant of standby 0.5 and the combined efficiency of pump and prime mover = 0.75. The unit cost of energy = $0.0005k_m$ units. The annual and daily averaging factors are 0.8 and 0.4, respectively; $k_p/k_m = 1.6$ units. Adopt $r = 0.05$ per year.

REFERENCES

Samra, S., and Essery, C. (2003). *NSW Reference Rates Manual for Valuation of Water Supply, Sewerage and Stormwater Assets.* Ministry of Energy and Utilities, NSW, Australia.

Sharma, R.K. (1979). *Optimisation of Water Supply, Zones.* Thesis presented to the University of Roorkee, Roorkee, India, in the partial fulfillment of the requirements for the degree of Master of Engineering.

Sharma, A.K. (1989). *Water Distribution Network Optimisation.* Thesis presented to the University of Roorkee, Roorkee, India, in the fulfillment of the requirements for the degree of Doctor of Philosophy.

Sharma, A.K. and Swamee, P.K. (2006). Cost considerations and general principles in the optimal design of water distribution systems. 8th Annual International Symposium on Water Distribution Systems Analysis, Cincinnati, OH, 27–30 August 2006.

Swamee, P.K. (1996). Design of multistage pumping main. *J. Transport. Eng.* 122(1), 1–4.

Swamee, P.K., and Kumar, V. (2005). Optimal water supply zone size. *J. Water Supply: Research and Technology–AQUA* 54(3), 179–187.

5

GENERAL PRINCIPLES OF NETWORK SYNTHESIS

A pipe network should be designed in such a way to minimize its cost, keeping the aim of supplying the fluid at requisite quantity and prescribed pressure head. The maximum savings in cost are achieved by selecting proper geometry of the network. Usually, water

Design of Water Supply Pipe Networks. By Prabhata K. Swamee and Ashok K. Sharma
Copyright © 2008 John Wiley & Sons, Inc.

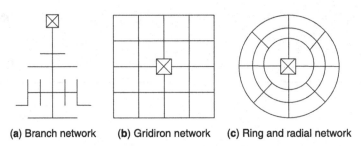

(a) Branch network   (b) Gridiron network   (c) Ring and radial network

Figure 5.1. Water supply network configurations.

distribution lines are laid along the streets of a city. Therefore, optimal design of a water supply system should determine the pattern and the length of the street system in the planning of a city. The water networks have either branched or looped geometry. Branched networks (Fig. 5.1a) are not preferred due to reliability and water quality considerations. Two basic configurations of a looped water distribution system shown are gridiron pattern (Fig. 5.1b) and ring and radial pattern (Fig. 5.1c). The gridiron and radial pattern are the best arrangement for a water supply system as all the mains are looped and interconnected. Thus, in the event of any pipe break, the area can be supplied from other looped mains. However, such distribution systems may not be feasible in areas where ground elevations vary greatly over the service area. Moreover, it is not possible to find the optimal geometric pattern for an area that minimizes the cost.

In Chapter 4, cost functions of various components of a pipe network have been formulated that can be used in the synthesis of a water supply system based on cost considerations. In the current form, disregarding reliability, we restrict ourselves to the cost of the network only. Thus, minimization of cost is the objective of the design. In such a problem, the cost function is the objective function of the system.

The objective function F is a function of *decision variables* (which are commonly known as *design variables*) like pipe diameters and pumping heads, which can be written as

$$F = F(D_1, D_2, D_3, \ldots D_i \ldots D_{i_L}, h_{01}, h_{02}, h_{03}, \ldots h_{0k} \ldots h_{0n_L}), \qquad (5.1)$$

where D_i = diameter of pipe link i, h_{0k} = input head or source point (through pumping stations or through elevated reservoirs), i_L = number of pipe links in a network, and n_L = number of input source points.

5.1. CONSTRAINTS

The problem is to minimize the objective function F. By selecting all the link diameters and the input heads to zero, the objective function can be reduced to zero. This is not an acceptable situation as there will be no pipe network, and the objective of fluid transport will not be achieved. In order to exclude such a solution, additional conditions of transporting the fluid at requisite pressure head have to be prescribed (Sharma and Swamee,

2006). Furthermore, restrictions of minimum diameter and maximum average velocity have to be observed. The restriction of minimum diameter is from practical considerations, whereas the restriction of maximum average velocity avoids excessive velocities that are injurious to the pipe material. These restrictions are called *safety constraints.* Additionally, certain relationships, like the summation of discharges at a nodal point should be zero, and so forth, have to be satisfied in a network. Such restrictions are called *system constraints.* These constraints are discussed in detail in the following sections.

5.1.1. Safety Constraints

The minimum diameter constraint can be written as

$$D_i \geq D_{min} \quad i = 1, 2, 3, \ldots i_L, \tag{5.2}$$

where D_{min} = the minimum prescribed diameter. The value of D_{min} depends on the pipe material, operating pressure, and size of the city. The minimum head constraint can be written as

$$h_j \geq h_{min} \quad j = 1, 2, 3, \ldots j_L, \tag{5.3a}$$

where h_j = nodal head, and h_{min} = minimum allowable pressure head. For water supply network, h_{min} depends on the type of the city. The general consideration is that the water should reach up to the upper stories of low-rise buildings in sufficient quality and pressure, considering firefighting requirements. In case of high-rise buildings, booster pumps are installed in the water supply system to cater for the pressure head requirements. With these considerations, various codes recommend h_{min} ranging from 8 m to 20 m for residential areas. However, these requirements vary from country to country and from state to state. The designers are advised to check local design guidelines before selecting certain parameters. To minimize the chances of leakage through the pipe network, the following maximum pressure head constraint is applied:

$$h_j \leq h_{max} \quad j = 1, 2, 3, \ldots j_L, \tag{5.3b}$$

where h_{max} = maximum allowable pressure head at a node. The maximum velocity constraint can be written as

$$\frac{4Q_i}{\pi D_i^2} \leq V_{max} \quad i = 1, 2, 3, \ldots i_L, \tag{5.4}$$

where Q_i = pipe discharge, and V_{max} = maximum allowable velocity. The maximum allowable velocity depends on the pipe material. The minimum velocity constraint can also be considered if there is any issue with the sediment deposition in the pipelines.

5.1.2. System Constraints

The network must satisfy Kirchhoff's current law and the voltage law, stated as

1. The summation of the discharges at a node is zero; and
2. The summation of the head loss along a loop is zero.

Kirchhoff's current law can be written as:

$$\sum_{i \in N_P(j)} S_i Q_i = q_j \quad j = 1, 2, 3, \dots j_L, \tag{5.5}$$

where $N_P(j) =$ the set of pipes meeting at the node j, and $S_i = 1$ for flow direction toward the node, -1 for flow direction away from the node.
Kirchhoff's voltage law for the loop k can be written as:

$$\sum_{i \in I_k(k)} S_{k,i}\left(h_{f_1} + h_{mi}\right) = 0 \quad k = 1, 2, 3, \dots k_L, \tag{5.6}$$

where $S_{k,i} = 1$ or -1 depending on whether the flow direction is clockwise or anticlockwise, respectively, in the link i of loop k; $h_{fi} =$ friction loss; $h_{mi} =$ form loss; and $I_k(k) =$ the set of the pipe links in the loop k.

5.2. FORMULATION OF THE PROBLEM

The synthesis problem thus boils down to minimization of Eq. (5.1) subject to the constraints given by Eqs. (5.2), (5.3a, b), (5.4), (5.5), and (5.6). The objective function is nonlinear in D_i. Similarly, the nodal head constraints Eqs. (5.3a, b), maximum velocity constraints Eq. (5.4), and the loop constraints Eq. (5.6) are also nonlinear. Such a problem cannot be solved mathematically; however, it can be solved numerically. Many numerical algorithms have been devised from time to time to solve such problems.

For nonloop systems, it is easy to eliminate the *state variables* (pipe discharges and nodal heads) from the problem. Thus, the problem is greatly simplified and reduced in size. These simplified problems are well suited to Lagrange multiplier method and geometric programming method to yield closed form solutions. In the Chapters 6 and 7, closed form optimal design of nonloop systems, like water transmission lines and water distribution lines, is described.

5.3. ROUNDING OFF OF DESIGN VARIABLES

The calculated pipe diameter, pumping head, and the pumping horsepower are continuous in nature, thus can never be provided in actual practice as the pipe and the pumping plant of requisite sizes and specifications are not manufactured commercially.

The designer has to select a lower or higher size out of the commercially available pipe sizes than the calculated size. If a lower size is selected, the pipeline cost decreases at the expense of the pumping cost. On the other hand, if the higher size is selected, the pumping cost decreases at the expense of the pipeline cost. Out of these two options, one is more economical than the other. For a pumping main, both the options are evaluated, and the least-cost solution can be adopted.

As the available pumping horsepower varies in certain increments, one may select the pumping plant of higher horsepower. However, it is not required to revise the pipe diameter also as the cost of the pumping plant is insignificant in comparison with the pumping cost or the power cost. Similarly, if the number of pumping stages in a multi-stage pumping main involves a fractional part, the next higher number should be adopted for the pumping stages.

5.4. ESSENTIAL PARAMETERS FOR NETWORK SIZING

The selection of the design period of a water supply system, projection of water demand, per capita rate of water consumption, design peak factors, minimum prescribed pressure head in distribution system, maximum allowable pressure head, minimum and maximum pipe sizes, and reliability considerations are some of the important parameters required to be selected before designing any water system. A brief description of these parameters is provided in this section.

5.4.1. Water Demand

The estimation of water demand for the sizing of any water supply system or its component is the most important part of the design methodology. In general, water demands are generated from residential, industrial, and commercial developments, community facilities, firefighting demand, and account for system losses. It is difficult to predict water demand accurately as a number of factors affect the water demand (i.e., climate, economic and social factors, pricing, land use, and industrialization of the area). However, a comprehensive study should be conducted to estimate water demand considering all the site-specific factors. The residential forecast of future demand can be based on house count, census records, and population projections.

The industrial and commercial facilities have a wide range of water demand. This demand can be estimated based on historical data from the same system or from comparable users from other systems. The planning guidelines provided by engineering bodies/regulatory agencies should be considered along with known historical data for the estimation of water demand.

The firefighting demand can be estimated using Kuichling or Freeman's formula. Moreover, local guidelines or design codes also provide information for the estimation of water demand for firefighting. The estimation of system losses is difficult as it usually depends on a number of factors. The system losses are a function of the age of the system, minimum prescribed pressure, and maximum pressure in the system. Historical data can be used for the assessment of system losses. Similarly, water unaccounted for due to unmetered usage, sewer line flushing, and irrigation of public parks should also be considered in total water demand projections.

5.4.2. Rate of Water Supply

To estimate residential water demand, it is important to know the amount of water consumed per person per day for in-house (kitchen, bathing, toilet, and laundry) usage and external usage for garden irrigation. The average daily per capita water consumption varies widely, and as such, variations depend upon a number of factors.

Fair et al. (1981) indicated that per capita water usage varies widely due to the differences in (1) climatic conditions, (2) standard of living, (3) extent of sewer system, (4) type of commercial and industrial activity, (5) water pricing, (6) resort to private supplies, (7) water quality for domestic and industrial purposes, (8) distribution system pressure, (9) completeness of meterage, and (10) system management.

The Organization for Economic Co-operation and Development (OECD, 1999) has listed per capita household water consumption rates across OECD member countries. It can be seen that the consumption rates vary from just over 100 L per capita per day to more than 300 L per capita per day based on climatic and economic conditions. Similarly, Lumbrose (2003) has provided information on typical rural domestic water use figures for some of the African countries.

Water Services Association of Australia (WSAA, 2000) published the annual water consumption figure of 250.5 Kiloliters (KL)/year for the average household in *WSAA Facts*. This consumption comprises 12.5 KL for kitchen, 38.3 KL for laundry, 48.6 KL for toilet, 65 KL for bathroom, and 86 KL for outdoor garden irrigation. The parentage break-up of internal household water consumption of 164.5 KL is shown in Fig. 5.2a and also for total water consumption in Fig. 5.2b. The internal water consumption relates to usage in kitchen, bathroom, laundry, and toilet, and the external water consumption is mainly for garden irrigation including car washing. The sum of the two is defined as total water consumption.

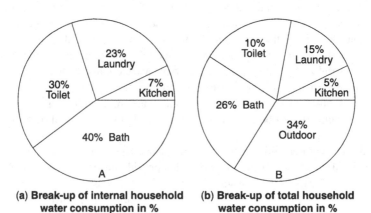

(a) Break-up of internal household water consumption in %   **(b) Break-up of total household water consumption in %**

Figure 5.2. Break-up of household water consumption for various usages.

Buchberger and Wells (1996) monitored the water demand for four single-family residences in City of Milford, Ohio, for a 1-year period. The year-long monitoring program recorded more than 600,000 signals per week per residence. The time series of daily per capita water demand indicated significant seasonal variation. Winter water demands were reported reasonably homogenous with average daily water demands of 250 L per capita and 203 L per capita for two houses.

The peak water demand σ per unit area $(\mathrm{m}^3/\mathrm{s}/\mathrm{m}^2)$ is an important parameter influencing the optimal cost of a pumping system. Swamee and Kumar (2005) developed an empirical relationship for the estimation of optimal cost F^* (per m^3/s of peak water supply) of a circular zone water supply system having a pumping station located at the center, and n equally spaced branches:

$$F^* = 1.2\,k_m \left(\frac{64\,k_T \rho f n^3}{3\,\pi^5\,k_m \sigma^3} \right)^{\frac{1}{6}}. \tag{5.6a}$$

5.4.3. Peak Factor

The water demand is not constant throughout the day and varies greatly over the day. Generally, the demand is lowest during the night and highest during morning or evening hours of the day. Moreover, this variation is very high for single dwellings and decreases gradually as population increases. The ratio of peak hourly demand to average hourly demand is defined as peak factor.

The variation in municipal water demand over the 24-hour daily cycle is called a diurnal demand curve. The diurnal demand patterns are different for different cities and are influenced by climatic conditions and economic development of the area. Two typical diurnal patterns are shown in Fig. 5.3. These curves are different in nature depicting the different pattern of diurnal water consumption. Pattern A indicates that two demand peaks occur, one in morning and the other in the evening hours of

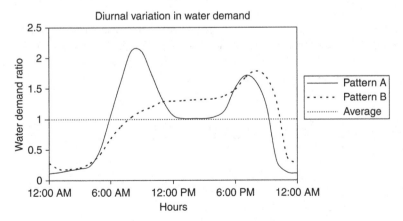

Figure 5.3. Diurnal variation curves.

the day. On the other hand, in pattern B only one peak occurs during the evening hours of the day.

Peaks of water demand affect the design of the water distribution system. High peaks of hourly demand can be expected in predominately residential areas; however, the hours of occurrence depend upon the characteristics of the city. In case of an industrial city, the peaks are not pronounced, thus the peak factors are relatively low.

Generally, the guidelines for suitable peak factor adoption are provided by local, state, or federal regulatory agencies or engineering bodies. However, it remains the designer's choice based on experience to select a suitable peak factor. To design the system for worst-case scenario, the peak factor can be based on the ratio of hourly demand of the maximum day of the maximum month to average hourly demand.

WSAA (2002) suggested the following peak factors for water supply system where water utilities do not specify an alternative mode.

Peak day demand over a 12-month period required for the design of a distribution system upstream of the balancing storage shall be calculated as:

$$\text{Peak day demand} = \text{Average day demand} \times \text{Peak day factor}$$

Peak day factor can be defined as the ratio of peak day demand or maximum day demand during a 12-month period over average day demand of the same period. Peak hour demand or maximum hour demand over a 24-hour period required for the design of a distribution system downstream of the balancing storage can be calculated as:

$$\text{Peak hour demand} = \text{Average hour demand (on peak day)} \times \text{Peak hour factor}$$

Thus, the peak hour factor can be defined as the ratio of peak hour demand on peak day over average hour demand over the same 24 hours. The peak day factor and peak hour factor are listed in Table 5.1. These values for population between 2000 and 10,000 can be interpolated using the data.

Peak factor for a water distribution design can also be estimated from the ratio of peak hourly demand on a maximum demand day during the year over the average hourly demand over the same period. The readers are advised to collect local information or guidelines for peak factor selection.

TABLE 5.1. Peak Day and Peak Hour Factors

Peak day factor
1.5 for population over 10,000
2 for population below 2000
Peak hour factor/peak factor
2 for population over 10,000
5 for population below 2000

On the other hand, annual averaging factor F_A and daily averaging factor F_D were considered for the estimation of annual energy in Eq. (4.9). F_A can be defined as a fraction of the year over which the system would supply water to the customers. It can be correlated with the reliability of pumping system. Similarly, the product of F_D and peak discharge should be equal to average discharge over the day. Thus, the F_D can be defined as the inverse of peak factor.

5.4.4. Minimum Pressure Requirements

The minimum design nodal pressures are prescribed to discharge design flows onto the properties. Generally, it is based on population served, types of dwellings in the area, and firefighting requirements. The information can be found in local design guidelines. As it is not economic to maintain high pressure in the whole system just to cater to the need of few highrise buildings in the area, the provision of booster pumps are specified. Moreover, water leakage losses increase with the increase in system pressure in a water distribution system.

5.4.5. Minimum Size of Distribution Main

The minimum size of pipes in a water distribution system is specified to ensure adequate flow rates and terminal pressures. It works as factor of safety against assumed population load on a pipe link and also provides a guarantee to basic firefighting capability. The minimum pipe sizes are normally specified based on total population of a city. Generally, a minimum size pipe of 100 mm for residential areas and 150 mm for commercial/industrial areas is specified. Local design guidelines should be referred to for minimum size specifications.

5.4.6. Maximum Size of Water Distribution

The maximum size of a distribution main depends upon the commercially available pipe sizes for different pipe material, which can be obtained from local manufacturers. The mains are duplicated where the design diameters are larger than the commercially available sizes.

5.4.7. Reliability Considerations

Generally, water distribution systems are designed for optimal configuration that could satisfy minimum nodal pressure criteria at required flows. The reliability considerations are rarely included in such designs. The reliability of water supply system can be divided into structural and functional forms. The structural reliability is associated with pipe, pump, and other system components probability of failure, and the functional reliability is associated with meeting nodal pressure and flow requirements.

The local regulatory requirements for system reliability must be addressed. Additional standby capacity of the important system components (i.e., treatment units and pumping plants) should be provided based on system reliability requirements.

Generally, asset-based system reliability is considered to guarantee customer service obligations.

In a water distribution system, pipe bursts, pump failure, storage operation failure, and control system failure are common system failures. Thus, the overall reliability of a system should be based on the reliability of individual components.

Su et al. (1987) developed a method for the pipe network reliability estimation. The probability of failure P_i of pipe i using Poisson probability distribution is

$$P_i = 1 - e^{-\beta_i}, \qquad (5.7a)$$

and $\beta_i = p_i L_i$, where β_i is the expected number of failures per year for pipe i, p_i is the expected number of failures per year per unit length of pipe i, and L_i is the length of pipe i.

The overall probability of failure of the system was estimated on the values of system and nodal reliabilities based on minimum cut-sets (MCs). A cut-set is a pipe (or combination of pipes) where, upon breakage, the system does not meet minimum system hydraulic requirements. The probability of failure P_s of the system in case of total minimum cut-sets T_{MC} with n pipes in jth cut-set:

$$P_s = \sum_{j=1}^{T_{MC}} P(MC_j), \qquad (5.7b)$$

where
$$P(MC_j) = \prod_{i=1}^{n} P_i = P_1 \times P_2 \times \cdots \times P_n. \qquad (5.7c)$$

The system reliability can be estimated as

$$R_s = 1 - P_s. \qquad (5.7d)$$

Swamee et al. (1999) presented an equation for the estimation of probability of breakage p in pipes in breaks/meter/year as:

$$p = \frac{0.0021\, e^{-4.35D} + 21.4D^8\, e^{-3.73D}}{1 + 10^5 D^8}, \qquad (5.7e)$$

where D is in meters. It can be seen from Eq. (5.7e) that the probability of breakage of a pipe link is a decreasing function of the pipe diameter D (m), whereas it is linearly proportional to its length.

5.4.8. Design Period of Water Supply Systems

Water supply systems are planned for a predecided time horizon generally called design period. In current design practices, disregarding the increase in water demand, the life of pipes, and future discount rate, the design period is generally adopted as 30 years on an *ad hoc* basis.

For a static population, the system can be designed either for a design period equal to the life of the pipes sharing the maximum cost of the system or for the perpetual existence of the supply system. Pipes have a life ranging from 60 years to 120 years depending upon the material of manufacture. Pipes are the major component of a water supply system having very long life in comparison with other components of the system. Smith et al. (2000) have reported the life of cast iron pipe as above 100 years. Alferink et al. (1997) investigated old poly (vinyl chloride) (PVC) water pipes laid 35 years ago and concluded that the new PVC pipes would continue to perform for considerably more than a 50-year lifetime. Plastic Pipe Institute (2003) has reported about the considerable supporting justification for assuming a 100-year or more design service life for corrugated polyethylene pipes. The exact information regarding the life of different types of pipes is not available. The PVC and asbestos cement (AC) pipes have not even crossed their life expectancy as claimed by the manufacturers since being used in water supply mains. Based on available information from manufacture's and user organizations, Table 5.2 gives the average life T_u of different types of pipes.

For a growing population or water demand, it is always economical to design the mains in staging periods and then strengthen the system after the end of every staging period. In the absence of a rational criterion, the design period of a water supply system is generally based on the designer's intuition disregarding the life of the component sharing maximum cost, pattern of the population growth or increase in water demand, and discount rate.

For a growing population, the design periods are generally kept low due to uncertainty in population prediction and its implications to the cost of the water supply systems. Hence, designing the water systems for an optimal period should be the main consideration. The extent to which the life-cycle cost can be minimized would depend upon the planning horizon (design period) of the water supply mains. As the pumping and transmission mains differ in their construction and functional requirements (Swamee and Sharma, 2000), separate analytical analysis is conducted for these two systems.

TABLE 5.2. Life of Pipes

Pipe Material	Life, T_u (Years)
Cast iron (CI)	120
Galvanized iron (GI)	120
Electric resistance welded (ERW)	120
Asbestos cement (AC)	60
Poly(vinyl chloride) (PVC)	60

Sharma and Swamee (2004) gave the following equation for the design period T of gravity flow systems:

$$T = T_u\left(1 + 2\alpha rT_u^2\right)^{0.375} \tag{5.8}$$

and the design period for pumping system as:

$$T = T_u\left(1 + 0.417\alpha rT_u^2 + 0.01\alpha^2 T_u^2\right)^{0.5}, \tag{5.9}$$

where r is discount rate factor, and α is rate of increase in water demand such that the initial water demand Q_0 increases to Q after time t as $Q = Q_0 e^{\alpha t}$.

Example 5.1. Estimate the design period for a PVC water supply gravity as well as pumping main, consider $\alpha = 0.04/\text{yr}$ and $r = 0.05$.

Solution. Using Eq. (5.8), the design period for gravity main is obtained as:

$$T = 60\left(1 + 2 \times 0.04 \times 0.05 \times 60^2\right)^{0.375} = 20.46\,\text{yr} \approx 20\,\text{yr}.$$

Similarly, using Eq. (5.9), the design period for a pumping main is obtained as:

$$T = 60\left(1 + 0.417 \times 0.04 \times 0.05 \times 60^2 + 0.01 \times 0.04^2 \times 60^2\right)^{0.5} = 29.77\,\text{yr} \approx 30\,\text{yr}.$$

Thus, the water supply gravity main should be designed initially for 20 years and then restrengthened after every 20 years. Similarly, the pumping main should be designed initially for 30 years and then restrengthened after every 30 years.

5.4.9. Water Supply Zones

Large water distribution systems are difficult to design, maintain, and operate, thus are divided into small subsystems called water supply zones. Each subsystem contains an input point (supply source) and distribution network. These subsystems are interconnected with nominal size pipe for interzonal water transfer in case of a system breakdown or to meet occasional spatial variation in water demands. It is not only easy to design subsystems but also economic due to reduced pipe sizes. Swamee and Sharma (1990) presented a method for splitting multi-input system into single-input systems based on topography and input pumping heads without cost considerations and also demonstrated reduction in total system cost if single-input source systems were designed separately. Swamee and Kumar (2005) developed a method for optimal zone sizes based on cost considerations for circular and rectangular zones. These methods are described in Chapter 12.

5.4.10. Pipe Material and Class Selection

Commercial pipes are manufactured in various pipe materials; for example, poly (vinyl chloride) (PVC), unplasticised PVC (uPVC), polyethylene (PE), asbestos cement (**AC**), high-density polyethylene (HDPE), mild steel (MS), galvanized iron (GI) and electric resistance welded (ERW). These pipes have different roughness heights, working pressure, and cost. The distribution system can be designed initially for any pipe material on an *ad hoc* basis, say CI, and then economic pipe material for each pipe link of the system can be selected. Such a pipe material selection should be based on maximum water pressure on pipes and their sizes, considering the entire range of commercial pipes, their materials, working pressures, and cost. A methodology for economic pipe material selection is described in Chapter 8.

EXERCISES

5.1. Describe constraints in the design problem formulation of a water distribution network.

5.2. Select the essential design parameters for the design of a water distribution system for a new development/subdivision having a design population of 10,000.

5.3. Estimate the design period of a gravity as well as a pumping main of CI pipe. Consider $\alpha = 0.03/\text{yr}$ and $r = 0.04$.

REFERENCES

Alferink, F., Janson, L.E., and Holloway, L. (1997). Old unplasticised poly(vinyl chloride) water pressure pipes Investigation into design and durability. *Plastics, Rubber and Composites Processing and Applications* 26(2), 55–58.

Buchberger, S.G., and Wells, J.G. (1996). Intensity, duration and frequency of residential water demands. *J. Water Resources Planning & Management* 122(1), 11–19.

Fair, G.M., Gayer, J.C., and Okun, D.A. (1981). *Elements of Water Supply and Wastewater Disposal*, Second ed. John Wiley & Sons, New York; reprint (1981), Toppan Printing Co.(s) Pte. Ltd., Singapore.

Lumbrose, D. (2003). *Handbook for the Assessment of Catchment Water Demand and Use.* HR Wallingford, Oxon, UK.

OECD. (1999). *The Price of Water, Trends in the OECD Countries.* Organisation for Economic Cooperation and Development, Paris, France.

Plastic Pipe Institute. (2003). *Design Service Life of Corrugated HDPE Pipe.* PPI report TR-43/2003, Washington, DC.

Sharma, A.K., and Swamee, P.K. (2004). Design life of water transmission mains for exponentially growing water demand. *J. Water Supply: Research and Technology-AQUA, IWA* 53(4), 263–270.

Sharma, A.K., and Swamee, P.K. (2006). Cost considerations and general principles in the optimal design of water distribution systems. 8th Annual International Symposium on Water Distribution Systems Analysis, Cincinnati, Ohio, 27–30 August 2006.

Smith, L.A., Fields, K.A., Chen, A.S.C., and Tafuri, A.N. (2000). *Options for Leak and Break Detection and Repair of Drinking Water Systems*. Battelle Press, Columbus, OH.

Su, Y., Mays, L.W., Duan, N., and Lansey, K.E. (1987). Reliability-based optimization model for water distribution systems. *J. Hydraul. Eng.* 114(12), 1539–1559.

Swamee, P.K., and Sharma, A.K. (1990). Decomposition of a large water distribution system. *J. Env. Eng.* 116(2), 296–283.

Swamee, P.K., Tyagi, A., and Shandilya, V.K. (1999). Optimal configuration of a well-field. *Ground Water* 37(3), 382–386.

Swamee, P.K., and Sharma, A.K. (2000). Gravity flow water distribution network design. *J. Water Supply: Research and Technology-AQUA* 49(4), 169–179.

Swamee, P.K., and Kumar, V. (2005). Optimal water supply zone size. *J. Water Supply: Research and Technology-AQUA* 54(3), 179–187.

WSAA. (2000). *The Australian Urban Water Industry WSAA Facts 2000*. Water Services Association of Australia, Melbourne, Australia.

WSAA. (2002). Water supply code of Australia. Melbourne retail water agencies edition, Version 1.0. Water Service Association of Australia, Melbourne, Australia.

6

WATER TRANSMISSION LINES

Water or any other liquid is required to be carried over long distances through pipelines. Like electric transmission lines transmit electricity, these pipelines transmit water. As defined in chapter 3, if the flow in a water transmission line is maintained by creating a pressure head by pumping, it is called a pumping main. On the other hand, if the flow in a water transmission line is maintained through the elevation difference, it is called a gravity main. There are no intermediate withdrawals in a water transmission line. This chapter discusses the design aspects of water transmission lines.

    The pumping and the gravity-sustained systems differ in their construction and functional requirements (Swamee and Sharma, 2000) as listed in Table 6.1.

Design of Water Supply Pipe Networks. By Prabhata K. Swamee and Ashok K. Sharma
Copyright © 2008 John Wiley & Sons, Inc.

TABLE 6.1. Comparison of Pumping and Gravity Systems

Item	Gravity System	Pumping System
1. Conveyance main	Gravity main	Pumping main
2. Energy source	Gravitational potential	External energy
3. Input point	Intake chamber	Pumping station
4. Pressure corrector	Break pressure tank	Booster
5. Storage reservoir	Surface reservoir	Elevated reservoir
6. Source of water	Natural water course	Well, river, lake, or dam

6.1. GRAVITY MAINS

A typical gravity main is depicted in Fig. 6.1. Because the pressure head h_0 (on account of water level in the collection tank) varies from time to time, much reliance cannot be placed on it. For design purposes, this head should be neglected. Also neglecting the entrance and the exit losses, the head loss can be written as:

$$h_L = z_0 - z_L - H. \tag{6.1}$$

Using Eqs. (2.22a) and (6.1), the pipe diameter is found to be

$$D = 0.66 \left\{ \varepsilon^{1.25} \left[\frac{LQ^2}{g(z_0 - z_L - H)} \right]^{4.75} + vQ^{9.4} \left[\frac{L}{g(z_0 - z_L - H)} \right]^{5.2} \right\}^{0.04}. \tag{6.2a}$$

Using Eqs. (4.4) and (6.2a), the capitalized cost of the gravity main works out as

$$F = 0.66^m L k_m \left\{ \varepsilon^{1.25} \left[\frac{LQ^2}{g(z_0 - z_L + H)} \right]^{4.75} + vQ^{9.4} \left[\frac{L}{g(z_0 - z_L + H)} \right]^{5.2} \right\}^{0.04m}. \tag{6.2b}$$

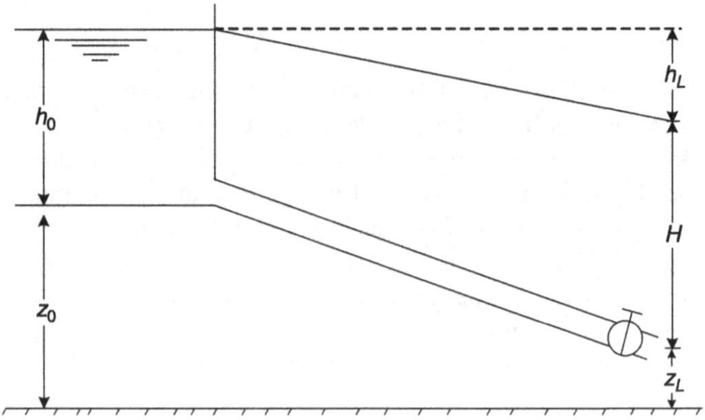

Figure 6.1. A gravity main.

The constraints for which the design must be checked are the minimum and the maximum pressure head constraints. The pressure head h_x at a distance x from the source is given by

$$h_x = z_0 + h_0 - z_x - 1.07 \frac{xQ^2}{gD^5} \left\{ \ln \left[\frac{\varepsilon}{3.7D} + 4.618 \left(\frac{vD}{Q} \right)^{0.9} \right] \right\}^{-2}, \qquad (6.3)$$

where z_x = elevation of the pipeline at distance x. The minimum pressure head can be negative (i.e., the pressure can be allowed to fall below the atmospheric pressure). The minimum allowable pressure head is $-2.5\,\text{m}$ (Section 2.2.9). This pressure head ensures that the dissolved air in water does not come out resulting in the stoppage of flow. In case the minimum pressure head constraint is violated, the alignment of the gravity main should be changed to avoid high ridges, or the main should pass far below the ground level at the high ridges.

If the maximum pressure head constraint is violated, one should use pipes of higher strength or provide break pressure tanks at intermediate locations and design the connected gravity mains separately. A break pressure tank is a tank of small plan area (small footprint) provided at an intermediate location in a gravity main. The surplus elevation head is nullified by providing a fall within the tank (Fig. 6.2). Thus, a break pressure tank divides a gravity main into two parts to be designed separately.

The design must be checked for the maximum velocity constraint. If the maximum velocity constraint is violated marginally, the pipe diameter may be increased to satisfy the constraint. In case the constraint is violated seriously, break pressure tanks may be provided at the intermediate locations, and the connecting gravity mains should be designed separately.

Example 6.1. Design a cast iron gravity main for carrying a discharge of $0.65\,\text{m}^3/\text{s}$ over a distance of 10 km. The elevation of the entry point is 175 m, whereas the elevation of the exit point is 140 m. The terminal head at the exit is 5 m.

Solution. Average roughness height ε for a cast iron as per Table 2.1 is 0.25 mm. The kinematic viscosity of water at $20°$ C is $1 \times 10^{-6}\,\text{m}^2/\text{s}$. Substituting, these values in

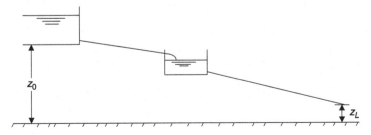

Figure 6.2. Location of break pressure tank.

Eq. (6.2a):

$$D = 0.66 \left\{ 0.00025^{1.25} \left[\frac{10{,}000 \times 0.65^2}{9.81 \times (175 - 140 + 5)} \right]^{4.75} \right.$$

$$\left. + 1 \times 10^{-6} \times 0.65^{9.4} \left[\frac{10{,}000}{9.81 \times (175 - 140 + 5)} \right]^{5.2} \right\}^{0.04},$$

$D = 0.69$ m. Adopt $D = 0.75$ m:

$$V = \frac{4 \times 0.65}{\pi \times 0.75^2} = 0.47 \text{ m/s},$$

which is within the permissible limits.

6.2. PUMPING MAINS

Determination of the optimal size of a pumping main has attracted the attention of engineers since the invention of the pump. Thresh (1901) suggested that in pumping mains, the average velocity should be about 0.6 m/s and in no case greater than 0.75 m/s. For the maximum discharge pumped, Q, this gives the pumping main diameter in SI units as $k\sqrt{Q}$; where $1.3 \leq k \leq 1.46$. On the other hand, the Lea formula (Garg, 1990) gives the range as $0.97 \leq k \leq 1.22$ in SI units. Using the Hazen–Williams equation, Babbitt and Doland (1949) and Turneaure and Russell (1955) obtained the economic diameter, whereas considering constant friction factor in the Darcy–Weisbach equation, Swamee (1993) found the pipe diameter.

A typical pumping main is shown in Fig. 6.3. The objective function to be minimized for a pumping main is

$$F = k_m L D^m + k_T \rho g Q h_0. \tag{6.4}$$

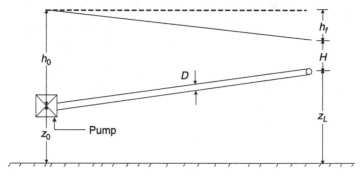

Figure 6.3. A pumping main.

The hydraulic constraint to be satisfied is

$$\frac{8fLQ^2}{\pi^2 gD^5} - h_0 - z_0 + H + z_L = 0. \tag{6.5}$$

Combining Eq. (6.4) with Eq. (6.5) through the Lagrange multiplier λ, the following merit function F_1 is obtained:

$$F_1 = k_m LD^m + k_T \rho g Q h_0 + \lambda \left(\frac{8fLQ^2}{\pi^2 gD^5} - h_0 - z_0 + H + z_L \right). \tag{6.6}$$

For optimality, the partial derivative of F_1 with respect to D, h_0, and λ should be zero.

6.2.1. Iterative Design Procedure

Assuming f to be constant, and differentiating partially F_1 with respect to D and simplifying, one gets

$$D = \left(\frac{40\lambda fQ^2}{\pi^2 gmk_m} \right)^{\frac{1}{m+5}}. \tag{6.7}$$

Differentiating F_1 partially with respect to h_0 and simplifying, one obtains

$$\lambda = k_T \rho g Q. \tag{6.8}$$

Eliminating λ between Eqs. (6.7) and (6.8), the optimal diameter D^* is obtained as

$$D^* = \left(\frac{40 k_T \rho f Q^3}{\pi^2 mk_m} \right)^{\frac{1}{m+5}}. \tag{6.9}$$

Substituting the optimal diameter in Eq. (6.5), the optimal pumping head h_0^* is obtained as:

$$h_0^* = H + z_L - z_0 + L \left[\left(\frac{8f}{\pi^2 g} \right)^m \left(\frac{mk_m}{5 k_T \rho g} \right)^5 Q^{-(5-2m)} \right]^{\frac{1}{m+5}}. \tag{6.10}$$

Substituting D^* and h_0^* in Eq. (6.4), the optimal cost F^* is found to be

$$F^* = k_m L \left(1 + \frac{m}{5} \right) \left(\frac{40 k_T \rho f Q^3}{\pi^2 mk_m} \right)^{\frac{m}{m+5}} + k_T \rho g Q (H + z_L - z_0). \tag{6.11}$$

Assuming an arbitrary value of f, the optimal diameter can be obtained by Eq. (6.9). Knowing the diameter, an improved value of f can be obtained by any of the Eqs. (2.6a–c). Using this value of f, an improved value of D^* can be obtained by Eq. (6.9). The process is repeated until the two successive values of D^* are very close. Knowing D^*, the values of h_0^* and F^* can be obtained by Eqs. (6.10) and (6.11), respectively.

It can be seen from Eq. (6.10) that the optimal pumping head is a decreasing function of the discharge, as m is normally less than 2.5. For $m = 2.5$, the optimal pumping head is independent of Q; and for $m > 2.5$, the optimal pumping head increases with the discharge pumped. However, at present there is no material for which $m \geq 2.5$.

Example 6.2. Design a ductile iron pumping main carrying a discharge of 0.25 m³/s over a distance of 5 km. The elevation of the pumping station is 275 m and that of the exit point is 280 m. The required terminal head is 10 m.

Solution. Adopting $k_T/k_m = 0.0131$, $m = 0.9347$, $\varepsilon = 0.25$ mm, and assuming $f = 0.01$ and using Eq. (6.9),

$$D^* = \left(\frac{40 \times 0.0131 \times 1000 \times 0.01 \times 0.25^3}{\pi^2 \times 0.9347} \right)^{\frac{1}{0.9347+5}} = 0.451 \text{m}.$$

Revising f as

$$f = 1.325 \left\{ \ln \left[\frac{0.25 \times 10^{-3}}{3.7 \times 0.451} + 4.618 \left(\frac{10^{-6} \times 0.451}{0.25} \right)^{0.9} \right] \right\}^{-2} = 0.01412.$$

The subsequent iteration yields $D^* = 0.478$ m using $f = 0.01412$. Based on revised pipe size, the friction factor is recalculated as $f = 0.01427$, and pipe size $D^* = 0.479$ m. Adopt 0.5 m as the diameter:

$$V = \frac{4 \times 0.25}{\pi \times 0.5^2} = 1.27 \text{ m/s},$$

which is within permissible limits.
Using Eq. (6.5), the optimal pumping head is 26.82 m, say 27 m.

6.2.2. Explicit Design Procedure

Eliminating f between Eqs. (2.6b) and (6.5), the constraint equation reduces to

$$z_0 + h_0 - H - z_L - 1.074 \frac{LQ^2}{gD^5} \left\{ \ln \left[\frac{\varepsilon}{3.7D} + 4.618 \left(\frac{vD}{Q} \right)^{0.9} \right] \right\}^{-2} = 0. \quad (6.12)$$

Using Eqs. (6.4) and (6.12), and minimizing the cost function, the optimal diameter is obtained. Relating this optimal diameter to the entry variables, the following empirical equation is obtained by curve fitting:

$$D^* = \left[\left(0.591 \frac{k_T \rho Q^3 \varepsilon^{0.263}}{mk_m} \right)^{\frac{40}{m+5.26}} + \left(0.652 \frac{k_T \rho Q^{2.81} v^{0.192}}{mk_m} \right)^{\frac{40}{m+4.81}} \right]^{0.025}. \quad (6.13a)$$

Putting $v = 0$ for a rough turbulent flow case, Eq. (6.13a) reduces to

$$D^* = \left(0.591 \frac{k_T \rho Q^3 \varepsilon^{0.263}}{m k_m} \right)^{\frac{1}{m+5.26}}. \qquad (6.13b)$$

Similarly, by putting $\varepsilon = 0$ in Eq. (6.13a), the optimal diameter for a smooth turbulent flow is

$$D^* = \left(0.652 \frac{k_T \rho Q^{2.81} v^{0.192}}{m k_m} \right)^{\frac{1}{m+4.81}}. \qquad (6.13c)$$

On substituting the optimal diameter from Eq. (6.13a) into Eq. (6.12), the optimal pumping head is obtained. Knowing the diameter and the pumping head, the optimal cost can be obtained by Eq. (6.4).

Example 6.3. Solve Example 6.2 using the explicit design procedure.

Solution. Substituting the values in Eq. (6.13a):

$$D^* = \left[\left(0.591 \frac{0.0131 \times 1000 \times 0.25^3 \times 0.00025^{0.263}}{0.9347} \right)^{\frac{40}{6.195}} \right.$$
$$\left. + \left(0.652 \frac{0.0131 \times 1000 \times 0.25^{2.81} \times (10^{-6})^{0.192}}{0.9347} \right)^{\frac{40}{5.745}} \right]^{0.025},$$

$D^* = 0.506$ m. Adopt 0.5 m diameter, the corresponding velocity is 1.27 m/s. It can be seen that Eq. (6.13a) slightly overestimates the diameter because in this case, both the roughness and the viscosity are approximately equally predominant.

6.3. PUMPING IN STAGES

Long-distance pipelines transporting fluids against gravity and frictional resistance involve multistage pumping. In a multistage pumping, the optimal number of pumping stages can be estimated by an enumeration process. Such a process does not indicate functional dependence of input parameters on the design variables (Swamee, 1996).

For a very long pipeline or for large elevation difference between the entry and exit points, the pumping head worked out using Eq. (6.10) is excessive and pipes withstanding such a high pressure may not be available, or the provision of high-pressure pipes may be uneconomical. In such a case, instead of providing a single pumping station, it is desirable to provide n pumping stations separated at a distance L/n. Provision of multiple pumping stations involves fixed costs associated at each pumping station.

This offsets the saving accrued by using low-pressure pipes. Thus, the optimal number of the pumping stages can be worked out to minimize the overall cost.

In current design practice, the number of pumping stages is decided arbitrarily, and the pumping main in between the two stages is designed as a single-stage pumping main. Thus, each of the pumping sections is piecewise optimal. Such a design will not yield an overall economy. The explicit optimal design equations for the design variables are described in this section.

The cost function F for a n stage pumping system is obtained by adding the pipe cost, pump and pumping cost, and the establishment cost E associated at each pumping station. Thus,

$$F = k_m\left(1 + \frac{h_0}{h_b}\right)LD^m + k_T\rho gQn(h_0 + h_c), \qquad (6.14)$$

where the allowable pressure head h_a has been taken as h_0; and the establishment cost was expressed as an extra pumping head h_c given by $E/(\rho g k_T Q)$.

Assuming a linear variation of the elevation profile, the elevation difference between the two successive pumping stations $= \Delta z/n$, where $\Delta z =$ the elevation difference between the inlet and the outlet levels. Using the Darcy–Weisbach equation for surface resistance, h_0 can be written as

$$h_0 = \frac{8fLQ^2}{\pi^2 gD^5 n} + H + \frac{\Delta z}{n}. \qquad (6.15)$$

Eliminating h_0 between Eqs. (6.14) and (6.15), one gets

$$F = k_n LD^m + \frac{k_m L\Delta zD^m}{nh_b} + \frac{8\,k_m L^2 fQ^2}{\pi^2 gh_b D^{5-m}n} + \frac{8\,k_T\rho fLQ^3}{\pi^2 D^5}$$

$$+ k_T\rho gQ(H + h_c)n + k_T\rho gQ\Delta z, \qquad (6.16)$$

where

$$k_n = k_m\left(1 + \frac{H}{h_b}\right). \qquad (6.17)$$

6.3.1. Long Pipeline on a Flat Topography

For a long pipeline on a relatively flat topography as shown in Fig. 6.4, Swamee (1996) developed a methodology for the determination of pumping main optimal diameter and the optimal number of pumping stations. The methodology is described below in which the multistage pumping main design is formulated as a geometric programming problem having a single degree of difficulty.

The total cost of a Multistage pumping system can be estimated using Eq. (6.16). The second term on the right-hand side of Eq. (6.16) being small can be neglected.

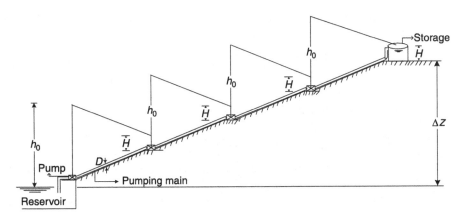

Figure 6.4. A typical multistage pumping main on flat topography.

Thus, Eq. (6.16) reduces to

$$F = k_n L D^m + \frac{8\, k_m L^2 f Q^2}{\pi^2 g h_b D^{5-m} n} + \frac{8\, k_T \rho f L Q^3}{\pi^2 D^5} + k_T \rho g Q (H + h_c) n + k_T \rho g Q \Delta z. \quad (6.18a)$$

The last term on the right-hand side of Eq. (6.18a) being constant will not enter into the optimization process; thus removing this term, Eq. (6.18a) changes to

$$F = k_n L D^m + \frac{8\, k_m L^2 f Q^2}{\pi^2 g h_b D^{5-m} n} + \frac{8\, k_T \rho f L Q^3}{\pi^2 D^5} + k_T \rho g Q (H + h_c) n. \quad (6.18b)$$

Thus, the design problem boils down to the minimization of a *posynomial* (positive polynomial) in the design variables D and n. This is a geometric programming problem having a single degree of difficulty.

Defining the weights w_1, w_2, w_3, and w_4 as

$$w_1 = \frac{k_n L D^m}{F} \quad (6.19a)$$

$$w_2 = \frac{8\, k_m L^2 f Q^2}{\pi^2 g h_b D^{5-m} n F} \quad (6.19b)$$

$$w_3 = \frac{8\, k_T \rho f L Q^3}{\pi^2 D^5 F} \quad (6.19c)$$

$$w_4 = \frac{k_T \rho g Q (H + h_c) n}{F}, \quad (6.19d)$$

and assuming f to be constant, the posynomial dual d of Eq. (6.18b) can be written as

$$d = \left(\frac{k_n LD^m}{w_1}\right)^{w_1} \left(\frac{8\,k_m L^2 fQ^2}{\pi^2 gh_b D^{5-m} nw_2}\right)^{w_2} \left(\frac{8\,k_T \rho fLQ^3}{\pi^2 D^5 w_3}\right)^{w_3} \left[\frac{k_T \rho gQ(H+h_c)n}{w_4}\right]^{w_4}. \quad (6.20)$$

The orthogonality conditions for Eq. (6.20) are

$$D: \quad mw_1^* - (5-m)w_2^* - 5w_3^* = 0 \quad\quad (6.21a)$$

$$n: \quad -w_2^* + w_4^* = 0, \quad\quad (6.21b)$$

and the normality condition for Eq. (6.20) is

$$w_1^* + w_2^* + w_3^* + w_4^* = 1, \quad\quad (6.21c)$$

where the asterisk indicates optimality. Solving Eqs. (6.21a–c) for w_1^*, w_2^*, and w_3^*, one gets

$$w_1^* = \frac{5}{m+5} - w_4^* \quad\quad (6.22a)$$

$$w_2^* = w_4^* \qu\quad (6.22b)$$

$$w_3^* = \frac{m}{m+5} - w_4^*. \quad\quad (6.22c)$$

Substituting w_1^*, w_2^*, and w_3^* from Eqs. (6.22a–c) into (6.20), the optimal dual d^* is

$$d^* = \frac{(m+5)k_n L}{5-(m+5)w_4^*} \left[\frac{5-(m+5)w_4^*}{m-(m+5)w_4^*}\frac{8\,k_T \rho fQ^3}{\pi^2 k_n}\right]^{\frac{m}{m+5}}$$

$$\times \left\{\frac{[m-(m+5)w_4^*][5-(m+5)w_4^*]}{(m+5)^2 w_4^{*2}}\frac{h_c+H}{h_b+H}\right\}^{w_4^*}, \quad\quad (6.23)$$

where w_1^* corresponds with optimality. Eliminating w_1^*, w_2^*, and w_3^*, and D, n, and F between Eqs. (6.19a–d) and Eqs.(6.22a–c), one gets the following quadratic equation in w_4^*:

$$\frac{(m+5)^2 w_4^{*2}}{[m-(m+5)w_4^*][5-(m+5)w_4^*]} = \frac{h_c+H}{h_b+H}. \quad\quad (6.24)$$

Equation (6.24) can also be obtained by equating the factor having the exponent w_4^* on the right-hand side of Eq. (6.23) to unity (Swamee, 1995). Thus, contrary to the optimization problem of zero degree of difficulty in which the weights are constants, in this problem of single degree of difficulty, the weights are functions of the parameters occurring in the objective function. The left-hand side of Eq. (6.24) is positive when $w_4^* < m/(m+5)$ or $w_4^* > 5/(m+5)$ (for which w_3^* is negative). Solving Eq. (6.24), the optimal weight was obtained as

$$w_4^* = \frac{10\,m}{(m+5)^2}\left\{1 + \left[1 - \frac{20\,m}{(m+5)^2}\frac{h_c - h_b}{h_c + H}\right]^{0.5}\right\}^{-1}. \tag{6.25a}$$

Expanding Eq. (6.25a) binomially and truncating the terms of the second and the higher powers, Eq. (6.25a) is approximated to

$$w_4^* = \frac{5\,m}{(m+5)^2}. \tag{6.25b}$$

Using Eqs. (6.23) and (6.24) and knowing $F^* = d^*$, one gets

$$F^* = \frac{(m+5)k_n L}{5 - (m+5)w_4^*}\left[\frac{5 - (m+5)w_4^*}{m - (m+5)w_4^*}\frac{8\,k_T \rho f Q^3}{\pi^2\,k_n}\right]^{\frac{m}{m+5}}. \tag{6.26}$$

Using Eqs. (6.19a), (6.22a), and (6.26), the optimal diameter D^* was obtained as

$$D^* = \left[\frac{5 - (m+5)w_4^*}{m - (m+5)w_4^*}\frac{8\,k_T \rho f Q^3}{\pi^2\,k_n}\right]^{\frac{1}{m+5}}. \tag{6.27}$$

Using Eqs. (6.19d) and (6.26), the optimal number of pumping stages n is

$$n^* = \frac{(m+5)w_4^*}{5 - (m+5)w_4^*}\frac{k_n L}{k_T \rho g Q(H + h_c)}\left[\frac{5 - (m+5)w_4^*}{m - (m+5)w_4^*}\frac{8\,k_T \rho f Q^3}{\pi^2\,k_n}\right]^{\frac{m}{m+5}}. \tag{6.28}$$

In Eqs. (6.26)–(6.28), the economic parameters occur as the ratio k_T/k_m. Thus, the inflationary forces, operating equally on K_T and K_m, have no impact on the design variables. However, technological innovations may disturb this ratio and thus will have a significant influence on the optimal design. Wildenradt (1983) qualitatively discussed these effects on pipeline design. The variation of f with D can be taken care of by the following iterative procedure:

1. Find w_4^* using Eq. (6.25a) or (6.25b)
2. Assume a value of f
3. Find D using Eq. (6.27)
4. Find f using Eq. (2.6a)
5. Repeat steps 3–5 until two successive D values are close
6. Find n using Eq. (6.28)
7. Find h_0 using Eq. (6.15)
8. Find F^* using Eq. (6.14)

The methodology provides D and n as continuous variables. In fact, whereas n is an integer variable, D is a set of values for which the pipe sizes are commercially available. Whereas the number of pumping stations has to be rounded up to the next higher integer, the optimal diameter has to be reduced to the nearest available size. In case the optimal diameter falls midway between the two commercial sizes, the costs corresponding with both sizes should be worked out by Eq. (6.18b), and the diameter resulting in lower cost should be adopted.

Example 6.4. Design a multistage cast iron pumping main for the transport of $0.4 \, \text{m}^3/\text{s}$ of water from a reservoir at 100 m elevation to a water treatment plant situated at an elevation of 200 m over a distance of 300 km. The water has $v = 1.0 \times 10^{-6} \, \text{m}^2/\text{s}$ and $\rho = 1000 \, \text{kg/m}^3$. The pipeline has $\varepsilon = 0.25 \, \text{mm}$, $m = 1.62$, and $h_b = 60 \, \text{m}$. The terminal head $H = 5 \, \text{m}$. The ratio $k_T/k_m = 0.02$, and $h_c = 150 \, \text{m}$.

Solution. For given $k_T/k_m = 0.02$, $H = 5 \, \text{m}$, and $h_b = 60 \, \text{m}$, calculate k_T/k_n applying Eq. (6.17) to substitute in Eqs. (6.27) and (6.28). Using Eq. (6.25), $w_4^* = 0.2106$. To start the algorithm, assume $f = 0.01$, and the outcome of the iterations is shown in Table 6.2.

Thus, a diameter of 0.9 m can be provided. Using Eq. (6.28), the number of pumping stations is 7.34, thus provide 8 pumping stations. Using Eq. (6.15), the pumping head is obtained as 31.30 m.

6.3.2. Pipeline on a Topography with Large Elevation Difference

Urban water supply intake structures are generally located at a much lower level than the water treatment plant or clear water reservoir to supply raw water from a river or lake. It is not economic to pump the water in a single stretch, as this will involve high-pressure pipes that may not be economic. If the total length of pumping main is divided into sublengths, the pumping head would reduce considerably, thus the resulting infrastructure would involve less cost. The division of the pumping main into submains on an *ad hoc* basis would generally result in a suboptimal solution. Swamee (2001) developed explicit equations for the optimal number of pumping stages, pumping main diameter, and the corresponding cost for a high-rise, multistage pumping system. This methodology involves the formulation of a geometric programming problem having a single degree of difficulty, which is presented in the following section. A typical multistage high-rise pumping main is shown in Fig. 6.5.

TABLE 6.2. Optimal Design Iterations

Iteration No.	Pipe Friction f	Pipe Diameter D (m)	No. of Pumping Stations n	Velocity V (m/s)	Reynolds No. $\mathbf{R}$
1	0.01	0.753	6.52	0.898	676,405
2	0.0163	0.811	7.35	0.775	628,281
3	0.0162	0.811	7.34	0.776	628,866

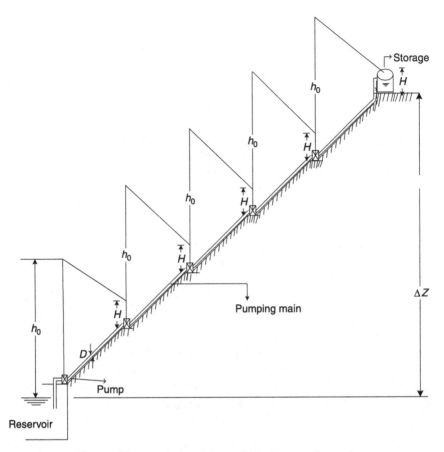

Figure 6.5. A typical multistage, high-rise pumping main.

Because there is a large elevation difference between the inlet and the outlet points, the third term in Eq. (6.16) involving h_b and f is much smaller than the term involving Δz. Thus, dropping the third term on the right-hand side of Eq. (6.16), one gets

$$F = k_n LD^m + \frac{k_m L \Delta z D^m}{h_b} + \frac{8 k_T \rho f L Q^3}{\pi^2 D^5} + k_T \rho g Q(H + h_c)n + k_T \rho g Q \Delta z. \qquad (6.29a)$$

As the last term on the right-hand side of Eq. (6.29a) is constant, it will not enter in the optimization process. Removing this term, Eq. (6.29a) reduces to

$$F = k_n LD^m + \frac{k_m L \Delta z D^m}{h_b} + \frac{8 k_T \rho f L Q^3}{\pi^2 D^5} + k_T \rho g Q(H + h_c)n. \qquad (6.29b)$$

As the cost function is in the form of a posynomial, it is a geometric programming formulation. Because Eq. (6.29b) contains four terms in two design variables, D and n, it has a single degree of difficulty. The weights w_1, w_2, w_3, and w_4 define contributions of various terms of Eq. (6.29b) in the following manner:

$$w_1 = \frac{k_n L D^m}{F} \tag{6.30a}$$

$$w_2 = \frac{k_m L \Delta z D^m}{n h_b F} \tag{6.30b}$$

$$w_3 = \frac{8 k_T \rho f L Q^3}{\pi^2 D^5 F} \tag{6.30c}$$

$$w_4 = \frac{k_T \rho g Q (H + h_c) n}{F}. \tag{6.30d}$$

Assuming f to be constant, the posynomial dual d of Eq. (6.29b) can be written as

$$d = \left(\frac{k_n L D^m}{w_1}\right)^{w_1} \left(\frac{k_m L \Delta z D^m}{n h_b w_2}\right)^{w_2} \left(\frac{8 k_T \rho f L Q^3}{\pi^2 D^5 w_3}\right)^{w_3} \left[\frac{k_T \rho g Q (H + h_c) n}{w_4}\right]^{w_4}. \tag{6.31}$$

Using Eq. (6.31), the orthogonality conditions in terms of optimal weights w_1^*, w_2^*, w_3^*, and w_4^* are given by

$$D: \quad m w_1^* + m w_2^* - 5 w_3^* = 0 \tag{6.32a}$$

$$n: \quad -w_2^* + w_4^* = 0, \tag{6.32b}$$

and the normality condition for Eq. (6.31) is written as

$$w_1^* + w_2^* + w_3^* + w_4^* = 1. \tag{6.32c}$$

Solving Eq. (6.32a–c) for optimal weights, w_1^*, w_2^*, and w_3^* are expressed as

$$w_1^* = \frac{5}{m + 5} - \frac{m + 10}{m + 5} w_4^* \tag{6.33a}$$

$$w_2^* = w_4^* \tag{6.33b}$$

$$w_3^* = \frac{m}{m + 5} - \frac{m}{m + 5} w_4^*. \tag{6.33c}$$

Substituting w_1^*, w_2^*, and w_3^* from Eq. (6.33a–c) into Eq. (6.31), the optimal dual d^* is

$$d^* = \frac{(m + 5) k_n L}{5 - (m + 10) w_4^*} \left[\frac{5 - (m + 10) w_4^*}{(1 - w_4^*)} \frac{8 k_T \rho f Q^3}{\pi^2 m k_n}\right]^{\frac{m}{m + 5}}$$

$$\times \left\{\left[\frac{5 - (m + 10) w_4^*}{(m + 5) w_4^*}\right]^2 \left[\frac{5 (1 - w_4^*)}{5 - (m + 10) w_4^*}\right]^{m/(m + 5)} \frac{h_c + H}{h_b + H} \frac{k_T \rho g Q \Delta z}{k_n L D_s^m}\right\}^{w_4^*}, \tag{6.34}$$

where w_4^* corresponds with optimality, and $D_s =$ the optimal diameter of a single-stage pumping main as given by Eq. (6.9), rewritten as

$$D_s = \left(\frac{40\,k_T \rho f Q^3}{\pi^2 m k_n}\right)^{\frac{1}{m+5}}.$$ (6.35)

Equating the factor having the exponent w_4^* on the right-hand side of Eq. (6.34) to unity (Swamee, 1995) results in

$$\left[\frac{(m+5)w_4^*}{5-(m+10)w_4^*}\right]^2 \left[\frac{5-(m+10)w_4^*}{5(1-w_4^*)}\right]^{m/(m+5)} = \frac{h_c + H}{h_b + H} \times \frac{k_T \rho g Q \Delta z}{k_n L D_s^m}.$$ (6.36a)

The following equation represents the explicit form of Eq. (6.36a):

$$w_4^* = 5\left[m + 10 + (m+5)\left(\frac{h_b + H}{h_c + H}\frac{k_n L D_s^m}{k_T \rho g Q \Delta z}\right)^{1/2}\right]^{-1}.$$ (6.36b)

The maximum error involved in the use of Eq. (6.36b) is about 1%. Using Eq. (6.34) and Eq. (6.35) with the condition at optimality $F^* = d^*$, one gets

$$F^* = \frac{(m+5)k_n L}{5-(m+10)w_4^*}\left[\frac{5-(m+10)w_4^*}{(1-w_4^*)}\frac{8\,k_T \rho f Q^3}{\pi^2 m k_n}\right]^{\frac{m}{m+5}},$$ (6.37)

where w_4^* is given by Eq. (6.36b). Using Eqs. (6.30a), (6.33a), and (6.37), the optimal diameter D^* was obtained as

$$D^* = \left[\frac{5-(m+10)w_4^*}{(1-w_4^*)}\frac{8\,k_T \rho f Q^3}{\pi^2 m k_n}\right]^{\frac{1}{m+5}}$$ (6.38)

Using Eqs. (6.30b), (6.33b), (6.30d), and (6.37), the optimal number of pumping stages is

$$n^* = \frac{\Delta z}{h_b + H}\frac{5-(m+10)w_4^*}{(m+5)w_4^*}.$$ (6.39)

The variation of f with D can be taken care of by the following iterative procedure:

1. Find w_4^* using Eq. (6.36b)
2. Assume a value of f
3. Find D^* using Eq. (6.38)
4. Find f using Eq. (2.6a)

TABLE 6.3. Optimal Design Iterations

Iteration No.	Pipe Friction f	Pipe Diameter D (m)	No. of Pumping Stations n	Velocity V (m/s)	Reynolds No. $\mathbf{R}$
1	0.01	0.402	3.12	2.36	937,315
2	0.01809	0.443	3.36	1.94	849,736
3	0.01779	0.442	3.36	1.95	852,089

5. Repeat steps 3–5 until two successive D^* values are close
6. Find n^* using Eq. (6.39)
7. Reduce D^* to the nearest higher and available commercial size
8. Reduce n^* to nearest higher integer
9. Find h_0 using Eq. (6.15)
10. Find F^* using Eq. (6.14)

Example 6.5. Design a multistage cast iron pumping main for carrying a discharge of 0.3 m^3/s from a river intake having an elevation of 200 m to a location at an elevation of 950 m and situated at a distance of 30 km. The pipeline has $\varepsilon = 0.25$ mm and $h_b = 60$ m. The terminal head $H = 5$ m. The ratio $k_T/k_m = 0.018$ units, and $E/k_m = 12,500$ units.

Solution. Now, $h_c = E/(\rho g k_T Q) = 236.2$ m. Using Eq. (6.36b), $w_4^* = 0.3727$. For starting the algorithm, f was assumed as 0.01 and the iterations were carried out. These iterations are shown in Table 6.3. Thus, a diameter of 0.5 m can be provided. Using Eq. (6.39), the number of pumping stages is found to be 3.36. Thus, providing 4 stages and using Eq. (6.15), the pumping head is obtained as 224.28 m.

6.4. EFFECT OF POPULATION INCREASE

The water transmission lines are designed to supply water from a source to a town's water distribution system. The demand of water increases with time due to the increase in population. The town water supply systems are designed for a predecided time span called the design period, and the transmission mains are designed for the ultimate discharge required at the end of the design period of a water supply system. Such an approach can be acceptable in the case of a gravity main. However, if a pumping main is designed for the ultimate water demand, it will prove be uneconomic in the initial years. As there exists a trade-off between pipe diameters and pumping head, the smaller diameter involves less capital expenditure but requires high pumping energy cost as the flow increases with time. Thus, there is a need to investigate the optimal sizing of the water transmission main in a situation where discharge varies with time.

The population generally grows according to the law of decreasing rate of increase. Such a law yields an exponential growth model that subsequently saturates to a constant population. Because the per capita demand also increases with the growth of the population, the variation of the discharge will be exponential for a much longer duration.

Thus, the discharge can be represented by the following exponential equation:

$$Q = Q_0 e^{\alpha t}, \tag{6.40}$$

where Q = the discharge at time t, Q_0 = the initial discharge, and α = a rate constant for discharge growth.

The initial cost of pipe C_m can be obtained using Eq. (4.4). As it is not feasible to change the pumping plant frequently, it is therefore assumed that a pumping plant able to discharge ultimate flow corresponding with the design period T is provided at the beginning. Using Eq. (4.2c) with exponent $m_p = 1$, the cost of the pumping plant C_p is

$$C_p = \frac{(1+s_b)k_p}{1000\eta}\left[\rho g Q_0 e^{\alpha T}(H + z_L - z_0) + \frac{8\rho f L Q_0^3 e^{3\alpha T}}{\pi^2 D^5}\right]. \tag{6.41}$$

The energy cost is widespread over the design period. The investment made in the distant future is discounted for its current value. The future discounting ensures that very large investments are not economic if carried out at initial stages, which yield results in a distant future. The water supply projects have a similar situation. Denoting the discount rate by r, any investment made at time t can be discounted by a multiplier e^{-rt}.

Applying Eq. (4.9), the elementary energy cost dC_e for the time interval dt years is

$$dC_e = \frac{8.76 F_A F_D R_E}{\eta}\rho g Q h_0 e^{-rt} dt. \tag{6.42}$$

The cost of energy C_e is obtained as

$$C_e = \frac{8.76 F_A F_D R_E \rho g Q_0}{\eta}\int_0^T\left[e^{(\alpha-r)t}(H + z_L - z_0) + \frac{8 f L Q_0^2 e^{(3\alpha-r)t}}{\pi^2 g D^5}\right]dt. \tag{6.43}$$

Evaluating the integral, Eq. (6.43) is written as

$$C_e = \frac{8.76 F_A F_D R_E \rho g Q_0}{\eta}\left[(H + z_L - z_0)\frac{e^{(\alpha-r)T} - 1}{\alpha - 1} + \frac{8 f L Q_0^2}{\pi^2 g D^5}\frac{e^{(3\alpha-r)T} - 1}{3\alpha - 1}\right]. \tag{6.44}$$

Using Eqs. (6.41) and (6.44), the cost function is

$$F = k_m L D^m + k_{T1}\frac{8\rho f L Q_0^3}{\pi^2 D^5} + k_{T2}\rho g Q_0(H + z_L - z_0) \tag{6.45}$$

where

$$k_{T1} = \frac{(1+s_b)k_p e^{3\alpha T}}{1000\eta} + \frac{8.76 F_A F_D R_E}{\eta}\frac{e^{(3\alpha-r)T} - 1}{3\alpha - 1} \tag{6.46a}$$

$$k_{T2} = \frac{(1+s_b)k_p e^{\alpha T}}{1000\eta} + \frac{8.76 F_A F_D R_E}{\eta}\frac{e^{(\alpha-r)T} - 1}{\alpha - 1}. \tag{6.46b}$$

Using Eq. (6.9), the optimal diameter is expressed as

$$D^* = \left(\frac{40\, k_{T1} \rho f Q_0^3}{\pi^2 m k_m}\right)^{\frac{1}{m+5}}$$ (6.47)

Depending on the discharge, pumping head is a variable quantity that can be obtained by using Eqs. (6.5), (6.40), and (6.47) as

$$h_0 = \frac{8 f L Q_0^2 e^{2\alpha t}}{\pi^2 g}\left(\frac{\pi^2 m k_m}{40\, k_{T1} \rho f Q_0^3}\right)^{\frac{5}{m+5}} - z_0 + H + z_L.$$ (6.48)

It can be seen from Eq. (6.48) that the pumping head increases exponentially as the population or water demand increases. The variable speed pumping plants would be able to meet such requirements.

6.5. CHOICE BETWEEN GRAVITY AND PUMPING SYSTEMS

A pumping system can be adopted in any type of topographic configuration. On the other hand, the gravity system is feasible only if the input point is at a higher elevation than all the withdrawal points. If the elevation difference between the input point and the withdrawal point is very small, the required pipe diameters will be large, and the design will not be economic in comparison with the corresponding pumping system. Thus, there exists a critical elevation difference at which both gravity and pumping systems will have the same cost. If the elevation difference is greater than this critical difference, the gravity system will have an edge over the pumping alternative. Here, a criterion for adoption of a gravity main was developed that gives an idea about the order of magnitude of the critical elevation difference (Swamee and Sharma, 2000).

6.5.1. Gravity Main Adoption Criterion

The cost of gravity main F_g consists of the pipe cost only; that is,

$$F_g = k_m L D^m.$$ (6.49)

The head loss occurring in a gravity main is expressed as

$$h_f = z_0 - z_L - H = \frac{8 f L Q^2}{\pi^2 g D^5}.$$ (6.50)

Equation (6.50) gives the diameter of the gravity main as

$$D = \left[\frac{8 f L Q^2}{\pi^2 g (z_0 - z_L - H)}\right]^{\frac{1}{5}}.$$ (6.51)

Equations (6.49) and (6.51) yield

$$F_g = k_m L \left[\frac{8fLQ^2}{\pi^2 g(z_0 - z_L - H)} \right]^{\frac{m}{5}}. \qquad (6.52)$$

Similarly, the overall cost of the pumping main is expressed as

$$F_P = k_m LD^m + k_T \rho g Q h_0, \qquad (6.53a)$$

and the pumping head of the corresponding pumping main can be rewritten as

$$h_0 = H + z_L - z_0 + \frac{8fLQ^2}{\pi^2 g D^5}. \qquad (6.53b)$$

Using Eqs. (6.53a) and (6.53b) and eliminating h_0, the optimal pipe diameter and optimal pumping main cost can be obtained similar to Eqs. (6.9) and (6.11) as

$$D^* = \left(\frac{40 \, k_T \rho f Q^3}{\pi^2 m k_m} \right)^{\frac{1}{m+5}} \qquad (6.54a)$$

$$F^* = k_m L \left(1 + \frac{m}{5} \right) \left(\frac{40 \, k_T \rho f Q^3}{\pi^2 m k_m} \right)^{\frac{m}{m+5}} + k_T \rho g Q(H + z_L - z_0). \qquad (6.54b)$$

The second term on the right-hand side of Eq. (6.54b) is the cost of pumping against gravity. For the case where the elevation of entry point z_0 is higher than exit point z_L, this term is negative. Because the negative term is not going to reduce the cost of the pumping main, it is taken as zero. Thus, Eq. (6.54b) reduces to the following form:

$$F_p^* = k_m L \left(1 + \frac{m}{5} \right) \left(\frac{40 \, k_T \rho f Q^3}{\pi^2 m k_m} \right)^{\frac{m}{m+5}}. \qquad (6.55)$$

The gravity main is economic when $F_g < F_p^*$. Using Eqs. (6.52) and (6.55), the optimality criteria for a gravity main to be economic is derived as

$$z_0 - z_L - H > \frac{L}{g} \left(\frac{5}{m+5} \right)^{\frac{5}{m}} \left(\frac{8fQ^2}{\pi^2} \right)^{\frac{m}{m+5}} \left(\frac{m k_m}{5 \rho k_T Q} \right)^{\frac{5}{m+5}}. \qquad (6.56)$$

Equation (6.56) states that for economic viability of a gravity main, the left-hand side of inequality sign should be greater than the critical value given by its right-hand side. The critical value has a direct relationship with f and k_m. Thus, a gravity-sustained system becomes economically viable by using smoother and cheaper pipes. As $m < 2.5$, the critical elevation difference has an inverse relationship with Q. Therefore, for the same topography, it is economically viable to transport a large discharge gravitationally.

Equation (6.56) can be written in the following form for the critical discharge Q_c for which the costs of pumping main and gravity main are equal:

$$Q_c = \left[\frac{L}{g(z_0 - z_L - H)} \left(\frac{5}{m+5} \right)^{\frac{5}{m}} \left(\frac{8f}{\pi^2} \right)^{\frac{m}{m+5}} \left(\frac{mk_m}{5\rho k_T} \right)^{\frac{5}{m+5}} \right]^{\frac{m+5}{5-2m}}. \tag{6.57}$$

For a discharge greater than the critical discharge, the gravity main is economic. Thus, (6.57) also indicates that for a large discharge, a gravity main is economic.

Example 6.6. Explore the economic viability of a 10-km-long cast iron gravity main for carrying a discharge of 0.1 m^3/s. The elevation difference between the input and exit points $z_0 - z_L = 20$ m and the terminal head $H = 1$ m. Adopt $k_T/k_m = 0.0185$ units.

Solution. Adopt $m = 1.62$ (for cast iron pipes), $g = 9.8$ m/s^2, and $f = 0.01$. Consider the left-hand side (LHS) of Eq. (6.56), $z_0 - z_L - H = 19$ m. On the other hand, the right-hand side (RHS) of Eq. (6.56) works out to be 11.48 m. Thus, carrying the discharge through a gravity main is economic. In this case, using Eq. (6.52), $F_g = 1717.8k_m$, and using Eq. (6.55), $F_p^* = 2027.0k_m$. The critical discharge as computed by Eq. (6.57) is 0.01503 m^3/s. For the critical discharge, both the pumping main and the gravity main have equal cost. Thus, LHS and RHS of Eq. (6.56) equal 19 m; and further, Eqs. (6.52) and (6.55) give $F_g = F_p^* = 503.09k_m$.

EXERCISES

6.1. Design a cast iron gravity main for carrying a discharge of 0.3 m^3/s over a distance of 5 km. The elevation of the entry point is 180 m, whereas the elevation of the exit point is 135 m. The terminal head at the exit is 5 m.

6.2. Design a ductile iron pumping main carrying a discharge of 0.20 m^3/s over a distance of 8 km. The elevation of the pumping station is 120 m and that of the exit point is 150 m. The required terminal head is 5 m. Use iterative design procedure for pipe diameter calculation.

6.3. Design a ductile iron pumping main carrying a discharge of 0.35 m^3/s over a distance of 4 km. The elevation of the pumping station is 140 m and that of the exit point is 150 m. The required terminal head is 10 m. Estimate the pipe diameter and pumping head using the explicit design procedure.

6.4. Design a multistage cast iron pumping main for the transport of 0.4 m^3/s of water from a reservoir at 150 m elevation to a water treatment plant situated at an elevation of 200 m over a distance of 100 km. The water has $v = 1.0 \times 10^{-6}$ m^2/s and $\rho = 1000$ kg/m^3. The pipeline has $\varepsilon = 0.25$ mm, $m = 1.6$ and $h_b = 60$ m. The terminal head $H = 10$ m. The ratio $k_T/k_m = 0.025$, and $h_c = 160$ m.

6.5. Design a multistage cast iron pumping main for carrying a discharge of $0.3 \, \text{m}^3/\text{s}$ from a river intake having an elevation of 100 m to a location at an elevation of 1050 m and situated at a distance of 25 km. The pipeline has $\varepsilon = 0.25$ mm and $h_b = 60$ m. The terminal head $H = 4$ m. The ratio $k_T/k_m = 0.019$ units, and $E/k_m = 15{,}500$ units.

6.6. Explore the economic viability of a 20-km-long cast iron gravity main for carrying a discharge of $0.2 \, \text{m}^3/\text{s}$. The elevation difference between the input and exit points $z_0 - z_L = 35$ m and the terminal head $H = 5$ m. Adopt $k_T/k_m = 0.0185$ units.

REFERENCES

Babbitt, H.E., and Doland, J.J. (1949). *Water Supply Engineering*, 4th ed. McGraw-Hill, New York, pp. 171–173.

Garg, S.K. (1990). *Water Supply Engineering*, 6th ed. Khanna Publishers, Delhi, India, pp. 287.

Swamee, P.K. (1993). Design of a submarine oil pipeline. *J. Transp. Eng.* 119(1), 159–170.

Swamee, P.K. (1995). Design of sediment-transporting pipeline. *J. Hydraul. Eng.* 121(1), 72–76.

Swamee, P.K. (1996). Design of multistage pumping main. *J. Transp. Eng.* 122(1), 1–4.

Swamee, P.K. (2001). Design of high-rise pumping main. *Urban Water* 3(4), 317–321.

Swamee, P.K. and Sharma, A.K. (2000). Gravity flow water distribution network design. *Journal of Water Supply: Research and Technology-AQUA, IWA.* 49(4), 169–179.

Thresh, J.C. (1901). *Water and Water Supplies*. Rebman Ltd., London.

Turneaure, F.E., and Russell, H.L. (1955). *Public Water Supplies*. John Wiley & Sons, New York.

Wildenradt, W.C. (1983). Changing economic factors affect pipeline design variables. *Pipeline and Gas J.* 210(8), 20–26.

7

WATER DISTRIBUTION MAINS

A pipeline with the input point at one end and several withdrawals at intermediate points and also at the exit point is called a *water distribution main*. The flow in a distribution main is sustained either by gravity or by pumping.

7.1. GRAVITY-SUSTAINED DISTRIBUTION MAINS

In case of gravity-sustained systems, the input point can be a reservoir or any water source at an elevation higher than all other points of the system. Such systems are generally possible where the topographical (elevation) differences between the source (input) and withdrawal (demand) points are reasonably high. A typical gravity-sustained distribution main is depicted in Fig. 7.1. Swamee and Sharma (2000) developed a methodology for computing optimal pipe link diameters based on elevation difference between input and terminal withdrawal point, minimum pressure head requirement, water demand, and pipe roughness. The methodology is described in the following section.

Design of Water Supply Pipe Networks. By Prabhata K. Swamee and Ashok K. Sharma
Copyright © 2008 John Wiley & Sons, Inc.

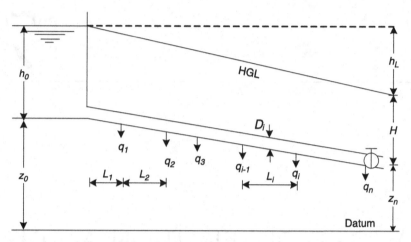

Figure 7.1. A gravity-sustained distribution main.

Denoting $n=$ the number of links, the cost function of such a system is expressed as

$$F = k_m \sum_{i=1}^{n} L_i D_i^m. \tag{7.1}$$

The system should satisfy the energy loss constraint; that is, the total energy loss is equal to the available potential head. Assuming the form losses to be small and neglecting water column h_0, the constraint is

$$\sum_{i=1}^{n} \frac{8 f_i L_i Q_i^2}{\pi^2 g D_i^5} - z_0 + z_n + H = 0, \tag{7.2}$$

where f_i can be estimated using Eq. (2.6c), rewritten as:

$$f_i = 1.325 \left\{ \ln \left[\frac{\varepsilon_i}{3.7 D_i} + 4.618 \left(\frac{v D_i}{Q_i} \right)^{0.9} \right] \right\}^{-2}. \tag{7.3}$$

Combining Eqs. (7.1) and (7.2) through Lagrange multiplier λ, the following merit function is formed:

$$F_1 = k_m \sum_{i=1}^{n} L_i D_i^m + \lambda \left[\sum_{i=1}^{n} \frac{8 f_i L_i Q_i^2}{\pi^2 g D_i^5} - z_0 + z_n + H \right]. \tag{7.4}$$

For optimality, partial derivatives of F_1 with respect to D_i ($i = 1, 2, 3, \ldots, n$) and λ should be zero. Considering f to be constant and differentiating F_1 partially with respect to D_i, equating it to zero, and simplifying, yields

$$D_i^* = \left(\frac{40\lambda f_i Q_i^2}{\pi^2 gmk_m}\right)^{\frac{1}{m+5}}.$$

(7.5a)

Putting $i = 1$ in Eq. (7.5a),

$$D_1^* = \left(\frac{40\lambda f_1 Q_1^2}{\pi^2 gmk_m}\right)^{\frac{1}{m+5}}.$$

(7.5b)

Using Eqs. (7.5a) and (7.5b),

$$D_i^* = D_1^* \left(\frac{f_i Q_i^2}{f_1 Q_1^2}\right)^{\frac{1}{m+5}}.$$

(7.5c)

Substituting D_i from Eq. (7.5c) into Eq. (7.2) and simplifying,

$$D_1^* = \left(f_1 Q_1^2\right)^{\frac{1}{m+5}} \left[\frac{8}{\pi^2 g(z_0 - z_n - H)} \sum_{p=1}^{n} L_p \left(f_p Q_p^2\right)^{\frac{m}{m+5}}\right]^{0.2},$$

(7.6a)

where p is an index for pipes in the distribution main.
Eliminating D_1 between Eqs. (7.5c) and (7.6a),

$$D_i^* = \left(f_i Q_i^2\right)^{\frac{1}{m+5}} \left[\frac{8}{\pi^2 g(z_0 - z_n - H)} \sum_{p=1}^{n} L_p \left(f_p Q_p^2\right)^{\frac{m}{m+5}}\right]^{0.2}.$$

(7.6b)

Substituting D_i from Eq. (7.6b) into Eq. (7.1), the optimal cost F^* works out to be

$$F^* = k_m \left[\frac{8}{\pi^2 g(z_0 - z_n - H)}\right]^{\frac{m}{5}} \left[\sum_{i=1}^{n} L_i \left(f_i Q_i^2\right)^{\frac{m}{m+5}}\right]^{\frac{m+5}{5}}.$$

(7.7)

Equation (7.6b) calculates optimal pipe diameters assuming constant friction factor. Thus, the diameters obtained using arbitrary values of f are approximate. These diameters can be improved by evaluating f using Eq. (7.3) and estimating a new set of diameters by Eq. (7.6b). The procedure can be repeated until the two successive solutions are close.

The design so obtained should be checked against the minimum and the maximum pressure constraints at all nodal points. In case these constraints are violated, remedial measures should be adopted. If the minimum pressure head constraint is violated, the distribution main has to be realigned at a lower level. In a situation where the distribution main cannot be realigned, pumping has to be restored to cater flows at required minimum pressure heads. Based on the local conditions, part-gravity and part-pumping systems can provide economic solutions. On the other hand, if maximum pressure constraint

TABLE 7.1. Data for Gravity-Sustained Distribution Main

Pipe i	Elevation z_i (m)	Length L_i (m)	Demand Discharge q_i (m³/s)	Pipe Discharge Q_i (m³/s)
0	100			
1	92	1500	0.01	0.065
2	94	200	0.015	0.055
3	88	1000	0.02	0.04
4	85	1500	0.01	0.02
5	87	500	0.01	0.01

TABLE 7.2. Design Output for Gravity-Sustained System

Pipe i	1st Iteration f_i	D_i	2nd Iteration f_i	D_i	3rd Iteration f_i	D_i	4th Iteration f_i	D_i
1	0.010	0.279	0.0199	0.322	0.0199	0.321	0.0199	0.321
2	0.010	0.264	0.0203	0.305	0.0203	0.304	0.0203	0.304
3	0.010	0.237	0.0209	0.276	0.0209	0.275	0.0209	0.275
4	0.010	0.187	0.0225	0.221	0.0225	0.220	0.0225	0.220
5	0.010	0.148	0.0244	0.177	0.0244	0.177	0.0244	0.177

is violated, break pressure tanks or other devices to increase form losses should be considered. The design should also be checked against the maximum velocity constraint. In the case of marginal violation, the pipe diameter may be increased. If the violation is serious, the form losses should be increased by installing energy dissipation devices.

Example 7.1. Design a gravity-sustained distribution main with the data given in Table 7.1. The system layout can be considered similar to that of Fig. 7.1.

Solution. The ductile iron pipe cost parameters ($k_m = 480$, $m = 0.935$) are taken from Fig. 4.3 and roughness height ($\varepsilon = 0.25$ mm) of pipe from Table 2.1. Adopting $f_i = 0.01$ and using Eq. (7.6b), the pipe diameters and the corresponding friction factors were obtained as listed in Table 7.2. The pipe diameters were revised for new f_i values using Eq. (7.6b) again. The process was repeated until the two consecutive solutions were close. The design procedure results are listed in Table 7.2. The final cost of the system worked out to be \$645,728. These pipes are continuous in nature; the nearest commercial sizes can be finally adopted.

7.2. PUMPED DISTRIBUTION MAINS

Pumping distribution mains are provided for sustaining the flow if the elevation difference between the entry and the exit points is very small, also if the exit point level or an

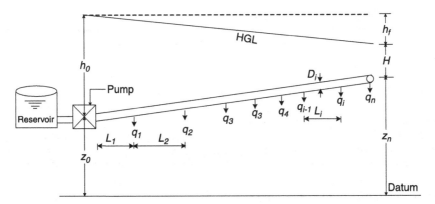

Figure 7.2. A pumped distribution main.

intermediate withdrawal point level is higher than the entry point level. Figure 7.2 depicts a typical pumping distribution main. It can be seen from Fig. 7.2 that a pumping distribution main consists of a pump and a distribution main with several withdrawal (supply) points. The source for the water can be a reservoir as shown in the figure. Swamee et al. (1973) developed a methodology for the pumping distribution mains design, which is highlighted in the following section.

The cost function of a pumping distribution main system is of the following form:

$$F = k_m \sum_{i=1}^{n} L_i D_i^m + k_T \rho g Q_1 h_0. \tag{7.8}$$

The head-loss constraint of the system is given by

$$\sum_{i=1}^{n} \frac{8 f_i L_i Q_i^2}{\pi^2 g D_i^5} - z_0 - h_0 + z_n + H = 0. \tag{7.9}$$

Combining Eqs. (7.8) and (7.9), the following merit function is formed:

$$F_1 = k_m \sum_{i=1}^{n} L_i D_i^m + k_T \rho g Q_1 h_0 + \lambda \left[\sum_{i=1}^{n} \frac{8 f_i L_i Q_i^2}{\pi^2 g D_i^5} - z_0 - h_0 + z_n + H \right]. \tag{7.10}$$

For minimum, the partial derivatives of F_1 with respect to D_i ($i = 1, 2, 3, \ldots, n$) and λ should be zero. Considering f_i to be constant and differentiating F_1 partially with respect to D_i, equating it to zero, and simplifying, one gets Eq. (7.5a). Differentiating Eq. (7.10) partially with respect to h_0 and simplifying, one obtains

$$\lambda = k_T \rho g Q_1. \tag{7.11a}$$

TABLE 7.3. Data for Pumped Distribution Main

Pipe i	Elevation Z_i (m)	Length L_i (m)	Demand Discharge q_i (m³/s)	Pipe Discharge Q_i (m³/s)
0	100			
1	102	1200	0.012	0.076
2	105	500	0.015	0.064
3	103	1000	0.015	0.049
4	106	1500	0.02	0.034
5	109	700	0.014	0.014

Substituting λ from Eq. (7.11a) into Eq. (7.5a),

$$D_i^* = \left(\frac{40 k_T \rho f_i Q_1 Q_i^2}{\pi^2 m k_m} \right)^{\frac{1}{m+5}}. \tag{7.11b}$$

Substituting D_i from Eq. (7.11b) into Eq. (7.9),

$$h_0^* = z_n + H - z_0 + \frac{8}{\pi^2 g} \left(\frac{\pi^2 m k_m}{40 \rho k_T Q_1} \right)^{\frac{5}{m+5}} \sum_{i=1}^{n} L_i \left(f_i Q_i^2 \right)^{\frac{m}{m+5}}. \tag{7.12}$$

Substituting D_i and h_0 from Eqs. (7.11b) and (7.12), and simplifying, the optimal cost as obtained from Eq. (7.8) is

$$F^* = \left(1 + \frac{m}{5} \right) k_m \sum_{i=1}^{n} L_i \left(\frac{40 k_T \rho f_i Q_1 Q_i^2}{\pi^2 m k_m} \right)^{\frac{m}{m+5}} + k_T \rho g Q_1 (z_n + H - z_0). \tag{7.13}$$

The optimal design values obtained by Eqs. (7.11b)–(7.13) assume a constant value of f_i. Thus, the design values are approximate. Knowing the approximate values of D_i, improved values of f_i can be obtained by using Eq. (7.3). The process should be repeated until the two solutions are close to the allowable limits.

Example 7.2. Design a pumped distribution main using the data given in Table 7.3. The terminal pressure head is 5 m. Adopt cast iron pipe for the design and layout similar to Fig. 7.2.

Solution. The cost parameters of a ductile iron pipe ($k_m = 480$, $m = 0.935$) are taken from Fig. 4.3 and roughness height of pipe ($\varepsilon = 0.25$ mm) from Table 2.1. The k_T/k_m ratio as 0.02 is considered in this example. Adopting $f_i = 0.01$ and using Eqs. (7.11b) and (7.3), the pipe diameters and the corresponding friction factors were obtained. Using the calculated friction factors, the pipe diameters were recalculated using Eq. (7.11b). The process was repeated until two solutions were close. The design output is listed in Table 7.4. The cost of the final system worked out to be $789,334, of which $642,843 is the cost of pipes.

TABLE 7.4. Design Iterations for Pumping Main

Pipe	1st Iteration		2nd Iteration		3rd Iteration		4th Iteration	
i	f_i	D_i	f_i	D_i	f_i	D_i	f_i	D_i
1	0.010	0.265	0.0203	0.299	0.0200	0.298	0.0200	0.298
2	0.010	0.250	0.0207	0.283	0.0203	0.282	0.0203	0.282
3	0.010	0.229	0.0212	0.260	0.0208	0.259	0.0208	0.259
4	0.010	0.202	0.0220	0.231	0.0216	0.230	0.0216	0.230
5	0.010	0.150	0.0241	0.174	0.0237	0.174	0.0237	0.174

EXERCISES

7.1. Design a ductile iron gravity-sustained water distribution main for the data given in Table 7.5, Use pipe cost parameters from Fig. 4.3, pipe roughness height from Table 2.1, and terminal head 5 m. Also calculate system cost.

7.2. Design a pumping main for the data in Table 7.6. The pipe cost parameters are $m = 0.9$ and $k_m = 500$ units. Use $k_T/k_m = 0.02$ units and terminal head as 5 m. The pipe roughness height is 0.25 mm. Calculate pipe diameters, pumping head, and cost of the system.

TABLE 7.5. Data for Gravity-Sustained Water Distribution Main

Pipe i	Elevation Z_i (m)	Length L_i (m)	Demand Discharge q_i (m^3/s)
0	100		
1	90	1000	0.012
2	85	500	0.015
3	83	800	0.02
4	81	1200	0.02
5	72	800	0.01
6	70	500	0.015

TABLE 7.6. Data for Pumping Distribution Main

Pipe i	Elevation Z_i (m)	Length L_i (m)	Demand Discharge q_i (m^3/s)
0	100		
1	105	1000	0.015
2	107	500	0.010
3	110	800	0.015
4	105	1200	0.025
5	118	800	0.015

REFERENCES

Swamee, P.K., Kumar, V., and Khanna, P. (1973). Optimization of dead-end water distribution systems. *J. Envir. Eng.* 99(2), 123–134.

Swamee, P.K., and Sharma, A.K. (2000). Gravity flow water distribution system design. *Journal of Water Supply: Research and Technology-AQUA, IWA* 49(4), 169–179.

8

SINGLE-INPUT SOURCE, BRANCHED SYSTEMS

A water distribution system is the pipe network that distributes water from the source to the consumers. It is the pipeline laid along the streets with connections to residential, commercial, and industrial taps. The flow and pressure in distribution systems are maintained either through gravitational energy gained through the elevation difference between source and supply point or through pumping energy.

Sound engineering methods and practices are required to distribute water in desired quantity, pressure, and reliably from the source to the point of supply. The challenge in such designs should be not only to satisfy functional requirements but also to provide economic solutions. The water distribution systems are designed with a number of objectives, which include functional, economic, reliability, water quality preservation, and future growth considerations. This chapter and other chapters on water distribution network design deal mainly with functional and economic objectives of the water

Design of Water Supply Pipe Networks. By Prabhata K. Swamee and Ashok K. Sharma
Copyright © 2008 John Wiley & Sons, Inc.

distribution. The future growth considerations are taken into account while projecting the design flows.

Water distribution systems receive water either from single- or multiple-input sources to meet water demand at various withdrawal points. This depends upon the size of the total distribution network, service area, water demand, and availability of water sources to be plugged in with the distribution system. A water distribution system is called a single-input source water system if it receives water from a single water source; on the other hand, the system is defined as a multi-input source system if it receives water from a number of water sources.

The water distribution systems are either branched or looped systems. Branched systems have a tree-like pipe configuration. It is like a tree trunk and branch structure, where the tree trunk feeds the branches and in turn the branches feed subbranches. The water flow path in branched system pipes is unique, thus there is only one path for water to flow from source to the point of supply (tap). The advantages and disadvantages of branched water distribution systems are listed in Table 8.1. The looped systems have pipes that are interconnected throughout the system such that the flow to a demand node can be supplied through several connected pipes. The flow direction in a looped system can change based on spatial or temporal variation in water demand, thus the flow direction in the pipe can vary based on the demand pattern. Hence, unlike the branched network, the flow directions in looped system pipes are not unique.

The water distribution design methods based on cost optimization have two approaches: (a) continuous diameter approach as described in previous chapters and (b) discrete diameter approach or commercial diameter approach. In the continuous diameter approach, the pipe links are calculated as continuous variables, and once the solution is obtained, the nearest commercial sizes are adopted. On the other hand, in the discrete diameter approach, commercially available pipe diameters are directly applied in the design methodology. In this chapter, discrete diameter approach will be introduced for the design of a branched water distribution system.

A typical gravity-sustained, branched water distribution system and a pumping system is shown in Fig. 8.1.

TABLE 8.1. Advantages and Disadvantages of Branched Water Distribution Systems

Advantages	Disadvantages
• Lower capital cost • Operational ease	• No redundancy in the system • One direction of flow to the point of use—main breaks put all customers out of service downstream of break
• Suitable for small rural areas of large lot sizes; low-density developments	• Water quality may deteriorate due to dead end in the system—may require periodic flushing in low-demand area • Less reliable—fire protection at risk • Less likely to meet increase in water demand

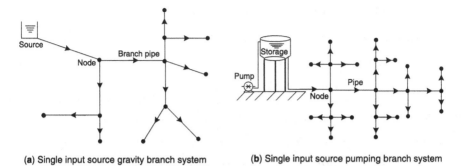

(a) Single input source gravity branch system   **(b)** Single input source pumping branch system

Figure 8.1. Branched water distribution system.

8.1. GRAVITY-SUSTAINED, BRANCHED SYSTEM

The gravity-sustained water distribution systems are generally suitable for areas where sufficient elevation difference is available between source (input) point and demand points across the system to generate sufficient gravitational energy to flow water at required quantity and pressure. Thus, in such systems the minimum available gravitational energy should be equal to the sum of minimum prescribed terminal head plus the frictional losses in the system. The objective of the design of such systems is to properly manipulate frictional energy losses so as to move the desired flows at prescribed pressure head through the system such that the system cost is minimum.

8.1.1. Radial Systems

Sometimes, radial water distribution systems are provided in hilly areas, based on the local development and location of water sources. A typical radial water distribution system is shown in Fig. 8.2. It can be seen from Fig. 8.2 that the radial system consists of a number of gravity-sustained water distribution mains (see Fig. 7.1). Thus, the radial water distribution system can be designed by designing each of its branches as a distribution main adopting the methodology described in Section 7.1.

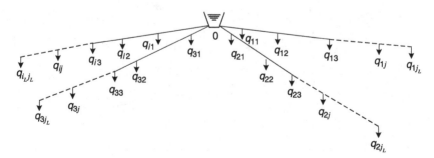

Figure 8.2. A radial, gravity water distribution system.

8.1.2. Branch Systems

The gravity-sustained systems are generally branched water distribution systems and are provided in areas with significant elevation differences and low-density-developments. A typical branched, gravity-sustained water distribution system is shown in Fig. 8.3. As described in the previous section, the design of such systems can be conducted using continuous diameter or discrete diameter approach. These approaches are described in the following sections.

8.1.2.1. Continuous Diameter Approach.

The method of the water distribution design is described by taking Fig. 8.3 as an example. The system data such as elevation, pipe length, nodal discharges, and cumulative pipe flows for Fig. 8.3 is given in Table 8.2.

The distribution system can be designed using the method described in Section 7.1 for gravity-sustained distribution mains and Section 3.9 for flow path development. The distribution system in Fig. 8.3 is decomposed into several distribution mains based on the number of flow paths. The total flow paths will be equal to the number of pipes in the system. Using the method for flow paths in Section 3.9, the pipe flow paths generated for Fig. 8.3 are tabulated in Table 8.3. The flow path for pipe 14 having pipes 14, 12, 7, 4 and pipe 1 is also highlighted in Fig. 8.3. The node $J_t(i)$ is the originating node of the flow path to which the pipe i is supplying the discharge.

Treating the flow path as a water distribution main and applying Eq. (7.6b), rewritten below, the optimal pipe diameters can be calculated:

$$D_i^* = \left(f_i Q_i^2\right)^{\frac{1}{m+5}} \left[\frac{8}{\pi^2 g(z_0 - z_n - H)} \sum_{p}^{N_t(i)} L_p \left(f_p Q_p^2\right)^{\frac{m}{m+5}}\right]^{0.2}, \qquad (8.1)$$

where $p = I_t(i,\ell)$, $\ell = 1, N_t(i)$ are the pipe in flow path of pipe i.

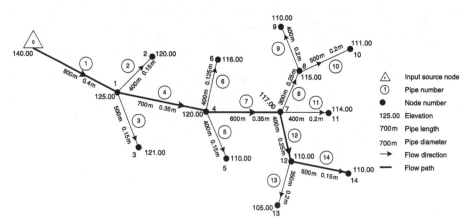

Figure 8.3. A branched, gravity water system (design based on continuous diameter approach).

TABLE 8.2. Design Data for Water Distribution System in Fig. 8.3.

Pipe/Node i/j	Elevation Z_j (m)	Length L_i (m)	Demand Discharge q_i (m³/s)	Pipe Discharge Q_i (m³/s)
0	140			
1	125	800	0.01	0.21
2	120	400	0.015	0.015
3	121	500	0.01	0.01
4	120	700	0.01	0.175
5	110	400	0.02	0.02
6	116	400	0.01	0.01
7	117	600	0.01	0.135
8	115	300	0.02	0.055
9	110	400	0.02	0.02
10	111	500	0.015	0.015
11	114	400	0.02	0.02
12	110	400	0.02	0.05
13	105	350	0.02	0.02
14	110	500	0.01	0.01

To apply Eq. (8.1) for the design of flow path of pipe 14 as a distribution main, the corresponding pipe flows, nodal elevations, and pipe lengths data are listed in Table 8.2. The total number of pipes in this distribution main is 5. The CI pipe cost parameters ($k_m = 480$, $m = 0.935$) similar to Fig. 4.3 and roughness ($\varepsilon = 0.25$ mm) from

TABLE 8.3. Total Water Distribution Mains

Pipe i	Flow Path Pipes Connecting to Input Point Node 0 and Generating Water Distribution Gravity Mains $I_t(i, \ell)$, $\ell = 1, N_t(i)$					$N_t(i)$	$J_t(i)$
	$\ell = 1$	$\ell = 2$	$\ell = 3$	$\ell = 4$	$\ell = 5$		
1	1					1	1
2	2	1				2	2
3	3	1				2	3
4	4	1				2	4
5	5	4	1			3	5
6	6	4	1			3	6
7	7	4	1			3	7
8	8	7	4	1		4	8
9	9	8	7	4	1	5	9
10	10	8	7	4	1	5	10
11	11	7	4	1		4	11
12	12	7	4	1		3	12
13	13	12	7	4	1	5	13
14	14	12	7	4	1	5	14

TABLE 8.4. Distribution Main Pipe Diameters

Pipe/ Node i/j	Elevation Z_j (m)	Length L_i (m)	Demand Discharge q_j (m³/s)	Pipe Discharge Q_i (m³/s)	Assumed Pipe f_i	Pipe Diameter D_i (m)	Calculated Pipe f_i
0	140						
1	125	800	0.01	0.21	0.0186	0.367	0.0186
4	120	700	0.01	0.175	0.0189	0.346	0.0189
7	117	600	0.01	0.135	0.0193	0.318	0.0193
12	110	400	0.02	0.05	0.0212	0.231	0.0212
14	110	500	0.01	0.01	0.0250	0.138	0.0250

Table 2.1 were considered in this example. The minimum terminal pressure of 5 m was maintained at nodes. The friction factor was improved iteratively until the two consecutive f values were close. The pipe diameters thus obtained are listed in Table 8.4. This gives the pipe diameters of pipes 1, 4, 7, 12, and 14.

Applying the similar methodology, the pipe diameters of all the flow paths were generated treating them as independent distribution mains (Table 8.3). The estimated pipe diameters are listed in Table 8.5.

It can be seen from Table 8.5 that different solutions are obtained for pipes common in various flow paths. To satisfy the minimum terminal pressures and maintain the desired flows, the maximum pipe diameters are selected in final solution. The maximum pipe sizes in various flow paths are highlighted in Table 8.5. Finally, continuous pipe sizes thus obtained are converted to nearest commercial pipe diameters for adoption. The commercial diameters adopted for the distribution system are listed in Table 8.5 and also shown in Fig. 8.3.

8.1.2.2. Discrete Pipe Diameter Approach. The conversion of continuous pipe diameters into discrete pipe diameters reduces the optimality of the solution. The consideration of commercial discrete pipe diameters directly in the design would eliminate such problem, and the solution thus obtained will be optimal. One of the methods for optimal system design using discrete pipe sizes is the application of linear programming (LP) technique. Karmeli et al. (1968) for the first time applied LP optimization approach for the optimal design of a branched water distribution system of single source.

In order to make LP application possible, it is assumed that each pipe link L_i consists of two commercially available discrete sizes of diameter D_{i1} and D_{i2} having lengths x_{i1} and x_{i2}, respectively. The pipe network system cost can be written as

$$F = \sum_{i=1}^{i_L} (c_{i1} x_{i1} + c_{i2} x_{i2}), \qquad (8.2)$$

TABLE 8.5. Water Distribution System Pipe Diameters

Pipe	Flow Path Pipes of Pipe (i) and Estimated Pipe Diameters of Pipes in Path														Maximum Pipe Diameter (m)	Adopted Pipe Size (m)
	1	2	3	4	5	6	7	8	9	10	11	12	13	14		
1	0.36	0.339	0.35	0.366	0.34	0.341	0.375	0.375	0.365	0.369	0.371	0.361	0.353	0.367	0.375	0.4
2		0.145													0.145	0.15
3			0.13												0.13	0.15
4				0.346	0.32	0.32	0.354	0.353	0.344	0.348	0.35	0.34	0.333	0.346	0.354	0.35
5					0.16										0.16	0.15
6						0.125									0.125	0.125
7							0.326	0.325	0.317	0.32	0.322	0.313	0.306	0.318	0.326	0.35
8								0.243	0.238	0.24					0.243	0.25
9									0.172						0.172	0.2
10										0.158					0.158	0.2
11											0.174				0.174	0.2
12												0.227		0.232	0.232	0.25
13													0.166		0.166	0.2
14														0.138	0.138	0.15

Note: The grey shaded portion indicates the maximum pipe sizes in various flow paths.

where c_{i1} and c_{i2} are the costs of 1-m length of the pipes (including excavation cost) of diameters D_{i1} and D_{i2}, respectively. The cost function F has to be minimized subject to the following constraints:

- The sum of lengths x_{i1} and x_{i2} is equal to the pipe link length L_i
- The pressure head at each node is greater than or equal to the prescribed minimum head H

The first constraint can be written as

$$x_{i1} + x_{i2} = L_i; \quad i = 1, 2, 3 \ldots i_L. \tag{8.3}$$

On the other hand, the second constraint gives rise to a head-loss inequality constraint for each pipe link i. The head loss h_{fi} in pipe link i, having diameters D_{i1} and D_{i2} of lengths x_{i1} and x_{i2}, respectively, is

$$h_{fi} = \frac{8 f_{i1} Q_i^2}{\pi^2 g D_{i1}^5} x_{i1} + \frac{8 f_{i2} Q_i^2}{\pi^2 g D_{i2}^5} x_{i2} + h_{mi}, \tag{8.4}$$

where f_{i1} and f_{i2} = friction factors for pipes of diameter D_{i1} and D_{i2}, respectively, and h_{mi} = form loss due to valves and fittings in pipe i. Considering the higher diameter of pipe link as the diameter of fittings, h_{mi} can be obtained as

$$h_{mi} = \frac{8 k_{fi} Q_i^2}{\pi^2 g D_{i2}^4} \tag{8.5}$$

where k_{fi} = form-loss coefficient for pipe link i. Starting from the originating node $J_t(i)$, which is the end of pipe link i, and moving in the direction opposite to the flow, one reaches the input point 0. The set of pipe links falling on this flow path is denoted by $I_t(i, \ell)$, where ℓ varies from 1 to $N_t(i)$. Summing up the head loss accruing in the flow path originating from $J_t(i)$, the head-loss constraint for the node $J_t(i)$ is written as

$$\sum_{p=I_t(i,\ell)} \left(\frac{8 f_{p1} Q_p^2}{\pi^2 g D_{p1}^5} x_{p1} + \frac{8 f_{p2} Q_p^2}{\pi^2 g D_{p2}^5} x_{p2} \right) \leq z_0 + h_0 - z_{j_t(i)} - H - \sum_{p=I_t(i,\ell)} \frac{8 k_{fp} Q_p^2}{\pi^2 g D_{p2}^4}$$

$$\ell = 1, 2, 3 \ N_t(i) \quad \text{For } i = 1, 2, 3 \ldots iL \tag{8.6}$$

where z_0 = the elevation of input source node, $z_{j_t(i)}$ = the elevation of node $J_t(i)$, and h_0 = the pressure head at input source node. Equations (8.2) and (8.3) and Inequation (8.6) constitute a LP problem. Unlike an equation containing an = sign, inequation is a mathematical statement that contains one of the following signs: $\leq$, $\geq$, $<$, and $>$. Thus, the LP problem involves $2i_L$ decision variables, consisting of i_L equality constraints and i_L inequality constraints. Taking lower and upper range of commercially available pipe sizes as D_{i1} and D_{i2} the problem is solved by using simplex algorithm as described in Appendix 1. Thus, the LP solution gives the minimum system cost and the corresponding pipe diameters.

For starting the LP algorithm, the uniform pipe material is selected for all pipe links; and using continuity conditions, the pipe discharges are computed. For known diameters D_{i1} and D_{i2} and discharge Q_i, the friction factors f_{i1} and f_{i2} are obtained by using Eq. (2.6c). Using Eqs. (8.2) and (8.3) and Inequation (8.6), the resulting LP problem is solved. The LP solution indicates preference of one diameter (lower or higher) in each pipe link. Knowing such preferences, the pipe diameter not preferred by LP is rejected and another diameter replacing it is introduced as D_{i1} or D_{i2}. The corresponding friction factor is also obtained subsequently. After completing the replacement process for $i = 1, 2, 3, \ldots i_L$, another LP solution is carried out to obtain the new preferred diameters. The process of LP and pipe size replacement is continued until D_{i1} and D_{i2} are two consecutive commercial pipe sizes. One more LP cycle now obtains the diameters to be adopted.

This can be explained using the following example of assumed commercial pipe sizes. As shown in Fig. 8.4, the available commercial pipe sizes for a pipe material are from D_1 to D_9. Selecting D_{i1} as D_1 and D_{i2} as D_9, the LP formulation can be developed using Eqs. (8.2) and (8.3) and constraint Inequation (8.6). If after the first iteration the D_{i1} is in the solution, then for the next iteration D_{i1} is kept as D_1 and D_{i2} is changed to D_8. If LP solution again results in providing the final pipe diameter D_{i1}, then for the next LP iteration D_{i1} is kept as D_1 and D_{i2} is changed to D_7. In the next LP iteration, the solution may indicate (say) pipe diameter as $D_{i2} = D_7$. Carrying out the next LP iteration with $D_{i1} = D_2$ and $D_{i2} = D_7$, if the final solution yields $D_{i2} = D_7$, then for the next iteration D_{i2} is kept as D_7 and D_{i1} is changed to D_3. Progressing in this manner, the next three formulations may be $(D_{i1} = D_3; D_{i2} = D_6)$, $(D_{i1} = D_4; D_{i2} = D_6)$, and $(D_{i1} = D_4; D_{i2} = D_5)$. In the third formulation, there is a tie between two consecutive diameters D_4 and D_5. Suppose the LP iteration indicates its preference to D_5 (as D_5 is in the final solution), then D_5 will be adopted as the pipe diameter for ith link. Thus, the algorithm will terminate at a point where the LP has to decide about its preference over two consecutive commercial diameters (Fig. 8.4). It can be concluded that to cover a range of only nine commercial sizes, eight LP iterations will be required to reach the final solution.

Starting the LP algorithm with D_{i1} and D_{i2} as lower and upper range of commercially available pipe sizes, the total number of LP iterations is very high resulting in large computation time. The LP iterations can be reduced if the starting diameters are taken close to the final solution. Using Eq. (8.1), the continuous optimal pipe diameters D_i^* can be calculated. Selecting the two consecutive commercially available sizes such that $D_{i1} \leq D_i^* \leq D_{i2}$, significant computational time can be saved. The branched water

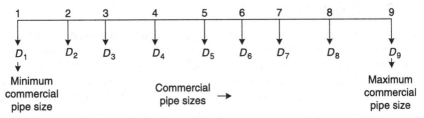

Figure 8.4. Application of commercial pipe sizes in LP formulation.

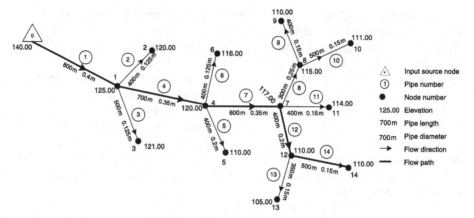

Figure 8.5. A branched, gravity water system (design based on discrete diameter approach).

distribution system shown in Fig. 8.1 was redesigned using the discrete diameter approach. The solution thus obtained is shown in Fig. 8.5.

It can be seen from Fig. 8.3 and Fig. 8.5 (depicting the solution by the two approaches) that the pipe diameters obtained from the discrete diameter approach are smaller in some of the branches than those obtained from the continuous diameter approach. This is because the discrete diameter approach delivers the solution by taking the system as a whole and no conversion from continuous to discrete diameters is required. Thus, the solution obtained by converting continuous sizes to nearest commercial sizes is not optimal.

8.2. PUMPING, BRANCHED SYSTEMS

The application of pumping systems is a must where topographic advantages are not available to flow water at desired pressure and quantity. In the pumping systems, the system cost includes the cost of pipes, pumps, pumping (energy), and operation and maintenance. The optimization of such systems is important due to high recurring energy cost. In the optimal design of pumping systems, there is an economic trade-off between the pumping head and pipe diameters. The design methodology for radial and branched water distribution systems is described in the following sections.

8.2.1. Radial Systems

Sometimes, water supply systems are conceived as a radial distribution network based on the local conditions and layout of the residential area. Radial systems have a central supply point and a number of radial branches with multiple withdrawals. These networks are ideally suited to rural water supply schemes receiving water from a single supply point. The radial system dealt with herein is a radial combination of i_L distribution lines with a single supply point at node 0. Each distribution line has j_L pipes.

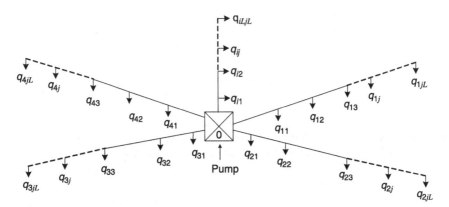

Figure 8.6. A radial, pumping water supply system.

A conceptual radial water distribution system is shown in Fig. 8.6. Let h_0 denote the pumping head at the supply point, H_i the minimum terminal head required at the last node of ith branch, q_{ij} the discharge withdrawal at jth node of ith branch, and L_{ij}, D_{ij}, h_{fij}, and Q_{ij} are the respective length, diameter, head loss, and discharge for jth pipe link of the ith branch.

The discharge Q_{ij} can be calculated by applying the continuity principle. The head loss can be expressed by the Darcy–Weisbach equation

$$h_{fij} = \frac{8 f_{ij} L_{ij} Q_{ij}^2}{\pi^2 g D_{ij}^5}, \tag{8.7}$$

where f_{ij} = the friction factor for jth pipe link of the ith branch expressed by

$$f_{ij} = 1.325 \left\{ \ln \left[\frac{\varepsilon_{ij}}{3.7 D_{ij}} + 4.618 \left(\frac{v D_{ij}}{Q_{ij}} \right)^{0.9} \right] \right\}^{-2}, \tag{8.8}$$

where ε_{ij} = average roughness height of the jth pipe link of the ith branch. Equating the total head loss in the ith branch to the combination of the elevation difference, pumping head, and the terminal head, the following equation is obtained:

$$\sum_{j=1}^{j_{Li}} \frac{8 f_{ij} L_{ij} Q_{ij}^2}{\pi^2 g D_{ij}^5} - h_0 - z_0 + z_{Li} + H_i = 0 \quad i = 1, 2, 3, \ldots, i_L, \tag{8.9}$$

where z_0 = the elevation of the supply point, z_{Li} = the elevation of the terminal point of the ith branch, and j_{Li} = the total number of the pipe links in the ith branch.

Considering the capitalized costs of pipes, pumps, and the pumping, the objective function is written as

$$F = k_m \sum_{i=1}^{i_L} \sum_{j=1}^{j_{Li}} L_{ij} D_{ij}^m + k_T \rho g Q_T h_0, \tag{8.10}$$

where Q_T = the discharge pumped. Combining Eqs. (8.9) and (8.10) through the Lagrange multipliers λ_i, the following merit function is formed

$$F_1 = k_m \sum_{i=1}^{i_L} \sum_{j=1}^{j_{Li}} L_{ij} D_{ij}^m + k_T \rho g Q_T h_0 + \sum_{i=1}^{i_L} \lambda_i \left(\sum_{j=1}^{j_{Li}} \frac{8 f_{ij} L_{ij} Q_{ij}^2}{\pi^2 g D_{ij}^5} - h_0 - z_0 + z_{Li} + H_i \right).$$
$$\tag{8.11}$$

Assuming f_{ij} to be constant, differentiating Eq. (8.11) partially with respect to D_{ij} for minimum, one gets

$$\lambda_i = \frac{\pi^2 g m k_m D_{ij}^{m+5}}{40 f_{ij} Q_{ij}^2}. \tag{8.12}$$

Putting $j = 1$ in Eq. (8.12) and equating it with Eq. (8.12) and simplifying, the following equation is obtained:

$$D_{ij}^* = D_{i1}^* \left(\frac{f_{ij} Q_{ij}^2}{f_{i1} Q_{i1}^2} \right)^{\frac{1}{m+5}}. \tag{8.13}$$

Combining Eqs. (8.9) and (8.13), the following equation is found:

$$D_{i1}^* = (f_{i1} Q_{i1}^2)^{\frac{1}{m+5}} \left[\frac{8}{\pi^2 g (h_0 + z_0 - z_{Li} - H_i)} \sum_{j=1}^{j_{Li}} L_{ij} (f_{ij} Q_{ij}^2)^{\frac{m}{m+5}} \right]^{\frac{1}{5}} \tag{8.14}$$

$$i = 1, 2, 3, \ldots, i_L.$$

Differentiating Eq. (8.11) partially with respect to h_0, the following equation is obtained:

$$k_T \rho g Q_T - \sum_{i=1}^{i_L} \lambda_i = 0. \tag{8.15}$$

Eliminating λ_i and D_{i1} between Eqs. (8.12), (8.14), and (8.15) for $j = 1$, the following equation is found:

$$\frac{40\,k_T\rho Q_T}{\pi^2 m k_m} = \sum_{i=1}^{i_L}\left[\frac{8}{\pi^2\,g(h_0 + z_0 - z_{Li} - H_i)}\sum_{j=1}^{j_{Li}} L_{ij}\left(f_{ij}Q_{ij}^2\right)^{\frac{m}{m+5}}\right]^{\frac{m+5}{5}}. \qquad (8.16)$$

Equation (8.16) can be solved for h_0 by trial and error. Knowing h_0, D_{i1} can be calculated by Eq. (8.14). Once a solution for D_{i1} and h_0 is obtained for assumed values of f_{ij}, it can be improved using Eq. (8.8), and the process is repeated until the convergence is arrived at. Knowing D_{i1}, D_{ij} can be obtained from Eq. (8.13).

For a flat area involving equal terminal head H for all the branches, Eq. (8.16) reduces to

$$h_0 = z_L + H - z_0 + \frac{8}{\pi^2 g}\left\{\frac{\pi^2 m k_m}{40\,k_T\rho Q_T}\sum_{i=1}^{i_L}\left[\sum_{j=1}^{j_{Li}} L_{ij}\left(f_{ij}Q_{ij}^2\right)^{\frac{m}{m+5}}\right]^{\frac{m+5}{5}}\right\}^{\frac{5}{m+5}}, \qquad (8.17)$$

whereas on substituting h_0 from Eq. (8.17) to Eq. (8.14) and using Eq. (8.13), D_{ij} is found as

$$D_{ij}^* = \left\{\frac{40\,k_T\rho f_{ij}Q_T Q_{ij}^2\left[\sum\limits_{p=1}^{j_{Li}} L_{ip}\left(f_{ip}Q_{ip}^2\right)^{\frac{m}{m+5}}\right]^{\frac{m+5}{5}}}{\pi^2 m k_m \sum\limits_{i=1}^{i_L}\left[\sum\limits_{p=1}^{j_{Li}} L_{ip}\left(f_{ip}Q_{ip}^2\right)^{\frac{m}{m+5}}\right]^{\frac{m+5}{5}}}\right\}^{\frac{1}{m+5}}. \qquad (8.18)$$

Using (8.10), (8.17), and (8.18), the optimal cost is obtained as

$$F^* = \left(1 + \frac{m}{5}\right)k_m\left(\frac{40\,k_T\rho\,Q_T}{\pi^2 m k_m}\right)^{\frac{m}{m+5}}\left\{\sum_{i=1}^{i_L}\left[\sum_{p=1}^{j_{Li}} L_{ip}\left(f_{ip}Q_{ip}^2\right)^{\frac{m}{m+5}}\right]^{\frac{m+5}{5}}\right\}^{\frac{5}{m+5}}$$

$$+ k_T\rho\,g Q_T(z_L + H - z_0). \qquad (8.19)$$

The effect of variation of f_{ij} can be corrected using Eq. (8.8) iteratively.

8.2.2. Branched, Pumping Systems

Generally, the rural water distribution systems are branched and dead-end systems. These systems typically consist of a source, pumping plant, water treatment unit, clear water reservoir, and several kilometers of pipe system to distribute the water. The design of such systems requires a method of analyzing the hydraulics of the network and also a method for obtaining the design variables pertaining to minimum system cost. Similar to gravity system design, the continuous diameter and discrete diameter approaches for the design of pumping systems are discussed in the following sections.

8.2.2.1. Continuous Diameter Approach.
In this approach, water distribution is designed by decomposing the entire network into a number of subsystems. Adopting a technique similar to the branched gravity system, the entire pumping network is divided into number of pumping distribution mains. Thus, in a pumping distribution system of i_L pipes, i_L distribution mains are generated. These distribution mains are the flow paths generated using the methodology described in Section 3.9. Such decomposition is essential to calculate optimal pumping head for the network. A typical branched pumping water distribution system is shown in Fig. 8.7.

These pumping distribution mains are listed in Table 8.6. Applying the method described in Section 7.2, the water distribution system can be designed.

Modify and rewrite Eq. (7.11b) for optimal pipe diameter as

$$D_i^* = \left(\frac{40\,k_T\rho f_i Q_T Q_i^2}{\pi^2 m k_m}\right)^{\frac{1}{m+5}},$$

(8.20)

where Q_T is the total pumping discharge. The data for the pipe network shown in Fig. 8.7 are given in Table 8.7. The optimal diameters were obtained applying Eq. (8.20) and using data from Table 8.7. Pipe parameters $k_m = 480$, $m = 0.935$, and $\varepsilon = 0.25\,\text{mm}$ were applied for this example. k_T/k_m ratio of 0.02 was adopted for this example.

To apply Eq. (8.20), the pipe friction f was considered as 0.01 in all the pipes initially, which was improved iteratively until the two consecutive solutions were

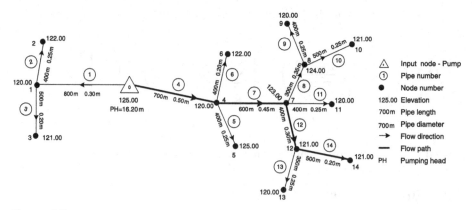

Figure 8.7. Branched, pumping water distribution system (design based on continuous diameter approach).

TABLE 8.6. Pipe Flow Paths as Pumping Distribution Mains

Pipe i	Flow Path Pipes Connecting to Input Point Node 0 and Generating Water Distribution Pumping Mains $I_t(i, \ell)$				$N_t(i)$	$J_t(i)$
	$\ell = 1$	$\ell = 2$	$\ell = 3$	$\ell = 4$		
1	1				1	1
2	2	1			2	2
3	3	1			2	3
4	4				1	4
5	5	4			2	5
6	6	4			2	6
7	7	4			2	7
8	8	7	4		3	8
9	9	8	7	4	4	9
10	10	8	7	4	4	10
11	11	7	4		3	11
12	12	7	4		3	12
13	13	12	7	4	4	13
14	14	12	7	4	4	14

TABLE 8.7. Pipe Diameters of Branched Pumping System

Pipe/ Node i/j	Elevation z_j (m)	Length L_i (m)	Nodal Demand Discharge q_i (m³/s)	Pipe Discharge Q_i (m³/s)	Estimated Pipe Diameter D_i (m)	Adopted Pipe Diameter (m)
0	125.00					
1	120.00	800	0.01	0.035	0.275	0.3
2	122.00	400	0.015	0.015	0.210	0.25
3	121.00	500	0.01	0.01	0.185	0.2
4	120.00	700	0.01	0.175	0.461	0.5
5	125.00	400	0.02	0.02	0.230	0.25
6	122.00	400	0.01	0.01	0.185	0.2
7	123.00	600	0.01	0.135	0.424	0.45
8	124.00	300	0.02	0.055	0.318	0.35
9	120.00	400	0.02	0.02	0.230	0.25
10	121.00	500	0.015	0.015	0.210	0.25
11	120.00	400	0.02	0.02	0.230	0.25
12	121.00	400	0.02	0.05	0.287	0.3
13	120.00	350	0.02	0.02	0.230	0.25
14	121.00	500	0.01	0.01	0.185	0.2

TABLE 8.8. Iterative Solution of Branched-Pipe Water Distribution System

Pipe i	1st Iteration		2nd Iteration		3rd Iteration	
	f_i	D_i	f_i	D_i	f_i	D_i
1	0.010	0.242	0.021	0.276	0.021	0.275
2	0.010	0.182	0.023	0.210	0.023	0.210
3	0.010	0.159	0.025	0.185	0.024	0.185
4	0.010	0.417	0.018	0.462	0.018	0.461
5	0.010	0.201	0.023	0.231	0.022	0.230
6	0.010	0.159	0.025	0.185	0.024	0.185
7	0.010	0.382	0.019	0.425	0.018	0.424
8	0.010	0.282	0.020	0.319	0.020	0.318
9	0.010	0.201	0.023	0.231	0.022	0.230
10	0.010	0.182	0.023	0.210	0.023	0.210
11	0.010	0.201	0.023	0.231	0.022	0.230
12	0.010	0.254	0.021	0.288	0.021	0.287
13	0.010	0.201	0.023	0.231	0.022	0.230
14	0.010	0.159	0.025	0.185	0.024	0.185

close. The output of these iterations is listed in Table 8.8. The calculated pipe diameters and adopted commercial sizes are listed in Table 8.7.

The pumping head required for the system can be obtained using Eq. (7.12), which is rewritten as:

$$h_0^* = z_n + H - z_0 + \frac{8}{\pi^2 g}\left(\frac{\pi^2 m k_m}{40\rho k_T Q_1}\right)^{\frac{5}{m+5}} \sum_{p=I_t(i,\ell)}^{n} L_p\left(f_p Q_p^2\right)^{\frac{m}{m+5}}. \tag{8.21}$$

Equation (8.21) is applied for all the water distribution mains (flow paths), which is equal to the total number of pipes in the distribution system. Thus, the variable n in Eq. (8.21) is equal to $N_t(i)$ and pipe in the distribution main $p = I_t(i, \ell)$, $\ell = 1, N_t(i)$. The discharge Q_1 is the flow in the last pipe $I_t(i,N_t(i))$ of a flow path for pipe i, which is directly connected to the source. The distribution main originates at pipe i as listed in Table 8.6. Thus, applying Eq. (8.21), the pumping heads for all the distribution mains as listed in Table 8.6 were calculated. The computation for pumping head is shown by taking an example of distribution main originating at pipe 13 with pipes in distribution main as 4, 7, 12, and 13 (Table 8.6). The originating node for this distribution main is node 13. The terminal pressure H across the network as 15 m was maintained. The pumping head required for distribution main originating at pipe 13 was calculated as 13.77 m. Repeating the process for all the distribution mains in the system, it was found that the maximum pumping head 17.23 m was required for distribution main originating at pipe 8 and node 8. The pumping heads for various distribution mains are listed in Table 8.9, and the maximum pumping head for the distribution main

TABLE 8.9. Pumping Heads for Distribution Mains

Pipe/ Node i/j	Elevation Z_j (m)	Length L_i (m)	Pipe Discharge Q_i (m³/s)	Continuous Pipe Diameters (m)	Pumping Head for Continuous Sizes h_0 (m)	Adopted Discrete Diameters D_i (m)	Pumping Head for Discrete Sizes h_0 (m)
0	125						
1	120	800	0.035	0.275	11.09	0.3	10.70
2	122	400	0.015	0.210	13.51	0.25	12.87
3	121	500	0.01	0.185	12.56	0.2	12.01
4	120	700	0.175	0.461	11.54	0.5	11.01
5	125	400	0.02	0.230	17.00	0.25	16.32
6	122	400	0.01	0.185	13.92	0.2	13.26
7	123	600	0.135	0.424	15.76	0.45	14.91
8	124	300	0.055	0.318	17.23	0.35	16.20
9	120	400	0.02	0.230	13.69	0.25	12.50
10	121	500	0.015	0.210	14.76	0.25	13.42
11	120	400	0.02	0.230	13.22	0.25	12.22
12	121	400	0.02	0.287	14.37	0.3	13.62
13	120	350	0.02	0.230	13.77	0.25	12.88
14	121	500	0.01	0.185	14.84	0.2	13.93

Note: The grey shaded portion indicates maximum pumping head for system.

starting from pipe 8 is also highlighted in this table. Thus, the required pumping head for the system is 17.23 m if continuous pipe sizes were provided.

As the continuous pipe diameters are converted into discrete diameters, Eq. (7.9) is applied directly to calculate pumping head and is rewritten below:

$$h_0^* = z_n + H - z_0 + \sum_{p=I_t(i,\ell)}^{N_t(i)} \frac{8 L_p f_p Q_p^5}{\pi^2 g D_p^5}.$$ (8.22)

Based on the finally adopted discrete diameters, the friction factor f was recalculated using Eq. (2.6c). Using Eq. (8.22), the pumping head for all the distribution mains was calculated and listed in Table 8.9. It can be seen that the maximum pumping head for the system was 16.20 m, which is again highlighted.

The adopted commercial sizes and pumping head required is shown in Fig. 8.7. It can be seen that a different solution is obtained if the commercial sizes are applied.

8.2.2.2. Discrete Diameter Approach.

In the discrete diameter approach of the design, the commercial pipe sizes are considered directly in the synthesis of water distribution systems. A method for the synthesis of a typical branched pumping system having i_L number of pipes, single input source, pumping station, and reservoir is presented in this section. The LP problem for this case is stated below.

The cost function F to be minimized includes the cost of pipes, pump, pumping (energy), and storage. The cost function is written as:

$$\min F = \sum_{i=1}^{i_L} (c_{i1} x_{i1} + c_{i2} x_{i2}) + \rho g \, k_T Q_T h_0,$$ (8.23)

subject to

$$x_{i1} + x_{i2} = L_i; \ i = 1, 2, 3 \ldots i_L,$$ (8.24)

$$\sum_{p=I_t(i,\ell)} \left[\frac{8 f_{p1} Q_p^2}{\pi^2 g D_{p1}^5} x_{p1} + \frac{8 f_{p2} Q_i^2}{\pi^2 g D_{p2}^5} x_{p2} \right] \leq z_0 + h_0 - z_{j_t(i)} - H - \sum_{p=I_t(i,\ell)} \frac{8 k_{fp} Q_p^2}{\pi^2 g D_{p2}^4}$$

$$\ell = 1, 2, 3, N_t(i) \qquad \text{For } i = 1, 2, 3 \ldots i_L \qquad (8.25)$$

Using Eq. (8.20), the continuous optimal pipe diameters D_i^* can be calculated. Two consecutive commercially available sizes such that $D_{i1} \leq D_i^* \leq D_{i2}$ are adopted to start the LP iterations. The branched water distribution system shown in Fig. 8.7 was redesigned using the discrete diameter approach. The solution thus obtained is shown in Fig. 8.8.

It can be seen that the pumping head is 20.15 m and the pipe sizes in the dead-end pipes are lower that the solution obtained with continuous diameter approach.

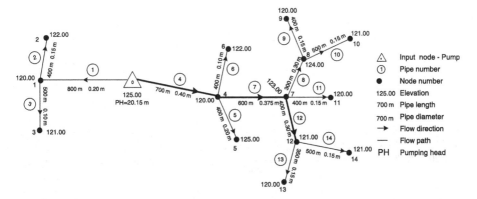

<u>Figure 8.8.</u> Branched pumping water distribution system (design based on discrete diameter approach).

Thus, in this case also two different solutions will be obtained by the two approaches.

8.3. PIPE MATERIAL AND CLASS SELECTION METHODOLOGY

Once the system design taking a particular pipe material is obtained, the economic pipe material and class (based on working pressure classification) should be selected for each pipe link. The selection of particular pipe material and class is based on local cost, working pressure, commercial pipe sizes availability, soil strata, and overburden pressure. Considering commercially available pipe sizes, per meter cost, and working pressure, Sharma (1989) and Swamee and Sharma (2000) developed a chart for the selection of pipe material and class based on local data. The modified chart is shown in Fig. 8.9 for demonstration purposes. Readers are advised to develop a similar chart for pipe material and class selection based on their local data and also considering regulatory requirements for pipe material usage. A simple computer program can then be written for the selection of pipe material and class. The pipe network analysis can be repeated for new pipe roughness height (Table 2.1) based on pipe material selection giving revised friction factor for pipes. Similarly, the pipe diameters are recalculated using revised analysis.

In water distribution systems, the effective pressure head h_{0i} at each pipe for pipe selection would be the maximum of the two acting on each node $J_1(i)$ or $J_2(i)$ of pipe i:

$$h_{0i} = h_0 + z_0 - \min\lfloor z_{J_1}(i), z_{J_2}(i)\rfloor, \qquad (8.26)$$

where h_0 is the pumping head in the case of pumping systems and the depth of the water column over the inlet pipe in reservoir in the case of gravity systems. For a commercial size of 0.30 m and effective pressure head h_{0i} of 80 m on pipe, the economic pipe

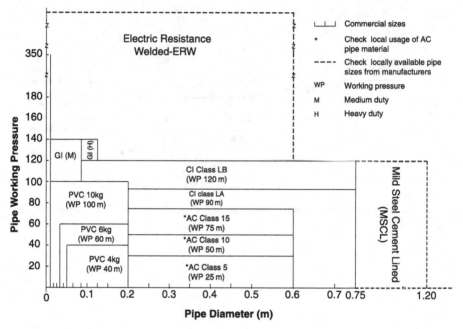

Figure 8.9. Pipe material selection based on available commercial sizes, cost, and working pressure.

material using Fig. 8.9 is CI Class A (WP 90 m). Similarly, pipe materials for each pipe for the entire network can be calculated.

EXERCISES

8.1. Describe advantages and disadvantages of branched water distribution systems. Provide examples for your description.

8.2. Design a radial, branched, gravity water distribution system for the data given below. The network can be assumed similar to that of Fig. 8.2. The elevation of the source point is 120.00 m. Collect local cost data required to design the system.

j	q_{1j} (m^3/s)	L_{1j} (m)	z_{1j} (m)	q_{2j} (m^3/s)	L_{2j} (m)	z_{2j} (m)	q_{3j} (m^3/s)	L_{3j} (m)	z_{3j} (m)
1	0.01	300	110	0.015	500	105	0.02	200	100
2	0.02	400	90	0.02	450	97	0.01	250	95
3	0.015	500	85	0.03	350	80	0.04	100	90
4	0.025	350	80	0.015	200	85	0.02	500	85

8.3. Design a gravity-flow branched system for the network given in Fig. 8.3. Modify the nodal demand to twice that given in Table 8.2 and similarly increase the pipe length by a factor of 2.

8.4. Design a radial, pumping water distribution system using the data given for Exercise 8.2. Consider the flat topography of the entire service area by taking elevation as 100.0 m. The system can be considered similar to that of Fig. 8.5.

8.5. Design a branched, pumping water distribution system similar to that shown in Fig. 8.6. Consider the nodal demand as twice that given in Table 8.7.

8.6. Develop a chart similar to Fig. 8.9 for locally available commercial pipe sizes.

REFERENCES

Karmeli, D., Gadish, Y., and Meyers, S. (1968). Design of optimal water distribution networks. *J. Pipeline Div.* 94(PL1), 1–10.

Sharma, A.K. (1989). *Water Distribution Network Optimisation.* Thesis presented to the University of Roorkee (presently Indian Institute of Technology, Roorkee), Roorkee, India, in the fulfillment of the requirements for the degree of Doctor of Philosophy.

Swamee, P.K., and Sharma, A.K. (2000). Gravity flow water distribution system design. *Journal of Water Supply Research and Technology-AQUA* 49(4), 169–179.

9

SINGLE-INPUT SOURCE, LOOPED SYSTEMS

Generally, town water supply systems are single-input source, looped pipe networks. As stated in the previous chapter, the looped systems have pipes that are interconnected throughout the system such that the flow to a demand node can be supplied through several connected pipes. The flow directions in a looped system can change based on spatial or temporal variation in water demand, thus unlike branched systems, the flow directions in looped network pipes are not unique.

The looped network systems provide redundancy to the systems, which increases the capacity of the system to overcome local variation in water demands and also ensures the distribution of water to users in case of pipe failures. The looped geometry is also favored from the water quality aspect, as it would reduce the water age. The pipe sizes and distribution system layouts are important factors for minimizing the water age. Due to the multidirectional flow patterns and also variations in flow patterns in the system over time, the water would not stagnate at one location resulting in reduced

Design of Water Supply Pipe Networks. By Prabhata K. Swamee and Ashok K. Sharma
Copyright © 2008 John Wiley & Sons, Inc.

TABLE 9.1. Advantages and Disadvantages of Looped Water Distribution Systems

Advantages	Disadvantages
• Minimize loss of services, as main breaks can be isolated due to multidirectional flow to demand points • Reliability for fire protection is higher due to redundancy in the system • Likely to meet increase in water demand—higher capacity and lower velocities • Better residual chlorine due to inline mixing and fewer dead ends • Reduced water age	• Higher capital cost • Higher operational and maintenance cost • Skilled operation

water age. The advantages and disadvantages of looped water distribution systems are given in Table 9.1.

It has been described in the literature that the looped water distribution systems, designed with least-cost consideration only, are converted into a tree-like structure resulting in the disappearance of the original geometry in the final design. Loops are provided for system reliability. Thus, a design based on least-cost considerations only defeats the basic purpose of loops provision in the network. In this chapter, a method for the design of a looped water distribution system is described. This method maintains the loop configuration of the network by bringing all the pipes of the network in the optimization problem formulation, although it is also based on least-cost consideration only.

Simple gravity-sustained and pumping looped water distribution systems are shown in Fig. 9.1. In case of pumping systems, the location of pumping station and reservoir can vary depending upon the raw water resource, availability of land for water works, topography of the area, and layout pattern of the town.

Analysis of a pipe network is essential to understand or evaluate a physical system, thus making it an integral part of the synthesis process of a network. In the case of a single-input system, the input source discharge is equal to the sum of withdrawals in the network. The discharges in pipes are not unique in looped water systems and are dependent on the pipe sizes and the pressure heads. Thus, the design of a looped

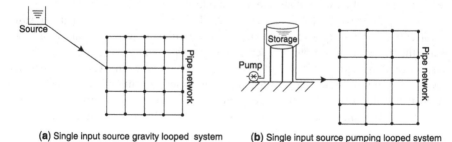

(a) Single input source gravity looped system   (b) Single input source pumping looped system

Figure 9.1. Looped water distribution systems.

network would require sequential application of analysis and synthesis techniques until a termination criterion is achieved. A pipe network can be analyzed using any of the analysis methods described in Chapter 3, however, the Hardy Cross method has been adopted for the analysis of water distribution network examples.

Similar to branched systems, the water distribution design methods based on cost optimization have two approaches: (a) continuous diameter approach and (b) discrete diameter approach or commercial diameter approach. In the continuous diameter approach, the pipe link sizes are calculated as continuous variables, and once the solution is obtained, the nearest commercial sizes are adopted. On the other hand, in the discrete diameter approach, commercially available pipe diameters are directly applied in the design method. The design of single-source gravity and pumping looped systems is described in this chapter.

9.1. GRAVITY-SUSTAINED, LOOPED SYSTEMS

The gravity-sustained, looped water distribution systems are suitable in areas where the source (input) point is at a higher elevation than the demand points. However, the area covered by the distribution network is relatively flat. The input source point is connected to the distribution network by a gravity-sustained transmission main. Such a typical water distribution system is shown in Fig. 9.2.

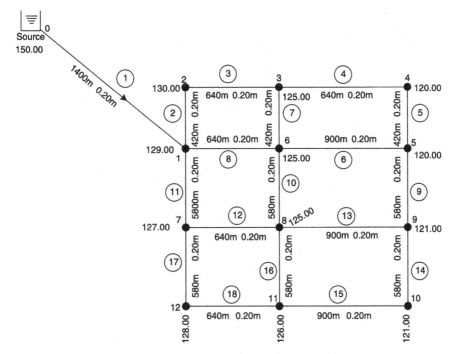

Figure 9.2. Gravity-sustained, looped water distribution system.

TABLE 9.2. Gravity-Sustained, Looped Water Distribution Network Data

Pipe/ Node i/j	Node 1 $J_1(i)$	Node 2 $J_2(i)$	Loop 1 $K_1(i)$	Loop 2 $K_2(i)$	Length L_i (m)	Form loss Coefficient k_{fi}	Population $P(i)$	Nodal Elevation $z(j)$ (m)
0								150
1	0	1	0	0	1400	0.5		129
2	1	2	1	0	420	0	200	130
3	2	3	1	0	640	0	300	125
4	3	4	2	0	900	0	450	120
5	4	5	2	0	580	0	250	120
6	5	6	2	4	900	0	450	125
7	3	6	1	2	420	0	200	127
8	1	6	1	3	640	0	300	125
9	5	9	4	0	580	0	250	121
10	6	8	3	4	580	0	250	121
11	1	7	3	0	580	0	250	126
12	7	8	3	5	640	0	300	128
13	8	9	4	6	900	0	450	
14	9	10	6	0	580	0	300	
15	10	11	6	0	900	0	450	
16	8	11	5	6	580	0	300	
17	7	12	5	0	580	0	300	
18	11	12	5	0	640	0	300	

The pipe network (Fig. 9.2) data are listed in Table 9.2. The data include the pipe number, both its nodes, loop numbers, form-loss co-efficient due to fittings in pipe, population load on pipe, and nodal elevations.

The pipe network shown in Fig. 9.2 has been analyzed for pipe discharges. Assuming peak discharge factor = 2.5, rate of water supply 400 liters/capita/day (L/c/d), the nodal discharges obtained using the method described in Chapter 3 (Eq. 3.29) are listed in Table 9.3. The negative nodal demand indicates the inflow into the distribution system at input source.

TABLE 9.3. Estimated Nodal Discharges

Node j	Discharge q_j (m³/s)	Node j	Discharge q_j (m³/s)
0	−0.06134	7	0.00491
1	0.00434	8	0.00752
2	0.00289	9	0.00579
3	0.00549	10	0.00434
4	0.00405	11	0.00607
5	0.00549	12	0.00347
6	0.00694		

TABLE 9.4. Looped Network Pipe Discharges

Pipe i	Discharge Q_i (m³/s)	Pipe i	Discharge Q_i (m³/s)	Pipe i	Discharge Q_i (m³/s)
1	0.06134	7	0.00183	13	0.00388
2	0.01681	8	0.01967	14	0.00187
3	0.01391	9	0.00377	15	−0.00247
4	0.00658	10	0.00782	16	0.00415
5	0.00252	11	0.02052	17	0.00786
6	−0.00674	12	0.00774	18	−0.00439

The pipe discharges in a looped water distribution network are not unique and thus require some looped network analysis technique. The pipe diameters are to be assumed initially to analyze the network, thus considering all pipe sizes = 0.2 m and pipe material as CI, the network was analyzed using the Hardy Cross method described in Section 3.7. The estimated pipe discharges are listed in Table 9.4. As described in Chapter 3, the negative pipe discharge indicates that the discharge in pipe flows from higher magnitude node number to lower magnitude node number.

9.1.1. Continuous Diameter Approach

In this approach, the entire looped water distribution system is converted into a number of distribution mains. Each distribution main is then designed separately using the methodology described in Chapters 7 and 8. The total number of such distribution mains is equal to the number of pipes in the network system, as each pipe would generate a flow path forming a distribution main.

The flow paths for all the pipes of the looped water distribution network were generated using the pipe discharges (Table 9.4) and the network geometry data (Table 9.2). Applying the flow path selection method described in Section 3.9, the pipe flow paths along with their originating nodes $J_t(i)$ are listed in Table 9.5.

Treating the flow path as a water distribution main and applying Eq. (7.6b), rewritten below, the optimal pipe diameters can be calculated:

$$D_i^* = \left(f_i Q_i^2\right)^{\frac{1}{m+5}} \left[\frac{8}{\pi^2 g(z_0 - z_n - H)} \sum_{p=I_t(i,\ell)}^{n} L_p \left(f_p Q_p^2\right)^{\frac{m}{m+5}}\right]^{0.2}. \tag{9.1}$$

Applying Eq. (9.1), the design of flow paths of pipes as distribution mains was conducted using $n = N_t(i)$ and $p = I_t(i, \ell)$, $\ell = 1$, $N_t(i)$ the pipe in flow path of pipe i. The corresponding pipe flows, nodal elevations, and pipe lengths used are listed in Tables 9.2 and 9.4. The minimum terminal pressure head of 10 m was maintained at nodes. The friction factor was assumed 0.02 initially for all the pipes in the distribution main, which was improved iteratively until the two consecutive f values were close. The estimated pipe sizes for various flow paths as water distribution mains are listed in Table 9.6.

TABLE 9.5. Pipe Flow Paths Treated as Water Distribution Main

Pipe i	$\ell = 1$	$\ell = 2$	$\ell = 3$	$\ell = 4$	$\ell = 5$	$N_t(i)$	$J_t(i)$
1	1					1	1
2	2	1				2	2
3	3	2	1			3	3
4	4	3	2	1		4	4
5	5	4	3	2	1	5	5
6	6	8	1			3	5
7	7	3	2	1		4	6
8	8	1				2	6
9	9	6	8	1		4	9
10	10	8	1			3	8
11	11	1				2	7
12	12	11	1			3	8
13	13	10	8	1		4	9
14	14	13	10	8	1	5	10
15	15	18	17	11	1	5	10
16	16	10	8	1		4	11
17	17	11	1			3	12
18	18	17	11	1		4	11

Header spanning columns $\ell = 1$ through $J_t(i)$: Flow Path Pipes Connecting to Input Point Node 0 and Generating Water Distribution Gravity Mains $I_t(i, \ell)$

It can be seen from Table 9.5 and Table 9.6 that there are a number of pipes that are common in various flow paths (distribution mains); the design of each distribution main provides different pipe sizes for these common pipes. The largest pipe sizes are highlighted in Table 9.6, which are taken into final design.

The estimated pipe sizes and nearest commercial sizes adopted are listed in Table 9.7.

Based on the adopted pipe sizes, the pipe network should be analyzed again for another set of pipe discharges (Table 9.4). The pipe flow paths are regenerated using the revised pipe flows (Table 9.5). The pipe sizes are calculated for new set of distribution mains. The process is repeated until the two solutions are close. Once the final design is achieved, the economic pipe material can be selected using the method described in Section 8.3. The application of economic pipe material selection method is described in Section 9.1.2.

9.1.2. Discrete Diameter Approach

The conversion of continuous pipe diameters into commercial (discrete) pipe diameters reduces the optimality of the solution. Similar to the method described in Chapter 8, a method considering commercial pipe sizes directly in the looped network design process using LP optimization technique is given in this section. The important feature of this method is that all the looped network pipes are brought in the optimization problem

TABLE 9.6. Pipe Sizes of Water Distribution Mains (Pipe Flow Paths)

Pipe i	\multicolumn{18}{c}{Flow Path Pipes of Pipe (/) and Estimated Pipe Diameters of Pipes in Path of Pipe i}

Pipe i	1	2	3	4	5	6	7	8	9	10	11	12	13	14	15	16	17	18
1	0.242	0.256	0.247	0.241	0.243	0.2386	0.254	0.241	0.246	0.248	0.247	0.249	0.245	0.249	**0.265**	0.258	0.259	0.258
2		**0.169**	0.163	0.159	0.161		0.168											
3			0.154	0.15	0.149		**0.158**											
4				0.118	**0.119**													
5					**0.088**													
6						0.1178			**0.121**									
7							**0.083**											
8						0.1658		0.167	0.171	0.172			0.17	0.173	**0.186**	0.179		
9									**0.101**									
10										0.129			0.127	0.129		**0.134**		
11											0.174	0.173					**0.183**	0.182
12												**0.128**						
13													0.102	**0.103**				
14														**0.082**				
15															**0.098**			
16																**0.109**		
17															**0.137**		0.135	0.134
18															**0.114**			0.111

Note: The grey shaded portion indicates the largest pipe sizes.

TABLE 9.7. Estimated and Adopted Pipe Sizes

Pipe	Estimated Continuous Pipe Size (m)	Adopted Pipe Size (m)	Pipe	Estimated Continuous Pipe Size (m)	Adopted Pipe Size (m)
1	0.265	0.300	10	0.134	0.150
2	0.169	0.200	11	0.183	0.200
3	0.158	0.150	12	0.128	0.125
4	0.119	0.125	13	0.103	0.100
5	0.088	0.100	14	0.082	0.100
6	0.121	0.125	15	0.098	0.100
7	0.083	0.100	16	0.109	0.100
8	0.186	0.200	17	0.137	0.150
9	0.101	0.100	18	0.114	0.125

formulation keeping the looped configuration intact (Swamee and Sharma, 2000). The LP problem in the current case is

$$\min \ F = \sum_{i=1}^{i_L} (c_{i1}x_{i1} + c_{i2}x_{i2}), \tag{9.2}$$

subject to

$$x_{i1} + x_{i2} = L_i; \quad i = 1, 2, 3 \ldots i_L, \tag{9.3}$$

$$\sum_{p=I_t(i,\ell)} \left[\frac{8f_{p1}Q_p^2}{\pi^2 g D_{p1}^5} x_{p1} + \frac{8f_{p2}Q_p^2}{\pi^2 g D_{p2}^5} x_{p2} \right] \le z_0 + h_0 - z_{J_t(i)}$$

$$- H - \sum_{p=I_t(i,\ell)} \frac{8 k_{fp} Q_p^2}{\pi^2 g D_{p2}^4}$$

$$\ell = 1, 2, 3 \, N_t(i) \qquad \text{for } i = 1, 2, 3, \ldots i_L \tag{9.4}$$

Equations (9.2) and (9.3) and Ineq. (9.4) constitute a LP problem. Using Inequation (9.4), the head-loss constraints for all the originating nodes $Z_{Jt(i)}$ of pipe flow paths in a pipe network are developed. (Unlike an equation containing an $=$ sign, inequation is a mathematical statement that contains one of the following signs: $\le$, $\ge$, $<$, and $>$.) Thus, it will bring all the pipes of the network into the optimization process. This will give rise to the formulation of more than one head-loss constraint inequation for some of the nodes. Such head-loss constraint equations for the same node will have a different set of pipes $I_t(i,\ell)$ in their flow paths.

As described in Section 8.1.2.2, starting LP algorithm with D_{i1} and D_{i2} as lower and upper range of commercially available pipe sizes, the total number of LP iterations is very high resulting in large computation time. In this case also, the LP iterations can be reduced if the starting diameters are taken close to the final solution. Using Eq. (9.1), the continuous optimal pipe diameters D_i^* can be calculated. Selecting the two consecutive commercially available sizes such that $D_{i1} \le D_i^* \le D_{i2}$, significant

TABLE 9.8. Looped Pipe Distribution Network Design

Pipe i	Pipe Length L_i (m)	Initial Design with Assumed Pipe Material		Maximum Pressure in Pipe h_i (m)	Final Design with Optimal Pipe Material	
		D_i (m)	Pipe Material		D_i (m)	Pipe Material
1	1400	0.300–975	CI[‡]	21.0	0.300–975	AC Class 5*
		0.250–425	CI	21.0	0.250–425	AC Class 5
2	420	0.200	CI	21.0	0.200	PVC 40 m†
3	640	0.150	CI	25.0	0.150	PVC 40 m
4	900	0.125	CI	30.0	0.100	PVC 40 m
5	580	0.100	CI	30.0	0.100	PVC 40 m
6	900	0.080	CI	30.0	0.650	PVC 40 m
7	420	0.080	CI	25.0	0.065	PVC 40 m
8	640	0.150	CI	25.0	0.150	PVC 40 m
9	580	0.080	CI	30.0	0.065	PVC 40 m
10	580	0.150	CI	25.0	0.150	PVC 40 m
11	580	0.150	CI	23.0	0.150	PVC 40 m
12	640	0.080	CI	25.0	0.065	PVC 40 m
13	900	0.080	CI	29.0	0.080	PVC 40 m
14	580	0.050	CI	29.0	0.050	PVC 40 m
15	900	0.100	CI	29.0	0.080	PVC 40 m
16	580	0.100	CI	25.0	0.100	PVC 40 m
17	580	0.150	CI	23.0	0.150	PVC 40 m
18	640	0.125	CI	24.0	0.125	PVC 40 m

*Asbestos cement pipe 25-m working pressure.
†Polyvinyl chloride 40-m working pressure.
‡Cast Iron pipe class LA-60 m working pressure.

computational time can be saved. The looped water distribution system shown in Fig. 9.2 was redesigned using the discrete diameter approach. The solution thus obtained using initially CI pipe material is given in Table 9.8. Once the final design with the initially assumed pipe material is obtained, the economic pipe material is selected using Section 8.3. Considering $h_0 = 0$, the maximum pressure on pipes h_i was calculated by applying Eq. (8.26) and is listed in Table 9.8. Using Fig. 8.9 for design pipe sizes based on assumed pipe material and maximum pressure h_i on pipe, the economic pipe material obtained for various pipes is listed in Table 9.8. The distribution system is reanalyzed for revised flows and redesigned for pipe sizes for economic pipe materials. The process is repeated until the two consecutive solutions are close within allowable limits. The final solution is listed in Table 9.8 and shown in Fig. 9.3.

The variation of system cost with LP iterations is shown in Fig. 9.4. The first three iterations derive the solution with initially assumed pipe material. The economic pipe material is then selected and again pipe network analysis and synthesis (LP formulation) carried out. The final solution with economic pipes is obtained after six iterations.

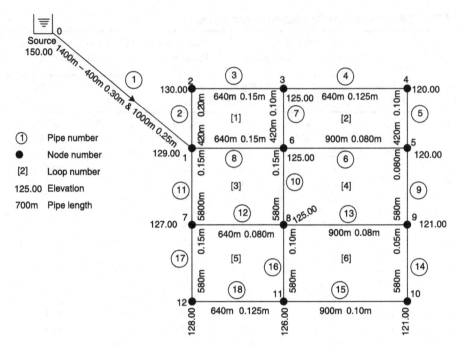

Figure 9.3. Looped water distribution network.

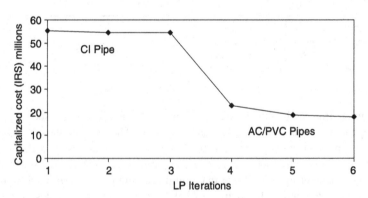

Figure 9.4. Number of LP iterations in system design.

9.2. PUMPING SYSTEM

The town water supply systems are generally single-input source, pumping, looped pipe networks. Pumping systems are provided where topography is generally flat or demand nodes are at higher elevation than the input node (source). In these circumstances, external energy is required to deliver water at required quantity and prescribed pressure.

The pumping systems include pipes, pumps, reservoirs, and treatment units based on the raw water quality. In case of bore water sources, generally disinfection may be sufficient. If the raw water is extracted from surface water, a water treatment plant will be required.

The system cost includes the cost of pipes, pumps, treatment, pumping (energy), and operation and maintenance. As described in Chapter 8, the optimization of such systems is therefore important due to the high recurring energy cost involved in it. This makes the pipe sizes in the system an important factor, as there is an economic trade-off between the pumping head and pipe diameters. Thus, there exists an optimum size of pipes and pump for every system, meaning that the pipe diameters are selected in such a way that the capitalized cost of the entire system is minimum. The cost of the treatment plant is not included in the cost function as it is constant for the desired degree of treatment based on raw water quality.

The design method is described using an example of a town water supply system shown in Fig. 9.5, which contains 18 pipes, 13 nodes, 6 loops, a single pumping source, and reservoir at node 0. The network data is listed in Table 9.9.

As the pipe discharges in looped water distribution networks are not unique, they require a looped network analysis technique. The Hardy Cross analysis method was applied similar as described in Section 9.1. Based on the population load on pipes, the nodal discharges were estimated using the method described in Chapter 3, Eq. (3.29). The rate of water supply 400 L/c/d and peak factor of 2.5 was considered for pipe flow estimation. The pipe discharges estimated for initially assumed pipes size 0.20 m of CI pipe material are listed in Table 9.10.

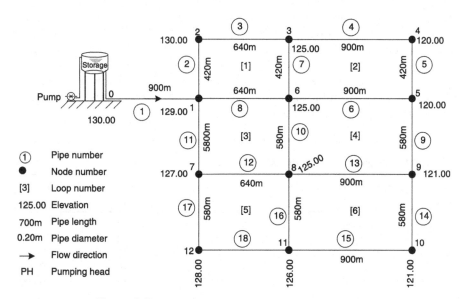

Figure 9.5. Looped, pumping water distribution system.

TABLE 9.9. Pumping, Looped Water Distribution Network Data

Pipe/ Node i/j	Node 1 $J_1(i)$	Node 2 $J_2(i)$	Loop 1 $K_1(i)$	Loop 2 $K_2(i)$	Length L_i (m)	Formloss coefficient k_{fi}	Population $P(i)$	Nodal elevation $z(j)$ (m)
0								130
1	0	1	0	0	900	0.5		129
2	1	2	1	0	420	0	200	130
3	2	3	1	0	640	0	300	125
4	3	4	2	0	900	0	450	120
5	4	5	2	0	580	0	250	120
6	5	6	2	4	900	0	450	125
7	3	6	1	2	420	0	200	127
8	1	6	1	3	640	0	300	125
9	5	9	4	0	580	0	250	121
10	6	8	3	4	580	0	250	121
11	1	7	3	0	580	0	250	126
12	7	8	3	5	640	0	300	128
13	8	9	4	6	900	0	450	
14	9	10	6	0	580	0	300	
15	10	11	6	0	900	0	450	
16	8	11	5	6	580	0	300	
17	7	12	5	0	580	0	300	
18	11	12	5	0	640	0	300	

9.2.1. Continuous Diameter Approach

The approach is similar to the gravity-sustained, looped network design described in Section 9.1.1. The entire looped water distribution system is converted into a number of distribution mains. Each distribution main is then designed separately using the methodology described in Chapter 7. The total number of such distribution mains is equal to the number of pipes in the distribution main as each pipe would generate a flow path forming a distribution main. Such conversion/decomposition is essential to calculate pumping head for the network.

TABLE 9.10. Looped Network Pipe Discharges

Pipe i	Discharge Q_i (m^3/s)	Pipe i	Discharge Q_i (m^3/s)	Pipe i	Discharge Q_i (m^3/s)
1	0.06134	7	0.00183	13	0.00388
2	0.01681	8	0.01967	14	0.00187
3	0.01391	9	0.00377	15	−0.00247
4	0.00658	10	0.00782	16	0.00415
5	0.00252	11	0.02052	17	0.00786
6	−0.00674	12	0.00774	18	−0.00439

TABLE 9.11. Pipe Flow Paths Treated as Water Distribution Main

Pipe i	Flow Paths Pipes Connecting to Input Point Node 0 and Generating Water Distribution Pumping Mains $I_t(i, \ell)$					$N_t(i)$	$J_t(i)$
	$\ell = 1$	$\ell = 2$	$\ell = 3$	$\ell = 4$	$\ell = 5$		
1	1					1	1
2	2	1				2	2
3	3	2	1			3	3
4	4	3	2	1		4	4
5	5	4	3	2	1	5	5
6	6	8	1			3	5
7	7	3	2	1		4	6
8	8	1				2	6
9	9	6	8	1		4	9
10	10	8	1			3	8
11	11	1				2	7
12	12	11	1			3	8
13	13	10	8	1		4	9
14	14	13	10	8	1	5	10
15	15	18	17	11	1	5	10
16	16	10	8	1		4	11
17	17	11	1			3	12
18	18	17	11	1		4	11

The flow paths for all the pipes of the looped water distribution network were generated using the network geometry data (Table 9.9) and pipe discharges (Table 9.10). Applying the flow path methodology described in Section 3.9, the pipe flow paths along with their originating nodes $J_t(i)$ are listed in Table 9.11.

The continuous pipe diameters can be obtained using Eq. (7.11b), which is modified and rewritten as:

$$D_i^* = \left(\frac{40 k_T f_i Q_T Q_i^2}{\pi^2 m k_m} \right)^{\frac{1}{m+5}}, \tag{9.5}$$

where Q_T is the total pumping discharge.

The optimal diameters were obtained applying Eq. (9.5) and pipe discharges from Table 9.10. Pipe and pumping cost parameters were similar to these adopted in Chapters 7 and 8. To apply Eq. (9.5), the pipe friction f was considered as 0.01 in all the pipes initially, which was improved iteratively until the two consecutive solutions were close. The calculated pipe diameters and adopted commercial sizes are listed in Table 9.12. The pumping head required for the system can be obtained using Eq. (7.12), which is rewritten as:

$$h_0^* = z_n + H - z_0 + \frac{8}{\pi^2 g} \left(\frac{\pi^2 m k_m}{40 \rho k_T Q_1} \right)^{\frac{5}{m+5}} \sum_{p=I_t(i,\ell)}^{n} L_p \left(f_p Q_p^2 \right)^{\frac{m}{m+5}}. \tag{9.6}$$

TABLE 9.12. Pumping, Looped Network Design: Continuous Diameter Approach

Pipe i	Length L_i (m)	Pipe Discharge Q_i (m³/s)	Calculated Pipe Diameter D_i (m)	Pipe Diameter Adopted D_i (m)	Elevation of Originating Node of Pipe Flow Path $z(J_t(i))$ (m)	Pumping Head with Calculated Pipe Sizes h_0 (m)	Pumping Head with Adopted Pipe Sizes h_0 (m)
0					130		
1	900	0.06134	0.269	0.300	129	13.0	11.4
2	420	0.01681	0.178	0.200	130	15.3	13.1
3	640	0.01391	0.167	0.200	125	12.2	8.9
4	900	0.00658	0.132	0.150	120	7.2	3.9
5	580	0.00252	0.097	0.100	120	7.2	3.9
6	900	0.00674	0.133	0.150	120	8.2	4.8
7	420	0.00183	0.088	0.100	125	12.9	9.1
8	640	0.01967	0.187	0.200	125	9.7	7.6
9	580	0.00377	0.110	0.125	121	9.3	5.9
10	580	0.00782	0.139	0.150	125	12.5	9.6
11	580	0.02052	0.189	0.200	127	12.9	10.8
12	640	0.00774	0.139	0.150	125	12.5	9.7
13	900	0.00388	0.111	0.125	121	10.3	6.4
14	580	0.00187	0.089	0.100	121	11.3	6.8
15	900	0.00247	0.097	0.100	121	6.9	4.8
16	580	0.00415	0.114	0.125	126	13.5	10.6
17	580	0.00786	0.139	0.150	128	13.9	11.8
18	640	0.00439	0.116	0.125	126	11.9	9.8

Equation (9.6) is applied for all the pumping water distribution mains (flow paths), which are equal to the total number of pipes in the distribution system. Thus, the variable n in Eq. (9.6) is equal to $N_t(i)$ and pipes p in the distribution main $I_t(i, \ell)$, $\ell = 1, N_t(i)$. The elevation z_n is equal to the elevation of originating node $J_t(i)$ of flow path for pipe i generating pumping distribution main. Q_1 is similar to that defined for Eq. (8.21). Thus, applying Eq. (9.6), the pumping heads for all the pumping mains listed in Table 9.11 were calculated. The minimum terminal pressure prescribed for Fig. 9.4 is 10 m. It can be seen from Table 9.12 that the pumping head for the network is 15.30 m if continuous pipe sizes are adopted. The flow path for pipe 2 provides the critical pumping head. The pumping head reduced to 13.1 m for adopted commercial sizes.

The adopted commercial sizes in Table 9.12 are based on the estimated pipe discharges, which are based on the initially assumed pipe diameters. Using the adopted commercial pipe sizes, the pipe network should be reanalyzed for new pipe discharges. This will again generate new pipe flow paths. The process of network analysis and pipe sizing should be repeated until the two solutions are close. The pumping head is estimated for the final pipe discharges and pipe sizes. Once the network design with initially assumed pipe material is obtained, the economic pipe material for each pipe link can be selected using the methodology described in Section 8.3. The process of network analysis and pipe sizing should be repeated for economic pipe material and pumping head estimated based on finally adopted commercial sizes. The pipe sizes and pumping head listed in Table 9.12 are shown in Fig. 9.6 for CI pipe material.

9.2.2. Discrete Diameter Approach

Similar to a branch system (Section 8.2.2.2), the cost function for the design of a looped system is formulated as

$$\min F = \sum_{i=1}^{i_L} (c_{i1}x_{i1} + c_{i2}x_{i2}) + \rho g \, k_T Q_T h_0, \tag{9.7}$$

subject to

$$x_{i1} + x_{i2} = L_i; \quad i = 1, 2, 3 \ldots i_L, \tag{9.8}$$

$$\sum_{p=I_t(i,\ell)} \left[\frac{8 f_{p1} Q_p^2}{\pi^2 g D_{p1}^5} x_{p1} + \frac{8 f_{p2} Q_p^2}{\pi^2 g D_{p2}^5} x_{p2} \right] \le z_0 + h_0 - z_{J_t(i)} - H$$

$$- \sum_{p=I_t(i,\ell)} \frac{8 k_{fp} Q_p^2}{\pi^2 g D_{p2}^4};$$

$$\ell = 1, 2, 3 \; N_t(i) \qquad \text{For } i = 1, 2, 3 \ldots i_L \tag{9.9}$$

As described in Section 9.1.2, the head-loss constraints Inequations (9.9) are developed for all the originating nodes of pipe flow paths. Thus, it will bring all the pipes of the network in the optimization process.

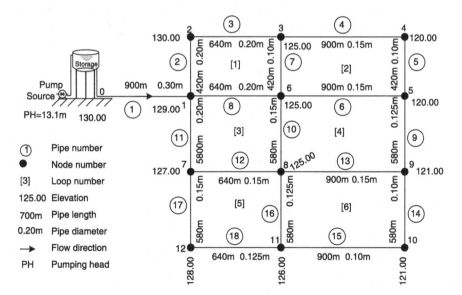

Figure 9.6. Pumping, looped water supply system: continuous diameter approach.

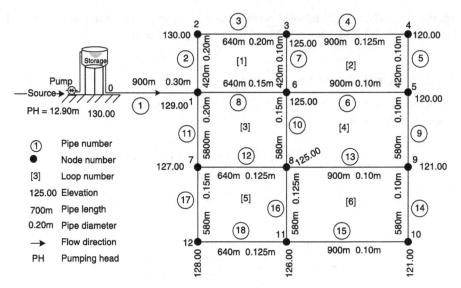

Figure 9.7. Pumping, looped water network design.

TABLE 9.13. Pumping, Looped Network Design

Pipe i	Length L_i (m)	Pipe Diameter D_i (m)	Pipe Material and Class	Pipe i	Length L_i (m)	Pipe Diameter D_i (m)	Pipe Material and Class
1	900	0.300	AC Class 5	10	580	0.150	PVC 40 m WP
2	420	0.200	PVC 40 m WP	11	580	0.150	PVC 40 m WP
3	640	0.200	PVC 40 m WP	12	640	0.150	PVC 40 m WP
4	900	0.150	PVC 40 m WP	13	900	0.125	PVC 40 m WP
5	580	0.100	PVC 40 m WP	14	580	0.100	PVC 40 m WP
6	900	0.150	PVC 40 m WP	15	900	0.100	PVC 40 m WP
7	420	0.100	PVC 40 m WP	16	580	0.125	PVC 40 m WP
8	640	0.200	PVC 40 m WP	17	580	0.150	PVC 40 m WP
9	580	0.125	PVC 40 m WP	18	640	0.125	PVC 40 m WP

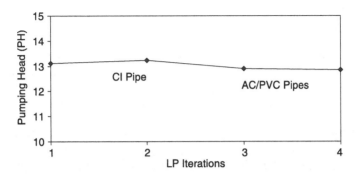

Figure 9.8. Variation of pumping head with LP iterations.

Equations (9.7) and (9.8) and Inequation (9.9) constituting a LP problem involve 2_{i_L} decision variables, i_L equality constraints, and i_L inequality constraints. As described in Section 9.1.2, the LP iterations can be reduced if the starting diameters are taken close to the final solution. Using Eq. (9.5), the continuous optimal pipe diameters D_i^* can be calculated. Selecting the two consecutive commercially available sizes such that $D_{i1} \leq D_i^* \leq D_{i2}$, significant computer time can be saved. Once the design for an initially assumed pipe material (CI) is obtained, economic pipe material is then selected applying the method described in Section 8.3. The network is reanalyzed and designed for new pipe material. The looped water distribution system shown in Fig. 9.6 was redesigned using the discrete diameter approach. The solution thus obtained is shown in Fig. 9.7 and listed in Table 9.13. The optimal pumping head was 12.90 m for 10-m terminal pressure head. The variation of pumping head with LP iterations is plotted in Fig. 9.8. From a perusal of Fig. 9.8, it can be seen that four LP iterations were sufficient using starting pipe sizes close to continuous diameter solution.

It can be concluded that the discrete pipe diameter approach provides an economic solution as it formulates the problem for the system as a whole, whereas piecemeal design is carried out in the continuous diameter approach and also conversion of continuous sizes to commercial sizes misses the optimality of the solution.

EXERCISES

9.1. Describe the advantages and disadvantages of looped water distribution systems. Provide examples for your description.

9.2. Design a gravity water distribution network by modifying the data given in Table 9.2. The length and population can be doubled for the new data set. Use continuous and discrete diameter approaches.

9.3. Create a single-loop, four-piped system with pumping input point at one of its nodes. Assume arbitrary data for this network, and design manually using discrete diameter approach.

9.4. Describe the drawbacks if the constraint inequations in LP formulation are developed only node-wise for the design of a looped pipe network.

9.5. Design a pumping, looped water distribution system using the data given in Table 9.9 considering the flat topography of the entire service area. Apply continuous and discrete diameter approaches.

REFERENCE

Swamee, P.K., and Sharma, A.K. (2000). Gravity flow water distribution system design. *Journal of Water Supply Research and Technology-AQUA* 49(4), 169–179.

10

MULTI-INPUT SOURCE, BRANCHED SYSTEMS

Sometimes, town water supply systems are multi-input, branched distribution systems because of insufficient water from a single source, reliability considerations, and development pattern. The multiple supply sources connected to a network also reduce the pipe sizes of the distribution system because of distributed flows. In case of multi-input source, branched systems, the flow directions in some of the pipes interconnecting the sources are not unique and can change due to the spatial or temporal variation in water demand.

Conceptual gravity-sustained and pumping multi-input source, branched water distribution systems are shown in Fig. 10.1. The location of input points/pumping stations and reservoirs can vary based on the raw water resources, availability of land for water works, topography of the area, and layout pattern of the town.

Because of the complexity in flow pattern in multi-input water systems, the analysis of pipe networks is essential to understand or evaluate a physical system, thus making it

Design of Water Supply Pipe Networks. By Prabhata K. Swamee and Ashok K. Sharma
Copyright © 2008 John Wiley & Sons, Inc.

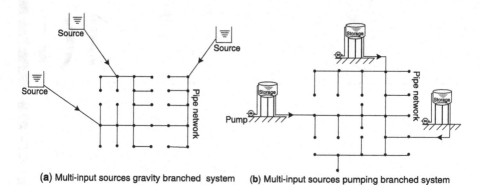

(a) Multi-input sources gravity branched system   (b) Multi-input sources pumping branched system

Figure 10.1. Multi-input sources branched water distribution systems.

an integral part of the synthesis process of a network. In case of a single-input system, the input source discharge is equal to the sum of withdrawals in the network. On the other hand, a multi-input network system has to be analyzed to obtain input point discharges based on their input point heads, nodal elevations, network configuration, and pipe sizes. Although some of the existing water distribution analysis models (i.e., Rossman, 2000) are capable of analyzing multi-input source systems, a water distribution network analysis model is developed specially to link with a cost-optimization model for network synthesis purposes. This analysis model has been described in Chapter 3. As stated earlier, in case of multi-input source, branched water systems, the discharges in some of the source-interconnecting pipes are not unique, which are dependent on the pipe sizes, locations of sources, their elevations, and availability of water from these sources. Thus, the design of a multi-input source, branched network would require sequential application of analysis and synthesis methods until a termination criterion is achieved.

As described in previous chapters, the water distribution design methods based on cost optimization have two approaches: (a) continuous diameter approach and (b) discrete diameter approach or commercial diameter approach. The design of multi-input source, branched network, gravity and pumping systems applying both the approaches is described in this chapter.

10.1. GRAVITY-SUSTAINED, BRANCHED SYSTEMS

The gravity-sustained, branched water distribution systems are suitable in areas where the source (input) points are at a higher elevation than the demand points. The area covered by the distribution network has low density and scattered development. Such a typical water distribution system is shown in Fig. 10.2.

The pipe network data are listed in Table 10.1. The data include the pipe number, both its nodes, form-loss coefficient due to fittings in pipe, population load on pipe, and nodal elevations. The two input points (sources) of the network are located at node 1 and node 17. Thus, the first source point $S(1)$ is located at node number 1 and

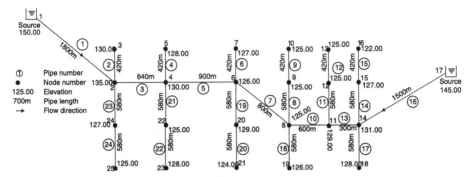

Figure 10.2. Multi-input sources gravity branched water distribution system.

TABLE 10.1. Multi-input Source, Gravity-Sustained Water Distribution Network Data

Pipe/Node i/j	Node 1 $J_1(i)$	Node 2 $J_2(i)$	Length L_i (m)	Form-Loss Coefficient $k_f(i)$	Population $P(i)$	Nodal Elevation $z(i)$ (m)
1	1	2	1800	0.5	0	150
2	2	3	420	0	200	135
3	2	4	640	0	300	130
4	4	5	420	0	200	130
5	4	6	900	0	500	128
6	6	7	420	0	200	126
7	6	8	800	0	300	127
8	8	9	580	0	250	125
9	9	10	420	0	200	125
10	8	11	600	0	300	125
11	11	12	580	0	300	129
12	12	13	420	0	200	125
13	11	14	300	0	150	125
14	14	15	580	0	300	131
15	15	16	420	0	200	127
16	14	17	1500	0.5	0	122
17	14	18	580	0	300	145
18	9	19	580	0	300	128
19	6	20	580	0	300	126
20	20	21	580	0	300	129
21	4	22	640	0	300	124
22	22	23	580	0	200	125
23	2	24	580	0	200	128
24	24	25	580	0	200	127
25	0	0	0	0	0	125

TABLE 10.2. Estimated Nodal Demand Discharges

Node j	Nodal Demand q_j (m³/s)	Node j	Nodal Demand q_j (m³/s)	Node j	Nodal Demand q_j (m³/s)	Node j	Nodal Demand q_j (m³/s)
1	0.0	8	0.0049	15	0.0029	22	0.0029
2	0.0041	9	0.0043	16	0.0012	23	0.0012
3	0.0012	10	0.0012	17	0.0	24	0.0023
4	0.0075	11	0.0043	18	0.0017	25	0.0012
5	0.0012	12	0.0029	19	0.0017		
6	0.0075	13	0.0012	20	0.0035		
7	0.0012	14	0.0043	21	0.0017		

the second source $S(2)$ at node 17. The pipe network has a total of 24 pipes, 25 nodes, and 2 sources. The nodal elevations are also provided in the data table.

The pipe network shown in Fig. 10.2 has been analyzed for pipe discharges. Assuming peak discharge factor = 2.5, rate of water supply 400 liters/capita/day (L/c/d), the nodal demand discharges obtained using the method described in Chapter 3 (Eq. 3.29) are estimated, which are listed in Table 10.2.

The pipe discharges in a multi-input source, water distribution network are not unique and thus require network analysis technique. The pipe diameters are to be assumed initially to analyze the network, thus considering all pipe sizes = 0.2 m and pipe material as CI, the network was analyzed using the method described in Sections 3.7 and 3.8. The estimated pipe discharges are listed in Table 10.3.

10.1.1. Continuous Diameter Approach

In this approach, the entire multi-input, branched water distribution system is converted into a number of gravity distribution mains. Each distribution main is then designed separately using the method described in Chapters 7 and 8. The total number of such distribution mains is equal to the number of pipes in the network system as each pipe would generate a flow path forming a distribution main.

TABLE 10.3. Multi-input Source, Gravity-Sustained Distribution Network Pipe Discharges

Pipe i	Discharge Q_i (m³/s)	Pipe i	Discharge Q_i (m³/s)	Pipe i	Discharge Q_i (m³/s)	Pipe i	Discharge Q_i (m³/s)
1	0.0341	7	−0.0012	13	−0.0217	19	0.0052
2	0.0012	8	0.0072	14	0.0041	20	0.0017
3	0.0254	9	0.0012	15	0.0012	21	0.0041
4	0.0012	10	−0.0134	16	−0.0319	22	0.0012
5	0.0127	11	0.0041	17	0.0017	23	0.0035
6	0.0012	12	0.0012	18	0.0017	24	0.0012

The flow paths for all the pipes of the looped water distribution network are generated using the pipe discharges (Table 10.3) and the network geometry data (Table 10.1). Applying the flow path method described in Section 3.9, the pipe flow paths along with their originating nodes $J_t(i)$ including the input sources $J_s(i)$ are identified and listed in Table 10.4. The pipe flow paths terminate at different input sources in a multi-input source network.

As listed in Table 10.4, the pipe flow paths terminate at one of the input points (sources), which is responsible to supply flow in that pipe flow path. Treating the flow path as a gravity water distribution main and applying Eq. (7.6b), which is modified and rewritten below, the optimal pipe diameters of gravity distribution mains can be calculated as

$$ D_i^* = \left(f_i Q_i^2\right)^{\frac{1}{m+5}} \left\{ \frac{8}{\pi^2 g \left[z_{Js(i)} - z_{Jt(i)} - H\right]} \sum_{p=I_t(i,l)}^{n} L_p \left(f_p Q_p^2\right)^{\frac{m}{m+5}} \right\}^{0.2}, \qquad (10.1) $$

TABLE 10.4. Pipe Flow Paths as Gravity-Sustained Water Distribution Mains

Pipe i	$\ell = 1$	$\ell = 2$	$\ell = 3$	$\ell = 4$	$\ell = 5$	$N_t(i)$	$J_t(i)$	$J_s(i)$
1	1					1	2	1
2	2	1				2	3	1
3	3	1				2	4	1
4	4	3	1			3	5	1
5	5	3	1			3	6	1
6	6	5	3	1		4	7	1
7	7	10	13	16		4	6	17
8	8	10	13	16		4	9	17
9	9	8	10	13	16	5	10	17
10	10	13	16			3	8	17
11	11	13	16			3	12	17
12	12	11	13	16		4	13	17
13	13	16				2	11	17
14	14	16				2	15	17
15	15	14	16			3	16	17
16	16					1	14	17
17	17	16				2	18	17
18	18	8	10	13	16	5	19	17
19	19	5	3	1		4	20	1
20	20	19	5	3	1	5	21	1
21	21	3	1			3	22	1
22	22	21	3	1		4	23	1
23	23	1				2	24	1
24	24	23	1			3	25	1

where $z_{Js(i)}$ = the elevation of input point source for pipe i, $z_{Jt(i)}$ = the elevation of originating node of flow path for pipe i, $n = N_t(i)$ number of pipe links in the flow path, and $p = I_t(i, \ell)$, $\ell = 1$, $N_t(i)$ are the pipe in flow path of pipe i. Applying Eq. (10.1), the design of flow paths of pipes as distribution mains is carried out applying the corresponding pipe flows, nodal elevations, and pipe lengths as listed in Table 10.1 and Table 10.3. The minimum terminal pressure of 10 m is maintained at nodes. The friction factor is assumed as 0.02 initially for all the pipes in the distribution main, which is improved iteratively until the two consecutive f values are close. The pipe sizes are calculated for various flow paths as gravity-flow water distribution mains using a similar procedure as described in Sections 8.1.2 and 9.1.1. The estimated pipe sizes and nearest commercial sizes adopted are listed in Table 10.5.

Using the set of adopted pipe sizes, shown in Table 10.5, the pipe network should be analyzed again for another set of pipe discharges (Table 10.3). The pipe flow paths (Table 10.4) are regenerated using the revised pipe flows. The pipe sizes are recalculated for a new set of gravity distribution mains. The process is repeated until the two solutions are close. Once the final design is achieved, the economic pipe material can be selected using the method described in Section 8.3.

10.1.2. Discrete Diameter Approach

As described in Chapters 8 and 9, the conversion of continuous pipe diameters into commercial (discrete) pipe diameters reduces the optimality of the solution. A method considering commercial pipe sizes directly in the design of a multi-input source water distribution system using LP optimization technique is described in this section.

TABLE 10.5. Multi-input Source, Gravity-Sustained System: Estimated and Adopted Pipe Sizes

Pipe	Estimated Continuous Pipe Size (m)	Adopted Pipe Size (m)	Pipe	Estimated Continuous Pipe Size (m)	Adopted Pipe Size (m)
1	0.225	0.250	13	0.190	0.200
2	0.064	0.065	14	0.100	0.100
3	0.200	0.200	15	0.061	0.065
4	0.065	0.065	16	0.219	0.250
5	0.159	0.200	17	0.076	0.080
6	0.067	0.065	18	0.080	0.080
7	0.069	0.065	19	0.117	0.125
8	0.131	0.150	20	0.076	0.080
9	0.068	0.065	21	0.103	0.100
10	0.161	0.200	22	0.067	0.065
11	0.100	0.100	23	0.090	0.100
12	0.066	0.065	24	0.061	0.065

For the current case, the LP problem is formulated as

$$\min F = \sum_{i=1}^{i_L} (c_{i1}x_{i1} + c_{i2}x_{i2}),\tag{10.2}$$

subject to

$$x_{i1} + x_{i2} = L_i; \quad i = 1, 2, 3 \dots i_L,\tag{10.3}$$

$$\sum_{p=I_t(i,\ell)} \left(\frac{8f_{p1}Q_p^2}{\pi^2 g D_{p1}^5}x_{p1} + \frac{8f_{p2}Q_p^2}{\pi^2 g D_{p2}^5}x_{p2} \right) \leq z_{Js(i)} + h_{Js(i)} - z_{J_t(i)} - H - \sum_{p=I_t(i,\ell)} \frac{8k_{fp}Q_p^2}{\pi^2 g D_{p2}^4}\tag{10.4}$$
$$\ell = 1, 2, 3 N_t(i) \qquad\qquad \text{for } i = 1, 2, 3, \dots, i_L$$

TABLE 10.6. Multi-input, Gravity-Sustained, Branched Pipe Distribution Network Design

Pipe i	Pipe Length L_i (m)	Initial Design with Assumed Pipe Material		Final Design with Optimal Pipe Material	
		D_i (m)	Pipe Material	D_i (m)	Pipe Material
1	1800	0.250	CI Class LA	0.250	AC Class 5*
2	420	0.050	CI Class LA	0.050	PVC 40 m WP†
3	640	0.200	CI Class LA	0.200	PVC 40 m WP
4	420	0.050	CI Class LA	0.050	PVC 40 m WP
5	900	0.200/460 0.150/440	CI Class LA	0.150	PVC 40 m WP
6	420	0.065	CI Class LA	0.065	PVC 40 m WP
7	800	0.065	CI Class LA	0.065	PVC 40 m WP
8	580	0.125	CI Class LA	0.125	PVC 40 m WP
9	420	0.065	CI Class LA	0.065	PVC 40 m WP
10	600	0.150	CI Class LA	0.150	PVC 40 m WP
11	580	0.100	CI Class LA	0.100	PVC 40 m WP
12	420	0.065	CI Class LA	0.065	PVC 40 m WP
13	300	0.200	CI Class LA	0.200	PVC 40 m WP
14	580	0.100	CI Class LA	0.100	PVC 40 m WP
15	420	0.050	CI Class LA	0.050	PVC 40 m WP
16	1500	0.250/800 0.200/700	CI Class LA	0.250/600 0.200/900 m	AC Class 5
17	580	0.080	CI Class LA	0.065	PVC 40 m WP
18	580	0.080	CI Class LA	0.080	PVC 40 m WP
19	580	0.125	CI Class LA	0.125	PVC 40 m WP
20	580	0.065	CI Class LA	0.065	PVC 40 m WP
21	640	0.100	CI Class LA	0.100	PVC 40 m WP
22	580	0.065	CI Class LA	0.065	PVC 40 m WP
23	580	0.080	CI Class LA	0.080	PVC 40 m WP
24	580	0.050	CI Class LA	0.050	PVC 40 m WP

*Asbestos cement pipe 25-m working pressure.
†Poly(vinyl chloride) 40-m working pressure.

where $z_{Js(i)}$ = elevation of input source node for flow path of pipe i, $z_{J_t(i)}$ = elevation of originating node of pipe i, and $h_{Js(i)}$ = water column (head) at input source node that can be neglected. Using Inequation (10.4), the head-loss constraint inequations for all the originating nodes of pipe flow paths are developed. This will bring all the pipes of the network into LP formulation.

Using Eq. (10.1), the continuous optimal pipe diameters D_i^* can be calculated. For LP iterations, the two consecutive commercially available sizes such that $D_{i1} \leq D_i^* \leq D_{i2}$, are selected. The selection between D_{i1} and D_{i2} is resolved by LP. Significant computational time is saved in this manner. The multi-input, branched water distribution system shown in Fig. 10.2 was redesigned using the discrete diameter approach. The solution thus obtained using initially CI pipe material is given in Table 10.6. Once the design with the initially assumed pipe material is obtained, the economic pipe material is selected using Section 8.3. The distribution system is reanalyzed for revised flows and redesigned for pipe sizes for economic pipe materials. The process is repeated until the two consecutive solutions are close within allowable limits. The final solution is also listed in Table 10.6 and shown in Fig. 10.3. The minimum diameter of 0.050 m was specified for the design.

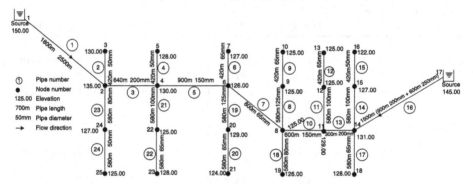

Figure 10.3. Multi-input gravity branched water distribution network design.

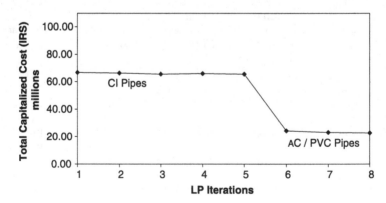

Figure 10.4. Number of LP iterations in multi-input branched gravity system design.

The total number of LP iterations required for the final design is shown in Fig. 10.4. The first five iterations derive the solution with initially assumed pipe material. The economic pipe material is then selected, and again pipe network analysis and synthesis (LP formulation) are carried out. The final solution with economic pipes is obtained after eight iterations.

10.2. PUMPING SYSTEM

Rural town water supply systems with bore water as source may have multi-input source, branched networks. These systems can be provided because of a variety of reasons such as scattered residential development, insufficient water from single source, and reliability considerations (Swamee and Sharma, 1988). As stated earlier, the pumping systems are essential where external energy is required to deliver water at required pressure and quantity. Such multi-input pumping systems include pipes, pumps, bores, reservoirs, and water treatment units based on the raw water quality.

The design method is described using an example of a conceptual town water supply system shown in Fig. 10.5, which contains 28 pipes, 29 nodes, 3 input sources as pumping stations, and reservoirs at nodes 1, 10, and 22. The network data is listed in Table 10.7. The minimum specified pipe size is 65 mm and the terminal head is 10 m.

The network data include pipe numbers, pipe nodes, pipe length, form-loss coefficient of valve fittings in pipes, population load on pipes, and nodal elevations.

Based on the population load on pipes, the nodal discharges were estimated using the method described in Chapter 3 (Eq. 3.29). The rate of water supply 400 L/c/d and a peak factor of 2.5 are assumed for the design. The nodal discharges thus estimated are listed in Table 10.8.

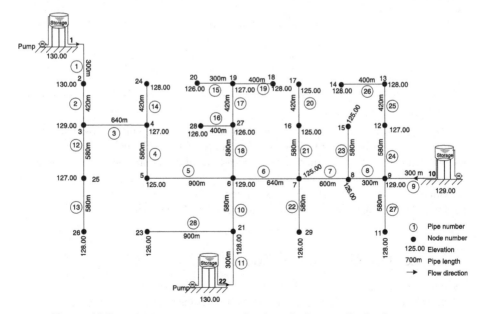

Figure 10.5. Multi-input sources pumping branched water distribution system.

TABLE 10.7. Multi-input, Branched, Pumping Water Distribution Network Data

Pipe/ Node i/j	Node 1 $J_1(i)$	Node 2 $J_2(i)$	Length L_i (m)	Form-Loss Coefficient $k_f(i)$	Population $P(i)$	Nodal Elevation $z(j)$ (m)
1	1	2	300	0.5	0	130
2	2	3	420	0	300	130
3	3	4	640	0	500	129
4	4	5	580	0	400	127
5	5	6	900	0	600	125
6	6	7	640	0	500	129
7	7	8	600	0	400	125
8	8	9	300	0	200	126
9	9	10	300	0.5	200	129
10	6	21	580	0	400	129
11	21	22	300	0.5	0	128
12	3	25	580	0	300	127
13	25	26	580	0	400	128
14	4	24	420	0	300	128
15	19	20	300	0	200	125
16	27	28	400	0	100	125
17	19	27	420	0	200	125
18	6	27	580	0	400	128
19	18	19	400	0	200	127
20	16	17	420	0	300	126
21	7	16	580	0	400	128
22	7	29	580	0	400	130
23	8	15	580	0	300	126
24	9	12	580	0	300	128
25	12	13	420	0	200	127
26	13	14	400	0	100	128
27	9	11	580	0	350	126
28	21	23	900	0	500	126
29						126

The pipe discharges in multi-input, branched water distribution networks are not unique and thus require a network analysis technique. The discharges in pipes interconnecting the sources are based on pipe sizes, location and elevation of sources, and overall topography of the area. The pipe network analysis was conducted using the method described in Chapter 3. The pipe discharges estimated for initially assumed pipes sizes equal to 0.20 m of CI pipe material are listed in Table 10.9.

10.2.1. Continuous Diameter Approach

The entire branched water distribution network is converted into a number of pumping distribution mains. Each pumping distribution main is then designed separately using the

TABLE 10.8. Nodal Water Demands

Node j	Nodal Demand $Q(j)$ (m³/s)	Node j	Nodal Demand $Q(j)$ (m³/s)	Node j	Nodal Demand $Q(j)$ (m³/s)
1	0.00000	11	0.00203	21	0.00405
2	0.00174	12	0.00289	22	0.00000
3	0.00637	13	0.00174	23	0.00289
4	0.00694	14	0.00058	24	0.00174
5	0.00579	15	0.00174	25	0.00405
6	0.00984	16	0.00405	26	0.00231
7	0.00984	17	0.00174	27	0.00405
8	0.00521	18	0.00116	28	0.00058
9	0.00492	19	0.00347	29	0.00231
10	0.00000	20	0.00116		

method described in Chapter 7. In case of a multi-input system, these flow paths will terminate at different sources, which will be supplying maximum flow into that flow path.

The flow paths for all the pipes of the branched network were generated using the network geometry data (Table 10.7) and pipe discharges (Table 10.9). Applying the flow path method described in Section 3.9, the pipe flow paths along with their originating nodes $J_t(i)$ and input source nodes $J_s(i)$ are listed in Table 10.10.

The continuous pipe diameters can be obtained using Eq. (7.11b), which is modified and rewritten for multi-input distribution system as

$$D_i^* = \left\{ \frac{40\, k_T \rho f_i Q_{T[Js(i)]} Q_i^2}{\pi^2 m k_m} \right\}^{\frac{1}{m+5}}, \tag{10.5}$$

where $Q_{T[Js(i)]}$ = the total pumping discharge at input source $J_s(i)$. The optimal diameters are obtained by applying Eq. (10.5) and using pipe discharges from Table 10.9. Pipe and pumping cost parameters were similar to those adopted in Chapters 7 and 8. The pipe friction factor f_i was considered as 0.01 for the entire set of pipe links initially,

TABLE 10.9. Network Pipe Discharges

Pipe i	Discharge Q_i (m³/s)	Pipe i	Discharge Q_i (m³/s)	Pipe i	Discharge Q_i (m³/s)	Pipe i	Discharge Q_i (m³/s)
1	0.02849	8	−0.02031	15	0.00116	21	0.00231
2	0.02675	9	−0.03246	16	0.00058	22	0.00174
3	0.01402	10	−0.02528	17	−0.00579	23	0.00521
4	0.00534	11	−0.03222	18	0.01042	24	0.00231
5	−0.00045	12	0.00637	19	−0.00116	25	0.00058
6	0.00458	13	0.00231	20	0.00174	26	0.00203
7	−0.01336	14	0.00174	21	0.00579	27	0.00289

TABLE 10.10. Pipe Flow Paths Treated as Water Distribution Mains

Pipe i	Flow Path Pipes Connecting to Input Point Nodes and Generating Water Distribution Pumping Mains $I_t(i, \ell)$					$N_t(i)$	$J_t(i)$	$J_s(i)$
	$\ell = 1$	$\ell = 2$	$\ell = 3$	$\ell = 4$	$\ell = 5$			
1	1					1	2	1
2	2	1				2	3	1
3	3	2	1			3	4	1
4	4	3	2	1		4	5	1
5	5	10	11			3	5	22
6	6	10	11			3	7	22
7	7	8	9			3	7	10
8	8	9				2	8	10
9	9					1	9	10
10	10	11				2	6	22
11	11					1	21	22
12	12	2	1			3	25	1
13	13	12	2	1		4	26	1
14	14	3	2	1		4	24	1
15	15	17	18	10	11	5	20	22
16	16	18	10	11		4	28	22
17	17	18	10	11		4	19	22
18	18	10	11			3	27	22
19	19	17	18	10	11	5	18	22
20	20	21	7	8	9	5	17	10
21	21	7	8	9		4	16	10
22	22	7	8	9		4	29	10
23	23	8	9			3	15	10
24	24	9				2	12	10
25	25	24	9			3	13	10
26	26	25	24	9		4	14	10
27	27	9				2	11	10
28	28	11				2	23	22

which was improved iteratively until the two consecutive solutions are close. The calculated pipe diameters and adopted commercial sizes are listed in Table 10.11.

The pumping head required for the system can be obtained using Eq. (7.12), which is modified and rewritten as

$$h^*_{Js(i)} = z_{Jt(i)} + H - z_{Js(i)} + \frac{8}{\pi^2 g} \left\{ \frac{\pi^2 m k_m}{40 \rho k_T Q_{T[Js(i)]}} \right\}^{-\frac{5}{m+5}} \sum_{p=I_t(i,\ell)}^{N_t(i)} L_p \left(f_p Q_p^2 \right)^{\frac{m}{m+5}} \quad (10.6)$$

$$i = 1, 2, 3, \ldots, i_L$$

where $h^*_{Js(i)}$ = the optimal pumping head for pumping distribution main generated from flow path of pipe link i. Equation (10.6) is applied for all the pumping water distribution

TABLE 10.11. Multi-input, Pumping, Branched Network Design

Pipe i	Calculated Pipe Diameter D_i (m)	Pipe Diameter Adopted D_i (m)	Pipe i	Calculated Pipe Diameter D_i (m)	Pipe Diameter Adopted D_i (m)
1	0.192	0.200	15	0.071	0.080
2	0.188	0.200	16	0.057	0.065
3	0.154	0.150	17	0.119	0.125
4	0.113	0.125	18	0.144	0.150
5	0.053	0.065	19	0.071	0.080
6	0.111	0.125	20	0.081	0.080
7	0.154	0.150	21	0.119	0.125
8	0.176	0.200	22	0.089	0.100
9	0.203	0.200	23	0.081	0.080
10	0.190	0.200	24	0.114	0.125
11	0.205	0.200	25	0.088	0.100
12	0.120	0.125	26	0.057	0.065
13	0.087	0.100	27	0.085	0.080
14	0.079	0.080	28	0.096	0.100

mains (flow paths), which are equal to the total number of pipe links in the distribution system. The total number of variables p in Eq. (10.6) is equal to $N_t(i)$ and pipe links in the distribution main $I_t(i, \ell)$, $\ell = 1, 2, 3, \ldots, N_t(i)$. The elevation $z_{J_t(i)}$ is equal to the elevation of the originating node of flow path for pipe i generating pumping distribution main, elevation $z_{J_s(i)}$ is the elevation of corresponding input source node, and $h_{0(Js(i))}^*$ is the optimal pumping head for pumping distribution main generated from flow path of pipe i.

Thus, applying Eq. (10.6), the pumping heads for all the pumping mains can be calculated using the procedure described in Sections 8.2.2 and 9.2. The pumping head at a source will be the maximum of all the pumping heads estimated for flow paths terminating at that source. The continuous pipe sizes and corresponding adopted commercial sizes listed in Table 10.11 are based on the pipe discharges calculated using initially assumed pipe diameters. The final solution can be obtained applying the procedure described in Sections 8.2.2 and 9.2.1.

10.2.2. Discrete Diameter Approach

The continuous pipe sizing approach reduces the optimality of the solution. The conversion of continuous pipe sizes to discrete pipe sizes can be eliminated if commercial pipe sizes are adopted directly in the optimal design process. A method for the design of a multi-input, branched water distribution network pumping system adopting commercial pipe sizes directly in the synthesis process is presented in this section. In this method, the configuration of the multi-sourced network remains intact.

The LP problem in this case is written as

$$\min F = \sum_{i=1}^{i_L} (c_{i1}x_{i1} + c_{i2}x_{i2}) + \rho g \, k_T \sum_{n=1}^{n_L} Q_{Tn}h_{0n}, \tag{10.7}$$

subject to,

$$x_{i1} + x_{i2} = L_i; \quad i = 1, 2, 3 \ldots i_L, \tag{10.8}$$

$$\sum_{p=I_t(i,\ell)} \left(\frac{8 f_{p1}Q_p^2}{\pi^2 g D_{p1}^5} x_{p1} + \frac{8 f_{p2}Q_p^2}{\pi^2 g D_{p2}^5} x_{p2} \right) \le z_{Js(i)} + h_{Js(i)} - z_{J_t(i)} - H - \sum_{p=I_t(i,\ell)} \frac{8 \, k_{fp} Q_p^2}{\pi^2 g D_{p2}^4}$$

$$\ell = 1, 2, 3 N_t(i) \quad \text{for } i = 1, 2, 3 \ldots i_L \tag{10.9}$$

where Q_{Tn} = the nth input point pumping discharge, and h_{on} = the corresponding pumping head. The constraint Ineqs. (10.9) are developed for all the originating nodes of pipe flow paths to bring all the pipes into LP problem formulation. The starting solution can be obtained using Eq. (10.5). The LP problem can be solved using the method described in Section 9.2.2 giving pipe diameters and input points pumping heads.

The water distribution system shown in Fig. 10.5 was redesigned using the discrete diameter approach. The solution thus obtained is shown in Fig. 10.6. The minimum pipe size as 65 mm and terminal pressure 10 m were considered for this design.

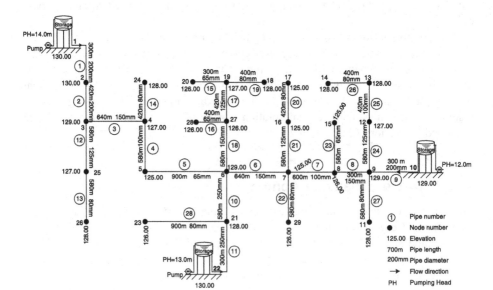

Figure 10.6. Pumping branched water network design.

TABLE 10.12. Multi-input, Branched, Pumping Network Design

Pipe I	Length L_i (m)	Pipe Diameter D_i (m)	Pipe Material and Class	Pipe i	Length L_i (m)	Pipe Diameter D_i (m)	Pipe Material and Class
1	300	0.200	PVC 40 m WP	15	300	0.065	PVC 40 m WP
2	420	0.200	PVC 40 m WP	16	400	0.065	PVC 40 m WP
3	640	0.150	PVC 40 m WP	17	420	0.125	PVC 40 m WP
4	580	0.100	PVC 40 m WP	18	580	0.150	PVC 40 m WP
5	900	0.065	PVC 40 m WP	19	400	0.080	PVC 40 m WP
6	640	0.150	PVC 40 m WP	20	420	0.080	PVC 40 m WP
7	600	0.100	PVC 40 m WP	21	580	0.125	PVC 40 m WP
8	300	0.150	PVC 40 m WP	22	580	0.080	PVC 40 m WP
9	300	0.200	PVC 40 m WP	23	580	0.065	PVC 40 m WP
10	580	0.250	PVC 40 m WP	24	580	0.125	PVC 40 m WP
11	300	0.250	PVC 40 m WP	25	420	0.100	PVC 40 m WP
12	580	0.125	PVC 40 m WP	26	400	0.080	PVC 40 m WP
13	580	0.080	PVC 40 m WP	27	580	0.080	PVC 40 m WP
14	420	0.080	PVC 40 m WP	28	900	0.080	PVC 40 m WP

TABLE 10.13. Input Points (Sources) Discharges and Pumping Heads

Input Source Point No.	Node	Pumping Head (m)	Pumping Discharge (m³/s)
1	1	14.00	0.0276
2	10	12.00	0.0242
3	22	13.00	0.0413

The final network design is listed in Table 10.12. The optimal pumping heads and corresponding input point discharges are listed in Table 10.13.

EXERCISES

10.1. Design a multi-input, gravity, branched system considering the system similar to that of Fig. 10.1A. Assume elevation of all the input source nodes at 100.00 m and the elevation of all demand nodes at 60.00 m. Consider minimum terminal head equal to 10 m, peak flow factor 2.5, water demand per person 400 L per day, population load on each distribution branch as 200 persons, and length of each distribution pipe link equal to 300 m. The length of transmission mains connecting sources to the distribution network is equal to 2000 m.

10.2. Design a multi-input, pumping, branched system considering the system similar to that of Fig. 10.1B. Assume elevation of all the input source nodes at 100.00 m and the demand nodes at 101 m. Consider minimum terminal head equal to 15 m, peak flow factor 2.5, water demand per person 400 L per day, population load on each distribution branch as 200 persons, and length of each distribution pipe link equal to 300 m. The length of pumping mains connecting sources to the distribution network is equal to 500 m.

REFERENCES

Rossman, L.A. (2000). *EPANET Users Manual, EPA/600/R-00/057.* US EPA, Cincinnati, OH.

Swamee, P.K., and Sharma, A.K. (1988). Branched Water Distribution System Optimization. *Proceedings of the National Seminar on Management of Water and Waste Water Systems.* Bihar Engineering College, Patna, Feb. 1988, pp. 5.9–5.28.

11

MULTI-INPUT SOURCE, LOOPED SYSTEMS

Generally, city water supply systems are multi-input source, looped pipe networks. The water supply system of a city receives water from various sources, as mostly it is not possible to extract water from a single source because of overall high water demand. Moreover, multi-input supply points also reduce the pipe sizes of the system because of distributed flows. Also in multi-input source systems, it is not only the pipe flow direction that can change because of the spatial or temporal variation in water demand but also the input point source supplying flows to an area or to a particular node.

The multi-input source network increases reliability against raw water availability from a single source and variation in spatial/temporal water demands. Conceptual gravity-sustained and pumping multi-input source, looped water distribution systems are shown in Fig. 11.1. The location of input points/pumping stations and reservoirs is dependent upon the availability of raw water resources and land for water works, topography of the area, and layout pattern of the city.

Design of Water Supply Pipe Networks. By Prabhata K. Swamee and Ashok K. Sharma
Copyright © 2008 John Wiley & Sons, Inc.

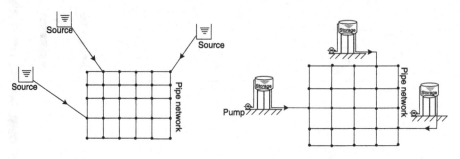

(a) Multi-input sources gravity looped system   (b) Multi-input sources pumping looped system

Figure 11.1. Multi-input source, looped water distribution system.

The analysis of multi-input, looped water systems is complex. It is, therefore, essential to understand or evaluate a physical system, thus making analysis of a network as an integral part of the synthesis process. As described in the previous chapter, some of the existing water distribution analysis models are capable in analyzing multi-input source systems. However, an analysis method was developed specially to link with a cost-optimization method for network synthesis purposes. This analysis method has been described in Chapter 3. In multi-input source, looped water supply systems, the discharges in pipes are not unique; these are dependent on the pipe sizes and location of sources, their elevations, and availability of water from the sources. Thus, as in the design of a multi-input source, branched networks, the looped network also requires sequential application of analysis and synthesis cycles. The design of multi-input sources, looped water distribution systems using continuous and discrete diameter approaches is described in this chapter.

11.1. GRAVITY-SUSTAINED, LOOPED SYSTEMS

Swamee and Sharma (2000) presented a method for the design of looped, gravity-flow water supply systems, which is presented in this section with an example. A typical gravity-sustained, looped system is shown in Fig. 11.2. The pipe network data of Fig. 11.2 are listed in Table 11.1. The pipe network has a total of 36 pipes, 24 nodes, 13 loops, and 2 sources located at node numbers 1 and 24.

The pipe network shown in Fig. 11.2 has been analyzed for pipe discharges. Assuming peak discharge factor = 2.5, rate of water supply 400 liters/capita/day (L/c/d), the nodal discharges are obtained using the method described in Chapter 3 (Eq. 3.29). These discharges are listed in Table 11.2.

For analyzing the network, all pipe link diameters are to be assumed initially as = 0.2 m and pipe material as CI. The network is then analyzed using the Hardy Cross method described in Section 3.7. The pipe discharges so obtained are listed in Table 11.3. The pipe flow discharge sign convention is described in Chapter 3.

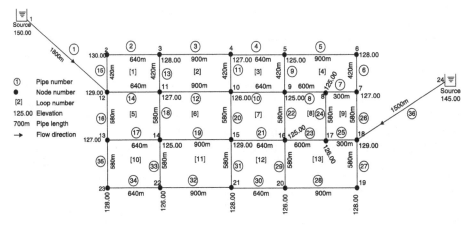

Figure 11.2. Multi-input source, gravity-sustained, looped water distribution system.

11.1.1. Continuous Diameter Approach

Similar to a multi-input, branched network, the entire multi-input, looped water distribution system is converted into a number of gravity distribution mains. Each distribution main is then designed separately using the method described in Chapters 7 and 8. Such distribution mains are equal to the number of pipe links in the network.

Using the data of Tables 11.1 and 11.3, the flow paths for all the pipe links of the network shown in Fig. 11.2 were generated applying the method described in Section 3.9. These flow paths are listed in Table 11.4.

The gravity water distribution mains (pipe flow paths) are designed applying Eq. (7.6b), rewritten as

$$D_i^* = \left(f_i Q_i^2\right)^{\frac{1}{m+5}} \left\{ \frac{8}{\pi^2 g [z_{Js(i)} - z_{Jt(i)} - H]} \sum_{p=I_t(i,\ell)}^{n} L_p \left(f_p Q_p^2\right)^{\frac{m}{m+5}} \right\}^{0.2}. \tag{11.1}$$

Using Tables 11.1 and 11.3, and considering $H = 10$ m and $f = 0.02$ for all pipe links, the pipe sizes are obtained by Eq. (11.1). The friction factor is improved iteratively until the two consecutive f values are close. The final pipe sizes are obtained using a similar procedure as described in Section 9.1.1 (Table 11.6). The calculated pipe sizes and adopted nearest commercial sizes are listed in Table 11.5.

Adopting the pipe sizes listed in Table 11.5, the pipe network was analyzed again by the Hardy Cross method to obtain another set of pipe discharges and the pipe flow paths. Any other analysis method can also be used. Using the new sets of the discharges, the flow paths were obtained again. These flow paths were used to recalculate the pipe sizes. This process was repeated until the two consecutive pipe diameters were close.

TABLE 11.1. Multi-input Sources, Gravity-Sustained, Looped Water Distribution
Network Data

Pipe/ Node i/j	Node 1 $J_1(i)$	Node 2 $J_2(i)$	Loop 1 $K_1(i)$	Loop 2 $K_2(i)$	Length L_i (m)	Form-Loss Coefficient k_{fi}	Population $P(i)$	Nodal Elevation $z(i)$ (m)
1	1	12	0	0	1800	0.5	0	150
2	2	3	1	0	640	0	300	130
3	3	4	2	0	900	0	500	128
4	4	5	3	0	640	0	300	127
5	5	6	4	0	900	0	500	125
6	6	7	4	0	420	0	200	128
7	7	8	4	9	300	0	150	127
8	8	9	4	8	600	0	250	125
9	5	9	3	4	420	0	200	125
10	9	10	3	7	640	0	300	126
11	4	10	2	3	420	0	200	127
12	10	11	2	6	900	0	500	129
13	3	11	1	2	420	0	200	127
14	11	12	1	5	640	0	300	125
15	2	12	1	0	420	0	200	129
16	12	13	5	0	580	0	300	125
17	13	14	5	10	640	0	400	126
18	11	14	5	6	580	0	300	129
19	14	15	6	11	900	0	500	128
20	10	15	6	7	580	0	500	126
21	15	16	7	12	640	0	300	128
22	9	16	7	8	580	0	200	126
23	16	17	8	13	600	0	300	128
24	8	17	8	9	580	0	200	145
25	17	18	9	13	300	0	150	
26	7	18	9	0	580	0	300	
27	18	19	13	0	580	0	300	
28	19	20	13	0	900	0	500	
29	16	20	12	13	580	0	300	
30	20	21	12	0	640	0	400	
31	15	21	11	12	580	0	350	
32	21	22	11	0	900	0	500	
33	14	22	10	11	580	0	300	
34	22	23	10	0	640	0	300	
35	13	23	10	0	580	0	300	
36	18	24	0	0	1500	0.5	0	

11.1.2. Discrete Diameter Approach

The important features of this method are (1) all the looped network pipe links are
brought into the optimization problem formulation keeping the looped configuration

TABLE 11.2. Estimated Nodal Demand Discharges

Node j	Discharge q_j (m³/s)	Node j	Discharge q_j (m³/s)	Node j	Discharge q_j (m³/s)	Node j	Discharge q_j (m³/s)
1	0	7	0.003761	13	0.005780	19	0.004629
2	0.002893	8	0.003472	14	0.008680	20	0.006944
3	0.005780	9	0.005496	15	0.009486	21	0.007234
4	0.00578	10	0.008680	16	0.006365	22	0.006365
5	0.005780	11	0.007523	17	0.003761	23	0.003472
6	0.004050	12	0.004629	18	0.004340	24	0

intact; and (2) the synthesis of the distribution system is conducted considering the entire system as a single entity.

In the current case, the LP formulation is stated as

$$\min F = \sum_{i=1}^{i_L}(c_{i1}x_{i1} + c_{i2}x_{i2}), \tag{11.2}$$

subject to

$$x_{i1} + x_{i2} = L_i; \quad i=1,2,3\ldots i_L, \tag{11.3}$$

$$\sum_{p=I_t(i,\ell)}\left(\frac{8f_{p1}Q_p^2}{\pi^2 g D_{p1}^5}x_{p1} + \frac{8f_{p2}Q_p^2}{\pi^2 g D_{p2}^5}x_{p2}\right) \le z_{Js(i)} + h_{Js(i)} - z_{J_t(i)} - H - \sum_{p=I_t(i,\ell)}\frac{8k_{fp}Q_p^2}{\pi^2 g D_{p2}^4}$$

$$\ell = 1,2,3\ N_t(i) \qquad \text{for } i = 1,2,3,\ldots i_L \tag{11.4}$$

The LP algorithm using commercial pipe sizes has been described in detail in Section 9.1.2. The starting solution is obtained by using Eq. (11.1) for optimal pipe diameters

TABLE 11.3. Multi-input Source, Gravity-Sustained Distribution Network Pipe Discharges

Pipe i	Discharge Q_i (m³/s)	Pipe i	Discharge Q_i (m³/s)	Pipe i	Discharge Q_i (m³/s)	Pipe i	Discharge Q_i (m³/s)
1	0.06064	10	0.0034401	19	0.002688	28	0.01158
2	0.01368	11	0.002315	20	0.003185	29	0.000446
3	0.006098	12	−0.006110	21	−0.004918	30	0.005084
4	−0.002003	13	0.001795	22	−0.001096	31	0.001243
5	−0.005688	14	−0.01939	23	−0.01282	32	−0.000905
6	−0.009731	15	−0.01657	24	−0.007981	33	0.003443
7	0.005434	16	−0.02034	25	−0.02456	34	−0.003827
8	0.009943	17	0.007243	26	−0.01893	35	0.007299
9	−0.002102	18	0.007559	27	0.01621	36	−0.06407

TABLE 11.4. Pipe Flow Paths Treated as Gravity-Sustained Water Distribution Main

Pipe i	Flow Path Pipes Connecting to Input Point Nodes (Sources) and Generating Water Distribution Gravity Mains $I_t(i, \ell)$						$N_t(i)$	$J_t(i)$	$J_s(i)$
	$\ell = 1$	$\ell = 2$	$\ell = 3$	$\ell = 4$	$\ell = 5$	$\ell = 6$			
1	1						1	12	1
2	2	15	1				3	3	1
3	3	2	15	1			4	4	1
4	4	5	6	26	36		5	4	24
5	5	6	26	26			4	5	24
6	6	26	36				3	6	24
7	7	26	36				3	8	24
8	8	24	25	36			4	9	24
9	9	8	24	25	36		5	5	24
10	10	8	24	25	36		5	10	24
11	11	3	2	15	1		5	10	1
12	12	14	1				3	10	1
13	13	2	15	1			4	11	1
14	14	1					2	11	1
15	15	1					2	2	1
16	16	1					2	13	1
17	17	16	1				3	14	1
18	18	14	1				3	14	1
19	19	18	14	1			4	15	1
20	20	12	14	1			4	15	1
21	21	23	25	36			4	15	24
22	22	23	25	36			4	9	24
23	23	25	36				3	16	24
24	24	25	36				3	8	24
25	25	36					2	17	24
26	26	36					2	7	24
27	27	36					2	19	24
28	28	27	36				3	20	24
29	29	23	25	36			4	20	24
30	30	28	27	36			4	21	24
31	31	21	23	25	36		5	21	24
32	32	34	35	16	1		5	21	1
33	33	17	16	1			4	22	1
34	34	35	16	1			4	22	1
35	35	16	1				3	23	1
36	36						1	18	24

D_i^* such that $D_{i1} \leq D_i^* \leq D_{i2}$. The multi-input, looped water distribution system shown in Fig. 11.2 was redesigned using the above-described LP formulation. The solution thus obtained using initially CI pipe material and then the economic pipe materials, is given in Table 11.6. The final solution is shown in Fig. 11.3.

TABLE 11.5. Multi-input Source, Gravity-Sustained System: Estimated and Adopted Pipe Sizes

Pipe	Calculated Continuous Pipe Size (m)	Adopted Pipe Size (m)	Pipe	Calculated Continuous Pipe Size (m)	Adopted Pipe Size (m)
1	0.235	0.250	19	0.084	0.100
2	0.143	0.150	20	0.091	0.100
3	0.109	0.100	21	0.102	0.102
4	0.076	0.080	22	0.058	0.080
5	0.108	0.100	23	0.141	0.150
6	0.130	0.125	24	0.112	0.125
7	0.109	0.100	25	0.176	0.200
8	0.134	0.150	26	0.167	0.200
9	0.074	0.080	27	0.155	0.150
10	0.093	0.100	28	0.138	0.150
11	0.078	0.080	29	0.043	0.050
12	0.109	0.100	30	0.104	0.100
13	0.072	0.080	31	0.064	0.065
14	0.152	0.150	32	0.058	0.065
15	0.153	0.150	33	0.091	0.100
16	0.166	0.200	34	0.091	0.100
17	0.117	0.125	35	0.113	0.125
18	0.115	0.125	36	0.245	0.250

The variation of system cost with LP iterations is shown in Fig. 11.4. The first three iterations pertain to initially assumed pipe material. Subsequently, the economic pipe material was selected and the LP cycles were carried out. A total of six iterations were needed to obtain a design with economic pipe material.

11.2. PUMPING SYSTEM

Generally, city water supply systems are multi-input, looped, pumping pipe networks. Multi-input systems are provided to meet the large water demand, which cannot be met mostly from a single source. Pumping systems are essential to supply water at required pressure and quantity where topography is flat or undulated. External energy is required to overcome pipe friction losses and maintain minimum pressure heads.

The design method is described using an example of a typical town water supply system shown in Fig. 11.5, which contains 37 pipes, 25 nodes, 13 loops, 3 input sources as pumping stations, and reservoirs at nodes 1, 24, and 25. The network data are listed in Table 11.7.

Considering the rate of water supply of 400 L/c/d and a peak factor of 2.5, the nodal discharges are worked out. These nodal discharges are listed in Table 11.8.

TABLE 11.6. Multi-input, Gravity-Sustained, Looped Pipe Distribution Network Design

Pipe i	Pipe Length L_i (m)	Initial Design with Assumed Pipe Material		Final Design with Optimal Pipe Material	
		D_i (m)	Pipe Material	D_i (m)	Pipe Material
1	1800	0.250	CI Class LA‡	0.250	AC Class 10*
2	640	0.150	CI Class LA	0.125	PVC 40 m WP†
3	900	0.125	CI Class LA	0.125	PVC 40 m WP
4	640	0.065	CI Class LA	0.065	PVC 40 m WP
5	900	0.100	CI Class LA	0.100	PVC 40 m WP
6	420	0.125	CI Class LA	0.100	PVC 40 m WP
7	300	0.125	CI Class LA	0.125	PVC 40 m WP
8	600	0.125	CI Class LA	0.125	PVC 40 m WP
9	420	0.050	CI Class LA	0.050	PVC 40 m WP
10	640	0.080	CI Class LA	0.080	PVC 40 m WP
11	420	0.050	CI Class LA	0.050	PVC 40 m WP
12	900	0.100	CI Class LA	0.100	PVC 40 m WP
13	420	0.050	CI Class LA	0.050	PVC 40 m WP
14	640	0.125	CI Class LA	0.125	PVC 40 m WP
15	420	0.150	CI Class LA	0.150	PVC 40 m WP
16	580	0.200	CI Class LA	0.200	PVC 40 m WP
17	640	0.150	CI Class LA	0.150	PVC 40 m WP
18	580	0.050	CI Class LA	0.050	PVC 40 m WP
19	900	0.080	CI Class LA	0.080	PVC 40 m WP
20	580	0.050	CI Class LA	0.050	PVC 40 m WP
21	640	0.150	CI Class LA	0.125	PVC 40 m WP
22	580	0.050	CI Class LA	0.050	PVC 40 m WP
23	600	0.200	CI Class LA	0.150	PVC 40 m WP
24	580	0.050	CI Class LA	0.050	PVC 40 m WP
25	300	0.200	CI Class LA	0.200	PVC 40 m WP
26	580	0.200	CI Class LA	0.150	PVC 40 m WP
27	580	0.200	CI Class LA	0.150	PVC 40 m WP
28	900	0.150	CI Class LA	0.150	PVC 40 m WP
29	580	0.050	CI Class LA	0.050	PVC 40 m WP
30	640	0.125	CI Class LA	0.125	PVC 40 m WP
31	580	0.050	CI Class LA	0.065	PVC 40 m WP
32	900	0.050	CI Class LA	0.050	PVC 40 m WP
33	580	0.080	CI Class LA	0.080	PVC 40 m WP
34	640	0.080	CI Class LA	0.080	PVC 40 m WP
35	580	0.100	CI Class LA	0.100	PVC 40 m WP
36	1500	0.250	CI Class LA	0.250	AC Class 10*

* Asbestos cement pipe 50 m working pressure.
† Poly(vinyl chloride) 40 m working pressure.
‡ Class LA based on pipe wall thickness = 60 m working pressure.

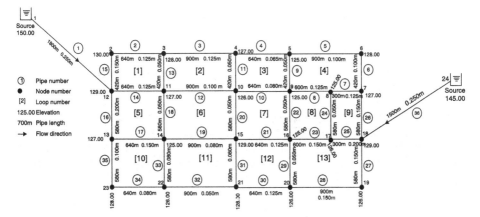

Figure 11.3. Multi-input, gravity-sustained looped water distribution network design.

Initially, pipes were assumed as 0.20 m of CI pipe material for the entire pipe network. The Hardy Cross analysis method was then applied to determine the pipe discharges. The pipe discharges so obtained are listed in Table 11.9.

11.2.1. Continuous Diameter Approach

As described in Section 10.2.1, the entire looped distribution system is converted into a number of distribution mains enabling them to be designed separately. The flow paths for all the pipe links are generated using Tables 11.7 and 11.9. Applying the method described in Section 3.9, the pipe flow paths along with their originating nodes $J_t(i)$ and input source nodes $J_s(i)$ are listed in Table 11.10.

Applying the method described in Section 10.2.1, the continuous pipe sizes using Eq. (10.5) and pumping head with the help of Eq. (10.6) can be calculated. The pipe

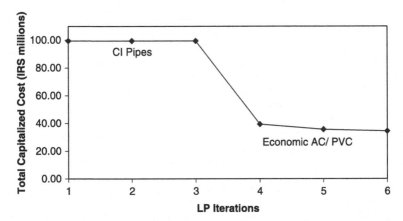

Figure 11.4. Number of LP iterations in gravity system design.

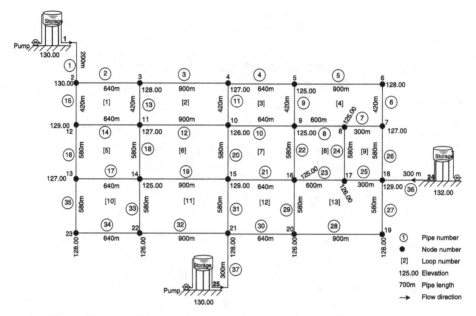

Figure 11.5. Multi-input source, pumping, looped water distribution system.

sizes and adopted nearest commercial sizes are listed in Table 11.11. The final solution can be obtained using the methods described in Sections 9.2.1 and 10.2.1.

11.2.2. Discrete Diameter Approach

The discrete diameter approach solves the design problem in its original form. In the current case, the LP formulation is

$$\min F = \sum_{i=1}^{i_L} (c_{i1}x_{i1} + c_{i2}x_{i2}) + \rho g k_T \sum_{n}^{n_L} Q_{Tn}h_{0n}, \qquad (11.5)$$

subject to

$$x_{i1} + x_{i2} = L_i, \quad i = 1,2,3 \ldots i_L, \qquad (11.6)$$

$$\sum_{p=I_t(i,\ell)} \left(\frac{8f_{p1}Q_p^2}{\pi^2 g D_{p1}^5} x_{p1} + \frac{8f_{p2}Q_p^2}{\pi^2 g D_{p2}^5} x_{p2} \right) \le z_{Js(i)} + h_{Js(i)} - z_{J_t(i)} - H$$

$$- \sum_{p=I_t(i,\ell)} \frac{8k_{fp}Q_p^2}{\pi^2 g D_{p2}^4}$$

$$\ell = 1, 2, 3 \, N_t(i) \qquad \text{For } i = 1,2,3 \ldots i_L \qquad (11.7)$$

Using Inequations (11.7), the head-loss constraint inequations for all the originating nodes of pipe flow paths are developed to bring all the looped network pipes into the LP formulation.

TABLE 11.7. Multi-input, Pumping, Looped Water Distribution Network Data

Pipe/Node i/j	Node 1 $J_1(i)$	Node 2 $J_2(i)$	Loop 1 $K_1(i)$	Loop 2 $K_2(i)$	Length L_i (m)	Form Loss Coefficient k_{fi}	Population $P(i)$	Nodal Elevation $z(i)$ (m)
1	1	2	0	0	200	0.5	0	130
2	2	3	1	0	640	0	300	130
3	3	4	2	0	900	0	500	128
4	4	5	3	0	640	0	300	127
5	5	6	4	0	900	0	500	125
6	6	7	4	0	420	0	200	128
7	7	8	4	9	300	0	150	127
8	8	9	4	8	600	0	250	125
9	5	9	3	4	420	0	200	125
10	9	10	3	7	640	0	300	126
11	4	10	2	3	420	0	200	127
12	10	11	2	6	900	0	500	129
13	3	11	1	2	420	0	200	127
14	11	12	1	5	640	0	300	125
15	2	12	1	0	420	0	200	129
16	12	13	5	0	580	0	300	125
17	13	14	5	10	640	0	400	126
18	11	14	5	6	580	0	300	129
19	14	15	6	11	900	0	500	128
20	10	15	6	7	580	0	500	126
21	15	16	7	12	640	0	300	128
22	9	16	7	8	580	0	200	126
23	16	17	8	13	600	0	300	128
24	8	17	8	9	580	0	200	132
25	17	18	9	13	300	0	150	130
26	7	18	9	0	580	0	300	
27	18	19	13	0	580	0	300	
28	19	20	13	0	900	0	500	
29	16	20	12	13	580	0	300	
30	20	21	12	0	640	0	400	
31	15	21	11	12	580	0	350	
32	21	22	11	0	900	0	500	
33	14	22	10	11	580	0	300	
34	22	23	10	0	640	0	300	
35	13	23	10	0	580	0	300	
36	18	24	0	0	300	0.5	0	
37	21	25	0	0	300	0.5	0	

The starting pipe sizes can be obtained using Eq. (10.5) for continuous optimal pipe diameters D_i^*, and the two consecutive commercially available sizes D_{i1} and D_{i2} are selected such that $D_{i1} \leq D_i^* \leq D_{i2}$. Following the LP method described in Section 9.2.2, the looped water distribution system shown in Fig. 11.5 was redesigned. The solution thus obtained is shown in Fig. 11.6. The design parameters such as

TABLE 11.8. Estimated Nodal Water Demands

Node j	Nodal Demand $Q(j)$ (m^3/s)	Node j	Nodal Demand $Q(j)$ (m^3/s)	Node j	Nodal Demand $Q(j)$ (m^3/s)
1	0	10	0.00868	19	0.00463
2	0.00289	11	0.00752	20	0.00694
3	0.00579	12	0.00463	21	0.00723
4	0.00579	13	0.00579	22	0.00637
5	0.00579	14	0.00868	23	0.00347
6	0.00405	15	0.00955	24	0
7	0.00376	16	0.00637	25	0
8	0.00347	17	0.00376		
9	0.00550	18	0.00434		

minimum terminal pressure of 10 m, minimum pipe diameter of 100 mm, rate of water supply per person as 400 L/day and peak flow discharge ratio of 2.5 were specified for the network.

The pipe sizes finally obtained from the algorithm are listed in Table 11.12 and the pumping heads including input source discharges are given in Table 11.13.

As stated in Chapter 9 for single-input, looped systems, the discrete pipe diameter approach provides an economic solution as it formulates the problem for the system as a whole, whereas piecemeal design is carried out in the continuous diameter approach and also conversion of continuous sizes to commercial sizes misses the optimality of the solution. A similar conclusion can be drawn for multi-input source, looped systems.

TABLE 11.9. Looped Network Pipe Discharges

Pipe i	Discharge Q_i (m^3/s)	Pipe i	Discharge Q_i (m^3/s)	Pipe i	Discharge Q_i (m^3/s)
1	0.04047	14	−0.00608	27	0.00789
2	0.01756	15	0.02002	28	0.00326
3	0.00683	16	0.00931	29	−0.00032
4	0.00057	17	0.00216	30	−0.00400
5	−0.00288	18	0.00120	31	−0.01405
6	−0.00694	19	−0.00248	32	0.01131
7	0.00454	20	−0.00400	33	−0.00284
8	0.00578	21	−0.00198	34	0.00211
9	−0.00233	22	−0.00397	35	0.00136
10	0.00191	23	−0.01200	36	−0.04793
11	0.00048	24	−0.00471	37	−0.03660
12	−0.00229	25	−0.02047		
13	0.00494	26	−0.01524		

TABLE 11.10. Pipe Flow Paths Treated as Water Distribution Main

Pipe i	$\ell=1$	$\ell=2$	$\ell=3$	$\ell=4$	$\ell=5$	$N_t(i)$	$J_t(i)$	$J_s(i)$
1	1					1	2	1
2	2	1				2	3	1
3	3	2	1			3	4	1
4	4	4	3	2	1	4	5	1
5	5	6	26	36		4	5	24
6	6	26	36			3	6	24
7	7	26	36			3	8	24
8	8	24	25	36		4	9	24
9	9	5	6	26	36	5	9	24
10	10	8	24	25	36	5	10	24
11	11	3	2	1		4	10	1
12	12	14	15	1		4	10	1
13	13	2	1			3	11	1
14	14	15	1			3	11	1
15	15	1				2	12	1
16	16	15	1			3	13	1
17	17	16	15	1		4	14	1
18	18	14	15	1		4	14	1
19	19	31	37			3	14	25
20	20	31	37			3	10	25
21	21	23	25	36		4	15	24
22	22	23	25	36		4	9	24
23	23	25	36			3	16	24
24	24	25	36			3	8	24
25	25	36				2	17	24
26	26	36				2	7	24
27	27	36				2	19	24
28	28	27	36			3	20	24
29	29	30	37			3	16	25
30	30	37				2	20	25
31	31	37				2	15	25
32	32	37				2	22	25
33	33	32	37			3	14	25
34	34	32	37			3	23	25
35	35	16	15	1		4	23	1
36	36					1	18	24
37	37					1	21	25

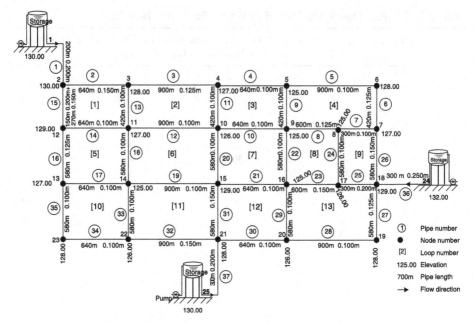

Figure 11.6. Pumping, looped water network design.

TABLE 11.11. Pumping, Looped Network Design

Pipe i	Calculated Pipe Diameter D_i (m)	Pipe Diameter Adopted D_i (m)	Pipe i	Calculated Pipe Diameter D_i (m)	Pipe Diameter Adopted D_i (m)
1	0.191	0.200	20	0.093	0.100
2	0.146	0.150	21	0.097	0.100
3	0.099	0.100	22	0.102	0.100
4	0.055	0.100	23	0.120	0.125
5	0.091	0.100	24	0.113	0.125
6	0.117	0.125	25	0.156	0.150
7	0.091	0.100	26	0.145	0.150
8	0.112	0.125	27	0.132	0.150
9	0.089	0.100	28	0.111	0.125
10	0.085	0.100	29	0.115	0.125
11	0.035	0.100	30	0.114	0.125
12	0.080	0.100	31	0.131	0.125
13	0.106	0.125	32	0.124	0.125
14	0.111	0.125	33	0.070	0.100
15	0.154	0.150	34	0.074	0.100
16	0.116	0.125	35	0.067	0.100
17	0.057	0.100	36	0.211	0.250
18	0.086	0.100	37	0.187	0.200
19	0.087	0.100			

TABLE 11.12. Multi-input, Looped, Pumping Network Design

Pipe i	Length L_i (m)	Pipe Diameter D_i (m)	Pipe Material and Class	Pipe i	Length L_i (m)	Pipe Diameter D_i (m)	Pipe Material and Class
1	200	0.200	PVC 40 m WP	20	580	0.100	PVC 40 m WP
2	640	0.150	PVC 40 m WP	21	640	0.100	PVC 40 m WP
3	900	0.125	PVC 40 m WP	22	580	0.100	PVC 40 m WP
4	640	0.100	PVC 40 m WP	23	600	0.150	PVC 40 m WP
5	900	0.100	PVC 40 m WP	24	580	0.100	PVC 40 m WP
6	420	0.125	PVC 40 m WP	25	300	0.200	PVC 40 m WP
7	300	0.100	PVC 40 m WP	26	580	0.150	PVC 40 m WP
8	600	0.125	PVC 40 m WP	27	580	0.125	PVC 40 m WP
9	420	0.100	PVC 40 m WP	28	900	0.100	PVC 40 m WP
10	640	0.100	PVC 40 m WP	29	580	0.100	PVC 40 m WP
11	420	0.100	PVC 40 m WP	30	640	0.100	PVC 40 m WP
12	900	0.100	PVC 40 m WP	31	580	0.150	PVC 40 m WP
13	420	0.100	PVC 40 m WP	32	900	0.150	PVC 40 m WP
14	640	0.100	PVC 40 m WP	33	580	0.100	PVC 40 m WP
15	420	0.200 (150)+ 0.150 (270)	PVC 40 m WP	34	640	0.100	PVC 40 m WP
16	580	0.125	PVC 40 m WP	35	580	0.100	PVC 40 m WP
17	640	0.100	PVC 40 m WP	36	300	0.250	AC C-5 25 m WP
18	580	0.100	PVC 40 m WP	37	300	0.200	PVC 40 m WP
19	900	0.100	PVC 40 m WP				

TABLE 11.13. Input Points (Sources) Discharges and Pumping Heads

Input Source Point No.	Node	Pumping Head (m)	Pumping Discharge (m³/s)
1	1	15.75	0.0404
2	24	11.50	0.0479
3	25	13.75	0.0365

EXERCISES

11.1. Describe the advantages of developing head-loss constraint inequations for all the originating nodes of pipe flow paths into LP problem formulation in multi-input, looped network.

11.2. Construct a two-input-source, gravity-sustained, looped network similar to Fig. 11.2 by increasing pipe lengths by a factor of 1.5. Design the system by increasing the population on each pipe link by a factor of 2 and keep the other parameters similar to the example in Section 11.1.

11.3. Construct a three-input-source, pumping, looped network similar to Fig. 11.5 by increasing pipe lengths by a factor of 1.5. Design the system by increasing the population on each pipe link by a factor of 2 and keep the other parameters similar to the example in Section 11.2.

REFERENCE

Swamee, P.K., and Sharma, A.K. (2000). Gravity flow water distribution network design. *Journal of Water Supply: Research and Technology-AQUA, IWA* 49(4), 169–179.

12

DECOMPOSITION OF A LARGE WATER SYSTEM AND OPTIMAL ZONE SIZE

Generally, urban water systems are large and have multi-input sources to cater for large population. To design such systems as a single entity is difficult. These systems are decomposed or split into a number of subsystems with single input source. Each subsystem is individually designed and finally interconnected at the ends for reliability considerations. Swamee and Sharma (1990) developed a method for decomposing multi-input large water distribution systems of predecided input source locations into subsystems of single input. The method not only eliminates the practice of decomposing or splitting large system by designer's intuition but also enables the designer to design a large water distribution system with a reasonable computational effort.

Design of Water Supply Pipe Networks. By Prabhata K. Swamee and Ashok K. Sharma
Copyright © 2008 John Wiley & Sons, Inc.

213

Estimating optimal zone size is difficult without applying optimization technique. Splitting of a large area into optimal zones is not only economic but also easy to design. Using geometric programming, Swamee and Kumar (2005) developed a method for optimal zone sizing of circular and rectangular geometry. These methods are described with examples in this chapter.

12.1. DECOMPOSITION OF A LARGE, MULTI-INPUT, LOOPED NETWORK

The most important factor encouraging decomposition of a large water distribution system into small systems is the difficulty faced in designing a large system as a single entity. The optimal size of a subsystem will depend upon the geometry of the network, spatial variation of population density, topography of the area, and location of input points. The computational effort required can be reckoned in terms of a number of multiplications performed in an algorithm. Considering that sequential linear programming is adopted as optimization technique, the computational effort required can be estimated for the design of a water distribution system.

The number of multiplications required for one cycle of linear programming (LP) algorithm is proportional to N^2 (N being the number of variables involved), and in general the number of iterations required are in proportion to N. Thus, the computational effort in an LP solution is proportional to N^3. If a large system is divided into M subsystems of nearly equal size, the computational effort reduces to $M(N/M)^3$ (i.e., N^3/M^2). Thus, a maximum reduction of the order M^2 can be obtained in the computational effort, which is substantial. On the other hand, the computer memory requirement reduces from an order proportional to N^2 to $M(N/M)^2$ (i.e., N^2/M). The large systems, which are fed by a large number of input points, could be decomposed into subsystems having an area of influence of each input point, and these subsystems can be designed independent of neighboring subsystems. Thus, the design of a very large network, which looked impossible on account of colossal computer time required, becomes feasible on account of independent design of the constituent subsystems.

12.1.1. Network Description

Figure 12.1 shows a typical water distribution network, which has been considered for presenting the method for decomposition. It consists of 55 pipe links, 33 nodes, and 3 input points. Three input points located at nodes 11, 22, and 28 have their influence zones, which have to be determined, and the pipe links have to be cut at points that are under the influence of two input points.

The network data are listed in Table 12.1. The data about pipes is given line by line, which contain ith pipe number, both nodal numbers, loop numbers, the length of pipe link, form-loss coefficient, and population load on pipe link. The nodal elevations corresponding with node numbers are also listed in this table. The nodal water demand due to industrial/commercial demand considerations can easily be included in the table.

The next set of data is about input points, which is listed in Table 12.2.

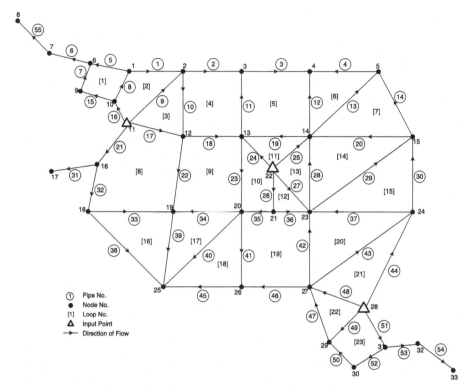

Figure 12.1. Multi-input looped network.

12.1.2. Preliminary Network Analysis

For the purpose of preliminary analysis of the network, all the pipe diameters D_i are assumed to be of 0.2 m and the total water demand is equally distributed among the input points to satisfy the nodal continuity equation. Initially, the pipe material is assumed as CI. The network is analyzed by applying continuity equations and the Hardy Cross method for loop discharge correction as per the algorithm described in Chapter 3. In the case of existing system, the existing pipe link diameters, input heads, and input source point discharges should be used for network decomposition. It will result in pipe discharges for assumed pipe diameters and input points discharges. The node pipe connectivity data generated for the network (Fig. 12.1) is listed in Table 12.3.

12.1.3. Flow Path of Pipes and Source Selection

The flow path of pipes of the network and the originating node of a corresponding flow path can be obtained by using the method as described in Chapter 3. The flow directions are marked in Fig. 12.1. A pipe receives the discharge from an input point at which the pipe flow path terminates. Thus, the source of pipe $I_s(i)$ is the input point number n at

TABLE 12.1. Pipe Network Data

Pipe/ Node i/j	First Node $J_1(i)$	Second Node $J_2(i)$	Loop 1 $K_1(i)$	Loop 2 $K_2(i)$	Pipe Length $L(i)$	Form-Loss Coefficient $k_f(i)$	Population Load $P(i)$	Nodal Elevation z_j
1	1	2	2	0	380	0.0	500	101.85
2	2	3	4	0	310	0.0	385	101.90
3	3	4	5	0	430	0.2	540	101.95
4	4	5	6	0	270	0.0	240	101.60
5	1	6	1	0	150	0.0	190	101.75
6	6	7	0	0	250	0.0	250	101.80
7	6	9	1	0	150	0.0	190	101.80
8	1	10	1	2	150	0.0	190	101.40
9	2	11	2	3	390	0.0	490	101.85
10	2	12	3	4	320	0.0	400	101.90
11	3	13	4	5	320	0.0	400	102.00
12	4	14	5	6	330	0.0	415	101.80
13	5	14	6	7	420	0.0	525	101.80
14	5	15	7	0	320	0.0	400	101.90
15	9	10	1	0	160	0.0	200	100.50
16	10	11	2	0	120	0.0	150	100.80
17	11	12	3	8	280	0.0	350	100.70
18	12	13	4	9	330	0.0	415	101.40
19	13	14	5	11	450	0.2	560	101.60
20	14	15	7	14	360	0.2	450	101.80
21	11	16	8	0	230	0.0	280	101.85
22	12	19	8	9	350	0.0	440	101.95
23	13	20	9	10	360	0.0	450	101.80
24	13	22	10	11	260	0.0	325	101.10
25	14	22	11	13	320	0.0	400	101.40
26	21	22	10	12	160	0.0	200	101.20
27	22	23	12	13	290	0.0	365	101.70
28	14	23	13	14	320	0.0	400	101.90
29	15	23	14	15	500	0.0	625	101.70
30	15	24	15	0	330	0.0	410	101.80
31	16	17	0	0	230	0.0	290	101.80
32	16	18	8	0	220	0.0	275	101.80
33	18	19	8	18	350	0.0	440	100.40
34	19	20	9	17	330	0.0	410	
35	20	21	10	19	220	0.0	475	
36	21	23	12	19	250	0.0	310	
37	23	24	15	20	370	0.0	460	
38	18	25	16	0	470	0.0	590	
39	19	25	16	17	320	0.0	400	
40	20	25	17	18	460	0.0	575	
41	20	26	18	19	310	0.0	390	
42	23	27	19	20	330	0.0	410	

(Continued)

TABLE 12.1. Continued

Pipe/ Node i/j	First Node $J_1(i)$	Second Node $J_2(i)$	Loop 1 $K_1(i)$	Loop 2 $K_2(i)$	Pipe Length $L(i)$	Form-Loss Coefficient $k_f(i)$	Population Load $P(i)$	Nodal Elevation z_j
43	24	27	20	21	510	0.0	640	
44	24	28	21	0	470	0.0	590	
45	25	26	18	0	300	0.0	375	
46	26	27	19	0	490	0.0	610	
47	27	29	22	0	230	0.0	290	
48	27	28	21	22	290	0.0	350	
49	28	29	22	23	190	0.0	240	
50	29	30	23	0	200	0.0	250	
51	28	31	23	0	160	0.0	200	
52	30	31	23	0	140	0.0	175	
53	31	32	0	0	200	0.0	110	
54	32	33	0	0	200	0.0	200	
55	7	8	0	0	200	0.0	250	

which the corresponding pipe flow path terminates. The flow path pipes $I_t(i, \ell)$ for each pipe i, the total number of pipes in the track $N_t(i)$, originating node of pipe track $J_t(i)$, and the input source (point) of pipe $I_s(i)$ are listed in Table 12.4. The originating node of a pipe flow path is the node to which the pipe flow path supplies the discharge. Using Table 12.3 and Table 12.4, one may find the various input points from which a node receives the discharge. These input points are designated as $I_n(j, \ell)$. The index ℓ varyies 1 to $N_n(j)$, where $N_n(j)$ is the total number of input points discharging at node j. The various input sources discharging to a node are listed in Table 12.5.

12.1.4. Pipe Route Generation Connecting Input Point Sources

A route is a set of pipes in the network that connects two different input point sources. Two different pipe flow paths leading to two different input points originating from a common node can be joined to form a route. The procedure is illustrated by considering the node $j = 26$. The flow directions in pipes based on initially assumed pipe sizes are shown in Fig. 12.1. Referring to Table 12.3, one finds that node 26 is connected to pipes 41, 45, and 46. Also from Table 12.4, one finds that the pipes 41 and 45 are connected to input point 1, whereas pipe 46 is connected to input point 3. It can be seen that flow path

TABLE 12.2. Input Point Nodes

Input Point n	Input Point Node $S(n)$
1	11
2	22
3	28

TABLE 12.3. Node Pipe Connectivity

Node j	Pipes Connected at Node j $I_P(j, \ell)$						Total Pipes $N(j)$
	$\ell = 1$	$\ell = 2$	$\ell = 3$	$\ell = 4$	$\ell = 5$	$\ell = 6$	
1	1	5	8				3
2	1	2	9	10			4
3	2	3	11				3
4	3	4	12				3
5	4	13	14				3
6	5	6	7				3
7	6	55					2
8	55						1
9	7	15					2
10	8	15	16				3
11	9	16	17	21			4
12	10	17	18	22			4
13	11	18	19	23	24		5
14	12	13	19	20	25	28	6
15	15	20	29	30			4
16	21	31	32				3
17	31						1
18	32	33	38				3
19	22	33	34	39			4
20	23	34	35	40	41		5
21	26	35	36				3
22	24	25	26	27			4
23	27	28	29	36	37	42	6
24	30	37	43	44			4
25	38	39	40	45			4
26	41	45	46				3
27	42	43	46	47	48		5
28	44	48	49	51			4
29	47	49	50				3
30	50	52					2
31	51	52	53				3
32	53	54					2
33	54						1

$I_t(45, \ell)$, $(\ell = 1, 8) = 45, 41, 23, 18, 10, 1, 8, 16$ is not originating from node 26. Whether a pipe flow path is originating from a node or not can be checked by finding the flow path originating node $J_t(i)$ from Table 12.4. For example, $J_t(i = 45)$ is 25. Thus, pipe 45 will not be generating a route at node 26. Hence, only the flow paths of pipes 41 and 46 will generate a route.

The flow path $I_t(41, \ell)$, $(\ell = 1,7) = 41, 23, 18, 10, 1, 8, 16$ ending up at input point 1 is reversed as 16, 8, 1, 10, 18, 23, 41, and combined with another flow path $I_t(46, \ell)$,

TABLE 12.4. Pipe Flow Paths, Originating Nodes, and Pipe Input Source Nodes

	Pipes in Flow Path of Pipe i $I_t(i, \ell)$								Total Pipes	Originating	Source Node
Pipe i	$\ell=1$	$\ell=2$	$\ell=3$	$\ell=4$	$\ell=5$	$\ell=6$	$\ell=7$	$\ell=8$	$N_t(i)$	Node $J_t(i)$	$I_s(i)$
1	1	8	16						3	2	1
2	2	1	8	16					4	3	1
3	3	2	1	8	16				5	4	1
4	4	13	20	29	27				5	4	2
5	5	8	16						3	6	1
6	6	5	8	16					4	7	1
7	7	15	16						3	6	1
8	8	16							2	1	1
9	9								1	2	1
10	10	1	8	16					4	12	1
11	11	18	10	1	8	16			6	3	1
12	12	20	29	27					4	4	2
13	13	14	20	29	27				4	5	2
14	14	29	27						3	5	2
15	15	16							2	9	1
16	16								1	10	1
17	17								1	12	1
18	18	10	1	8	16				5	13	1
19	19	20	29	27					4	13	2
20	20	29	27						3	14	2
21	21								1	16	1
22	22	10	1	8	16				5	19	1
23	23	18	10	1	8	16			6	20	1
24	24								1	13	2
25	25								1	13	2
26	26								1	21	2
27	27								1	23	2
28	28	27							2	14	2
29	29	27							2	15	2
30	30	43	47	49					4	15	3
31	31	21							2	17	1
32	32	21							2	18	1
33	33	32	21						3	19	1
34	34	23	18	10	1	8	16		7	19	1
35	35	26							2	20	2
36	36	26							2	23	2
37	37	43	47	49					4	23	3
38	38	32	31						3	25	1
39	39	22	10	1	8	16			6	25	1
40	40	23	18	10	1	8	16		7	25	1
41	41	23	18	10	1	8	16		7	26	1
42	42	47	49						3	23	3

(Continued)

TABLE 12.4. Continued

Pipe i	Pipes in Flow Path of Pipe i $I_t(i, \ell)$								Total Pipes $N_t(i)$	Originating Node $J_t(i)$	Source Node $I_s(i)$
	$\ell=1$	$\ell=2$	$\ell=3$	$\ell=4$	$\ell=5$	$\ell=6$	$\ell=7$	$\ell=8$			
43	43	47	49						3	24	3
44	44								1	24	3
45	45	41	23	18	10	1	8	16	8	25	1
46	46	47	49						3	26	3
47	47	49							2	27	3
48	48								1	27	3
49	49								1	29	3
50	50	52	51						3	29	3
51	51								1	31	3
52	52	51							2	30	3
53	53	51							2	32	3
54	54	53	51						3	33	3
55	55	6	5	8	16				5	8	1

TABLE 12.5. Nodal Input Point Sources

Node j	Input Sources $I_n(j, \ell)$		Total Sources $N_n(j)$	Node j	Input Sources $I_n(j, \ell)$		Total Sources $N_n(j)$
	$\ell=1$	$\ell=2$			$\ell=1$	$\ell=2$	
1	1		1	18	1		1
2	1		1	19	1		1
3	1		1	20	1	2	2
4	1	2	2	21	2		1
5	2		1	22	2		1
6	1		1	23	2	3	2
7	1		1	24	3		1
8	1		1	25	1		1
9	1		1	26	1	3	2
10	1		1	27	3		1
11	1		1	28	3		1
12	1		1	29	3		1
13	1	2	2	30	3		1
14	2		1	31	3		1
15	2	3	2	32	3		1
16	1		1	33	3		1
17	1		1				

$(\ell,1,3) = 46, 47, 49$ ending up at input point 3, the following route is obtained:

$$I_R(r,\ell),[\ell = 1,N_R(r)] = 16, 8, 1, 10, 18, 23, 41, 46, 47, 49, \qquad (12.1)$$

where r is the sequence in which various routes are generated, $N_R(r)$ = total pipes in the route (10 in the above route), and $I_R(r, \ell)$ is the set of pipes in the route.

The route r connects the two input points $M_1(r)$ and $M_2(r)$. These input points can be found from the initial and the final pipe numbers of the route r. The routes generated by the algorithm are shown in Table 12.6.

The routes emerging from or terminating at the input point source 1 can be found by scanning Table 12.6 for $M_1(r)$ or $M_2(r)$ to be equal to 1. These routes are shown in Table 12.7.

12.1.5. Weak Link Determination for a Route Clipping

A weak link is a pipe in the route through which a minimum discharge flows if designed separately as a single distribution main having input points at both ends.

Input point 1 can be separated from rest of the network if the process of generation of Table 12.7 and cutting of routes at suitable points is repeated untill the input point is separated. The suitable point can be the midpoint of the pipe link carrying the minimum discharge in that route.

For determination of the weak link, the route has to be designed by considering it as a separate entity from the remaining network. From the perusal of Table 12.7, it is clear that long routes are circuitous and thus are not suitable for clipping the pipe at the first instance when shorter routes are available. On the other hand, shorter routes more or less provide direct connection between the two input points. Selecting the first occurring route of minimum pipe links in Table 12.7, one finds that route for $r = 4$ is a candidate for clipping.

12.1.5.1. Design of a Route.   Considering a typical route (see Fig. 12.2a) consisting of i_L pipe links and $i_L + 1$ nodes including the two input points at the ends, one can find out the nodal withdrawals $q_1, q_2, q_3, \ldots, q_{i_L-1}$ by knowing the link populations. The total discharge Q_T is obtained by summing up these discharges, that is,

$$Q_T = q_1 + q_2 + q_3 + \cdots + q_{i_L-1}. \qquad (12.2)$$

The discharge Q_{T1} at input point 1 is suitably assumed initially, say $(Q_{T1} = 0.9Q_T)$, and the discharge Q_{T2} at input point 2 is:

$$Q_{T2} = Q_T - Q_{T1}. \qquad (12.3)$$

Considering the withdrawals to be positive and the input discharges to be negative, one may find the pipe discharges Q_i for any assumed value of Q_{T1}. This can be done by the

TABLE 12.6. Pipe Routes Between Various Input Point Sources

Route r	Pipes in Route r $I_R(r, \ell)$										Total Pipes $N_R(r)$	First Input Point of Route $M_1(r)$	Second Input Point of Route $M_2(r)$
	$\ell=1$	$\ell=2$	$\ell=3$	$\ell=4$	$\ell=5$	$\ell=6$	$\ell=7$	$\ell=8$	$\ell=9$	$\ell=10$			
1	16	8	1	2	3	4	13	20	29	27	10	1	2
2	16	8	1	2	3	12	20	29	27		9	1	2
3	16	8	1	10	18	19	20	29	27		9	1	2
4	16	8	1	10	18	24					6	1	2
5	27	29	30	43	47	49					6	2	3
6	16	8	1	10	18	23	35	26			8	1	2
7	27	37	43	47	49						5	2	3
8	27	42	47	49							4	2	3
9	26	36	37	43	47	49					6	2	3
10	26	36	42	47	49						5	2	3
11	16	8	1	10	18	23	41	46	47	49	10	1	3

TABLE 12.7. Routes Connected with Input Point Source 1

Route r	Pipes in Route r $I_R(r, \ell)$										Total Pipes $N_R(r)$	First Input Point of Route $M_1(r)$	Second Input Point of Route $M_2(r)$
	$\ell=1$	$\ell=2$	$\ell=3$	$\ell=4$	$\ell=5$	$\ell=6$	$\ell=7$	$\ell=8$	$\ell=9$	$\ell=10$			
1	16	8	1	2	3	4	13	20	29	27	10	1	2
2	16	8	1	2	3	12	20	29	27		9	1	2
3	16	8	1	10	18	19	20	29	27		9	1	2
4	16	8	1	10	18	24					6	1	2
5	16	8	1	10	18	23	35	26			8	1	2
6	16	8	1	10	18	23	41	46	47	49	10	1	3

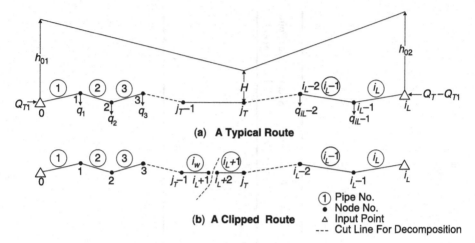

Figure 12.2. Pipe route connecting two input point sources.

application of continuity equation at various nodal points. The nodal point j_T that receives discharges from both the ends (connecting pipes) can be thus determined.

Thus, the route can be separated at j_T and two different systems are produced. Each one is designed separately by minimizing the system cost. For the design of the first system, the following cost function has to be minimized:

$$F_1 = \sum_{i=1}^{j_T} k_m L_i D_i^m + \rho g k_T Q_{T1} h_{01}, \qquad (12.4)$$

subject to the constraint

$$h_0 + z_0 - z_{j_T} - \sum_{i=1}^{j_T} \frac{8}{\pi^2 g D_i^5} f_i L_i Q_i^2 = H, \qquad (12.5)$$

where h_0 is the pumping head required at input point 0. The optimal diameter D_i^* is obtained by using Eq. (7.11b), which is rewritten as

$$D_i^* = \left(\frac{40 \rho k_T f_i Q_{T1} Q_i^2}{\pi^2 m k_m} \right)^{\frac{1}{m+5}}. \qquad (12.6)$$

The corresponding pumping head h_{01}^* is obtained using Eq. (7.12), which is also rewritten as

$$h_{01}^* = z_{j_T} + H - z_0 + \frac{8}{\pi^2 g} \left(\frac{\pi^2 m k_m}{40 \rho k_T Q_{T1}} \right)^{\frac{-5}{m+5}} \sum_{i=1}^{j_T} L_i \left(f_i Q_i^2 \right)^{\frac{m}{m+5}}. \qquad (12.7)$$

Substituting D_i and h_{01}^* from Eqs. (12.6) and (12.7) into Eq. (12.4), the minimum objective function

$$F_1^* = \left(1 + \frac{m}{5}\right) k_m \sum_{i=1}^{j_T} L_i \left(\frac{40 \, k_T \rho f_i Q_{T1} Q_i^2}{\pi^2 m k_m}\right)^{\frac{m}{m+5}} + k_T \rho g Q_{T1} \left(z_{j_T} + H - z_0\right). \quad (12.8)$$

Similarly, the design parameters of the second system are

$$D_i^* = \left(\frac{40 \rho k_T f_i Q_{T2} Q_i^2}{\pi^2 m k_m}\right)^{\frac{1}{m+5}} \quad (12.9)$$

$$h_{02}^* = z_{j_T} + H - z_{i_L} + \frac{8}{\pi^2 g} \left(\frac{\pi^2 m k_m}{40 \rho k_T Q_{T2}}\right)^{-\frac{5}{m+5}} \sum_{i=j_T+1}^{i_L} L_i \left(f_i Q_i^2\right)^{\frac{m}{m+5}} \quad (12.10)$$

$$F_2^* = \left(1 + \frac{m}{5}\right) k_m \sum_{i=j_T+1}^{i_L} L_i \left(\frac{40 \, k_T \rho f_i Q_{T2} Q_i^2}{\pi^2 m k_m}\right)^{\frac{m}{m+5}} + k_T \rho g Q_{T2} \left(z_{j_T} + H - z_{i_L}\right) \quad (12.11)$$

Thus, the optimal cost of the route for an arbitrary distribution of input point discharge is found to be

$$\text{Optimal system cost} = \text{Cost of first system} \quad + \quad \text{Cost of second system} \quad (12.12)$$
$$\text{having input head } h_{01} \qquad \text{having input head } h_{02}$$

which can be denoted as

$$F^* = F_1^* + F_2^*. \quad (12.13)$$

For an assumed value of Q_{T1}, F^* can be obtained for known values of k_m, m, k_T, ρ, f, and H. By varying Q_{T1}, the optimal value of F^* can be obtained. This minimum value corresponds with the optimal route design. For the optimal design, the minimum discharge flowing in a pipe link can be obtained. This link is the weakest link i_w in the system. This route can be clipped at the midpoint of this link. Thus, the system can be converted into two separate systems by introducing two nodes $i_L + 1$ and $i_L + 2$ at the midpoint of the weakest pipe link and redesignating the newly created pipe link to be $i_L + 1$ (see Fig. 12.2b). The newly introduced nodes may have mean elevations of their adjacent nodes. The population load is also equally divided on both pipes i_w and $i_L + 1$.

Following the procedure for route 4, the configuration of the network in Fig. 12.1 modifies to Fig. 12.3. This modification changes the withdrawals at the end points of the clipped link. It also affects the Tables 12.1, 12.3, 12.4, 12.5, 12.6, and 12.7.

Using the modified tables of the network geometry, the routes can now be regenerated, and the route connected to the input point 1 and having minimum number of pipe links is clipped by the procedure described earlier, and Tables 12.1, 12.3, 12.4, 12.5, 12.6 and 12.7 are modified again. Tables 12.4, 12.5, 12.6, and 12.7 are based on the revised flow estimations. The pipe flow analysis is described in Chapter 3. This

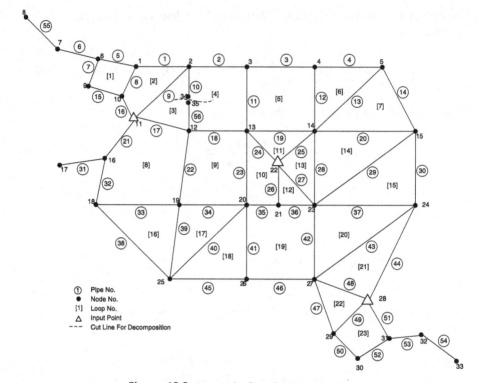

Figure 12.3. Network after clipping one pipe.

procedure is repeated untill the system fed by input point 1 is separated from the remaining network. This can be ascertained from the updated Table 12.5. The system fed by input point 1 is separated if at a node two or more input points supply flow, none of these input points should be input point 1. That is,

$$\text{For } N_n(j) > 1 \text{ Considering } j = 1 \text{ to } j_L{:}I_n(j, \ell) \neq 1 \text{ for } \ell = 1 \text{ to } N_n(j). \qquad (12.14)$$

Otherwise, the Tables 12.1, 12.3, 12.4, 12.5, 12.6, and 12.7 are updated and the Criteria (12.14) is applied again. The procedure is repeated until the system is separated. Figure 12.4 shows the successive progress for the algorithm.

Once the network connected to point 1 is separated, the remaining part of the network is renumbered, and Tables 12.1, 12.2, 12.3, 12.4, 12.5, 12.6, 12.7 and pipe discharges table similar to Table 3.5 are regenerated taking the remaining part of the network as a newly formed system. The process of selecting the weakest link and its clipping is repeated until all the input points are separated (Fig. 12.5).

After the separation of each input point, all the subsystems are designed separately by renumbering the subsystem network, and finally the decision parameters are produced as per the original geometry of the network.

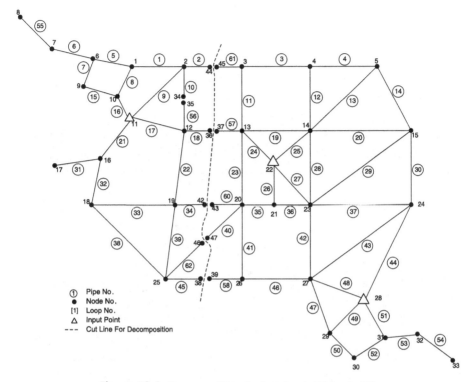

Figure 12.4. Decomposition for input point 1 (node 11).

12.1.6. Synthesis of Network

A multiple input system having i_L number of pipes after separation has to be synthesized separately for each subsystem of single input point. The network of subsystem 1 connected with input point 1 has to be synthesized first. All the pipes and nodes of this subsystem are renumbered such that the total number of pipes and nodes in this subsystem is i_{L1} and j_{L1}, respectively. The cost function F_1 for subsystem 1 is written as

$$F_1 = \sum_{i=1}^{i_{L1}} (c_{i1}x_{i1} + c_{i2}x_{i2}) + \rho g k_{T1} Q_{T1} h_{01}, \qquad (12.15a)$$

where Q_{T1} = the total water demand for subsystem 1. F_1 has to be minimized subject to the constraints as already described for pumping systems. The cost function and constraints constituting the LP problem can be solved using a simplex algorithm. The algorithm for selecting the starting basis has also been described in Chapters 10 and 11 for pumping systems. After selecting a suitable starting basis, the LP problem can

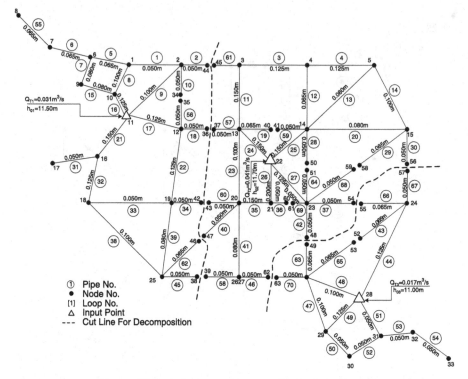

Figure 12.5. A decomposed water distribution system.

be solved. The process of synthesis is repeated for all the subsystems. The total system cost is

$$F = F_1 + F_2 + \cdots + F_{n_L}. \tag{12.15b}$$

The pipe link diameters, pumping and booster heads thus obtained for each subsystem are restored as per the original geometry of the network.

A three-input pumping system having a design population of 20,440 (see Fig. 12.1) has been separated into three subsystems (see Fig. 12.5) using the algorithm described herein. Each subsystem is synthesized separately, and decision parameters are produced as per the original geometry. The pipe link diameters, pumping heads, and input discharges are shown in Fig. 12.5. Thus, the decomposed subsystems can be designed separately as independent systems, and the weak links can then be restored at minimal prescribed diameters.

12.2. OPTIMAL WATER SUPPLY ZONE SIZE

Water distribution systems are generally designed with fixed configuration, but there must also be an optimal geometry to meet a particular water supply demand. A large area can be served by designing a single water supply system or it can be divided into a number of small zones each having an individual pumping and network system. The choice is governed by economic and reliability criteria. The economic criterion pertains to minimizing the water supply cost per unit discharge. The optimum zone size depends upon the network geometry, population density, topographical features, and the establishment cost E for a zonal unit. The establishment cost is described in Chapter 4.

Given an input point configuration and the network geometry, Section 12.1 describes an algorithm to decompose the water supply network into the zones under influence of each input point. However, in such decomposition, there is no cost consideration.

In this section, a method has been described to find the optimal area of a water supply zone. The area of a water supply network can be divided into various zones of nearly equal sizes. The pumping station (or input point) can be located as close to the center point as possible. It is easy to design these zones as separate entities and provide nominal linkage between the adjoining zones.

12.2.1. Circular Zone

12.2.1.1. Cost of Distribution System.
Considering a circular area of radius L, the area may be served by a radial distribution system having a pumping station located at the center and n equally spaced branches of length L as shown in Fig. 12.6. Assuming $\sigma =$ peak water demand per unit area $(\mathrm{m}^3/\mathrm{s}/\mathrm{m}^2)$, the peak discharge pumped in each branch is $\pi \sigma L^2/n$. Further, considering continuous withdrawal, the discharge withdrawn

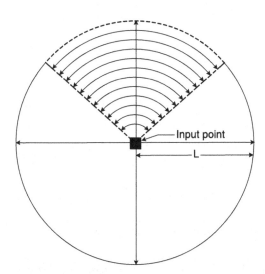

Figure 12.6. Circular zone.

in the length x is $\pi \sigma x^2/n$. Thus, the discharge Q flowing at a distance x from the center is the difference of these two expressions. That is,

$$Q = \frac{\pi \sigma L^2}{n}\left(1 - \xi^2\right), \tag{12.16}$$

where $\xi = x/L$. Initially considering continuously varying diameter, and using the Darcy–Weisbach equation with constant friction factor, the pumping head h_0 is

$$h_0 = \int_0^1 \frac{8fLQ^2}{\pi^2 gD^5}\,d\xi + z_L + H - z_0, \tag{12.17}$$

where D = branch pipe diameter, and z_0 and z_L = elevations of pumping station and the terminal end of the radial branch, respectively. For optimality, D should decrease with the increase in ξ, and finally at $\xi = 1$ the diameter should be zero. Such a variation of D is impractical, as D cannot be less than a minimum permissible diameter. Thus, it is necessary that the diameter D will remain constant throughout the pipe length, whereas the discharge Q will vary according to Eq. (12.16). Using Eq. (12.16), Eq. (12.17) is changed to

$$h_0 = \frac{64fL^5\sigma^2}{15gn^2D^5} + z_L + H - z_0. \tag{12.18}$$

The pumping cost F_p is written as

$$F_p = \pi k_T \rho g \sigma L^2 h_0, \tag{12.19}$$

Using Eq. (12.18), Eq. (12.19) is modified to

$$F_p = \frac{64\pi k_T \rho f \sigma^3 L^7}{15n^2 D^5} + \pi k_T \rho g \sigma L^2 (z_L + H - z_0). \tag{12.20}$$

The cost function F_m of the radial pipelines is written as

$$F_m = nk_m LD^m. \tag{12.21}$$

Adding Eqs. (12.20) and (12.21), the distribution system cost F_d is obtained as

$$F_d = nk_m LD^m + \frac{64\pi k_T \rho f \sigma^3 L^7}{15n^2 D^5} + \pi k_T \rho g \sigma L^2 (z_L + H - z_0). \tag{12.22}$$

For optimality, differentiating Eq. (12.22) with respect to D and equating it to zero and simplifying gives

$$D = \left(\frac{64\pi k_T \rho f \sigma^3 L^6}{3mn^3 k_m}\right)^{\frac{1}{m+5}}. \tag{12.23}$$

Using Eqs. (12.18) and (12.23), the pumping head works out to be

$$h_0 = \frac{64fL^5\sigma^2}{15gn^2}\left(\frac{3mn^3 k_m}{64\pi k_T \rho f \sigma^3 L^6}\right)^{\frac{5}{m+5}} + z_L + H - z_0. \tag{12.24}$$

Using Eqs. (12.19) and (12.24), the pumping cost is obtained as

$$F_p = \frac{64\pi k_T \rho f L^7 \sigma^3}{15n^2}\left(\frac{3mn^3 k_m}{64\pi k_T \rho f \sigma^3 L^6}\right)^{\frac{5}{m+5}} + \pi k_T \rho g \sigma L^2 (z_L + H - z_0). \tag{12.25}$$

Similarly, using Eqs. (12.21) and (12.23), the pipe cost is obtained as

$$F_m = nk_m L\left(\frac{64\pi k_T \rho f \sigma^3 L^6}{3mn^3 k_m}\right)^{\frac{m}{m+5}}. \tag{12.26}$$

Adding Eqs. (12.25) and (12.26), the cost of the distribution system is obtained as

$$F_d = nk_m L\left(1 + \frac{m}{5}\right)\left(\frac{64\pi k_T \rho f \sigma^3 L^6}{3mn^3 k_m}\right)^{\frac{m}{m+5}} + \pi k_T \rho g \sigma L^2 (z_L + H - z_0). \tag{12.27}$$

12.2.1.2. Cost of Service Connections. The cost of connections is also included in total system cost, which has been described in Chapter 4. The frequency and the length of the service connections will be less near the center and more toward the outskirts. Considering q_s as the discharge per ferrule through a service main of diameter D_s, the number of connections per unit length n_s at a distance x from the center is

$$n_s = \frac{2\pi\sigma x}{nq_s} \tag{12.28}$$

The average length L_s of the service main is

$$L_s = \pi x/n \tag{12.29}$$

The cost of the service connections F_s is written as

$$F_s = 2n \int_0^L k_s n_s L_s D_s^{m_s} \, dx, \tag{12.30}$$

where k_s and m_s = ferrule cost parameters. Using Eqs. (12.28) and (12.29), Eq. (12.30) is changed to

$$F_s = \frac{2\pi^2 \, k_s D_s^{m_s} \sigma L^3}{3 n q_s} \tag{12.31}$$

12.2.1.3. Cost per Unit Discharge of the System.　Adding Eqs. (12.27) and (12.31) and the establishment cost E, the overall cost function F_0 is

$$F_0 = n k_m L \left(1 + \frac{m}{5}\right) \left(\frac{64 \pi k_T \rho f \sigma^3 L^6}{3 m n^3 k_m}\right)^{\frac{m}{m+5}} + \frac{2\pi^2 \, k_s D_s^{m_s} \sigma L^3}{3 n q_s} + E$$
$$+ \pi k_T \rho g \sigma L^2 (z_L + H - z_0). \tag{12.32}$$

Dividing Eq. (12.32) by the discharge pumped $Q_T = \pi \sigma L^2$, the system cost per unit discharge F is

$$F = \left(1 + \frac{m}{5}\right) \frac{n k_m}{\pi \sigma L} \left(\frac{64 \pi k_T \rho f \sigma^3 L^6}{3 m n^3 k_m}\right)^{\frac{m}{m+5}} + \frac{2 \pi k_s D_s^{m_s} L}{3 n q_s} + \frac{E}{\pi \sigma L^2}$$
$$+ k_T \rho g (z_L + H - z_0). \tag{12.33}$$

12.2.1.4. Optimization.　As the last term of Eq. (12.33) is constant, it will not enter into the optimization process. Dropping this term, the objective function reduces to F_1 given by

$$F_1 = \left(1 + \frac{m}{5}\right) \frac{n k_m}{\pi \sigma L} \left(\frac{64 \pi k_T \rho f \sigma^3 L^6}{3 m n^3 k_m}\right)^{\frac{m}{m+5}} + \frac{2 \pi k_s D_s^{m_s} L}{3 n q_s} + \frac{E}{\pi \sigma L^2}. \tag{12.34}$$

The variable $n \geq 3$ is an integer. Considering n to be fixed, Eq. (12.34) is in the form of a *posynomial* (positive polynomial) in the design variable L. Thus, minimization of Eq. (12.34) reduces to a geometric programming with single degree of difficulty (Duffin et al., 1967). The contributions of various terms of Eq. (12.34) are described by the weights w_1, w_2, and w_3 given by

$$w_1 = \left(1 + \frac{m}{5}\right) \frac{nk_m}{\pi\sigma LF_1} \left(\frac{64\pi k_T \rho f \sigma^3 L^6}{3mn^3 k_m}\right)^{\frac{m}{m+5}} \tag{12.35}$$

$$w_2 = \frac{2\pi k_s D_s^{m_s} L}{3nq_s F_1} \tag{12.36}$$

$$w_3 = \frac{E}{\pi\sigma L^2 F_1}. \tag{12.37}$$

The dual objective function F_2 of Eq. (12.34) is

$$F_2 = \left[\left(1 + \frac{m}{5}\right) \frac{nk_m}{\pi\sigma Lw_1} \left(\frac{64\pi k_T \rho f \sigma^3 L^6}{3mn^3 k_m}\right)^{\frac{m}{m+5}}\right]^{w_1} \left(\frac{2\pi k_s D_s^{m_s} L}{3nq_s w_2}\right)^{w_2} \left(\frac{E}{\pi\sigma L^2 w_3}\right)^{w_3}. \tag{12.38}$$

The orthogonality condition for Eq. (12.38) is

$$\frac{5(m-1)}{m+5} w_1^* + w_2^* - 2w_3^* = 0, \tag{12.39}$$

whereas the normality condition of Eq. (12.38) is

$$w_1^* + w_2^* + w_3^* = 1, \tag{12.40}$$

Solving Eqs. (12.39) and (12.40) in terms of w_1^*, the following equations are obtained:

$$w_2^* = \frac{2}{3} - \frac{7m+5}{3(m+5)} w_1^* \tag{12.41}$$

$$w_3^* = \frac{1}{3} - \frac{2(5-2m)}{3(m+5)} w_1^*. \tag{12.42}$$

Substituting Eqs. (12.41) and (12.42) in Eq. (12.38) and using $F_1^* = F_2^*$, the optimal cost per unit discharge is

$$
F_1^* = \frac{2\pi(m+5)k_sD_s^{m_s}}{\left[2(m+5)-(7m+5)w_1^*\right]nq_s} \left[\frac{2(m+5)-(7m+5)w_1^*}{m+5-2(5-2m)w_1^*}\frac{3nq_sE}{2\pi^2\sigma k_sD_s^{m_s}}\right]^{\frac{1}{3}}
$$

$$
\times \left\{ \frac{nk_m}{15E}\left(\frac{64\pi k_T\rho f\sigma^3}{3mn^3 k_m}\right)^{\frac{m}{m+5}} \frac{m+5-2(5-2m)w_1^*}{w_1^*} \right.
$$

$$
\left. \times \left[\frac{2(m+5)-(7m+5)w_1^*}{m+5-2(5-2m)w_1^*}\frac{3nq_sE}{2\pi^2\sigma k_sD_s^{m_s}}\right]^{\frac{7m+5}{3(m+5)}} \right\}^{w_1^*} . \tag{12.43}
$$

Following Swamee (1995), Eq. (12.43) is optimal when the factor containing the exponent w_1^* is unity. Thus, denoting the parameter P by

$$
P = \frac{15E}{nk_m}\left(\frac{3mn^3 k_m}{64\pi k_T\rho f\sigma^3}\right)^{\frac{m}{m+5}}\left(\frac{2\pi^2\sigma k_sD_s^{m_s}}{3nq_sE}\right)^{\frac{7m+5}{3(m+5)}}, \tag{12.44}
$$

the optimality condition is

$$
P = \frac{m+5-2(5-2m)w_1^*}{w_1^*}\left[\frac{2(m+5)-(7m+5)w_1^*}{m+5-2(5-2m)w_1^*}\right]^{\frac{7m+5}{3(m+5)}}. \tag{12.45}
$$

For various w_1^*, corresponding values of P are obtained by Eq. (12.45). Using the data so obtained, the following equation is fitted:

$$
w_1^* = \frac{2(m+5)}{7m+5}\left[1+\left(\frac{P}{0.5+7m}\right)^{1.15}\right]^{-0.8}. \tag{12.46}
$$

The maximum error involved in the use of Eq. (12.46) is about 1.5%. Using Eqs. (12.43) and (12.45), the optimal objective function is

$$
F_1^* = \frac{2\pi(m+5)k_sD_s^{m_s}}{\left[2(m+5)-(7m+5)w_1^*\right]nq_s}\left[\frac{2(m+5)-(7m+5)w_1^*}{m+5-2(5-2m)w_1^*}\frac{3nq_sE}{2\pi^2\sigma k_sD_s^{m_s}}\right]^{\frac{1}{3}}, \tag{12.47}
$$

where w_1^* is given by Eq. (12.46). Combining Eqs. (12.36), (12.41), and (12.47), the optimal zone size L^* is

$$L^* = \left[\frac{2(m+5) - (7m+5)w_1^*}{m+5 - 2(5-2m)w_1^*} \frac{3nq_sE}{2\pi^2\sigma k_s D_s^{m_s}} \right]^{\frac{1}{3}}. \qquad (12.48)$$

Equation (12.48) reveals that the size L^* is a decreasing function of σ (which is proportional to the population density). Thus, a larger population density will result in a smaller circular zone size.

12.2.2. Strip Zone

Equations (12.28) and (12.29) are not applicable for $n = 2$ and 1, as for both these cases the water supply zone degenerates to a strip. Using Fig. 12.7 a,b, the pipe discharge is

$$Q = 2\sigma BL(1 - \xi), \qquad (12.49)$$

where B = half the zone width, and L = length of zone for $n = 1$ and half the zone length for $n = 2$. Using the Darcy–Weisbach equation, the pumping head is

$$h_0 = \frac{32f\sigma^2 B^2 L^3}{3\pi^2 g D^5} + z_L + H - z_0. \qquad (12.50)$$

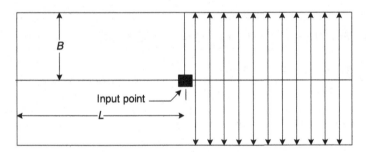

(a) Two branch strip zone - input point in the middle

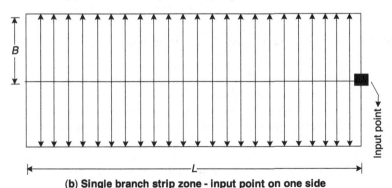

(b) Single branch strip zone - input point on one side

Figure 12.7. Strip zone.

For $n = 2$, the pumping discharge $Q_T = 4BL\sigma$. Thus, the pumping cost F_p is

$$F_p = 4\,k_T\rho g\sigma BL h_0 \tag{12.51}$$

Combining Eqs. (12.50) and (12.51), the following equation was obtained:

$$F_p = \frac{128\,k_T\rho f\sigma^3 B^3 L^4}{3\pi^2 n^2 D^5} + 4\,k_T\rho g\sigma BL(z_L + H - z_0). \tag{12.52}$$

The pipe cost function F_m of the two diametrically opposite radial pipelines is

$$F_m = 2k_m LD^m. \tag{12.53}$$

Summing up Eqs. (12.52) and (12.53), the distribution system cost F_d is

$$F_d = 2\,k_m LD^m + \frac{128\,k_T\rho f\sigma^3 B^3 L^4}{3\pi^2 n^2 D^5} + 4\,k_T\rho g\sigma BL(z_L + H - z_0) \tag{12.54}$$

Differentiating Eq. (12.54) with respect to D and equating it to zero and simplifying,

$$D = \left(\frac{320\,k_T\rho f\sigma^3 B^3 L^3}{3\pi^2 m k_m}\right)^{\frac{1}{m+5}} \tag{12.55}$$

Combining Eqs. (12.50), (12.51), and (12.55), pumping cost is

$$F_p = \frac{128\,k_T\rho f\sigma^3 B^3 L^4}{3\pi^2}\left(\frac{3\pi^2 m k_m}{320\,k_T\rho f\sigma^3 B^3 L^3}\right)^{\frac{5}{m+5}} + 4\,k_T\rho g\sigma BL(z_L + H - z_0). \tag{12.56}$$

Using Eqs. (12.53) and (12.55), the pipe cost is

$$F_m = 2\,k_m L\left(\frac{320\,k_T\rho f\sigma^3 B^3 L^3}{3\pi^2 m k_m}\right)^{\frac{m}{m+5}}. \tag{12.57}$$

Adding Eqs. (12.56) and (12.57), the cost of distribution system is

$$F_d = 2\,k_m L\left(1 + \frac{m}{5}\right)\left(\frac{320\,k_T\rho f\sigma^3 B^3 L^3}{3\pi^2 m k_m}\right)^{\frac{m}{m+5}} + 4\,k_T\rho g\sigma BL(z_L + H - z_0). \tag{12.58}$$

The number of ferrule connections $N_s = 4BL\sigma/q_s$. Thus, the cost of service connection is

$$F_s = \frac{4\sigma B^2 L k_s D_s^{m_s}}{q_s}. \tag{12.59}$$

Adding Eqs. (12.58) and (12.59) and E, the overall cost function was obtained as

$$F_d = 2\,k_m L\left(1 + \frac{m}{5}\right)\left(\frac{320\,k_T \rho f \sigma^3 B^3 L^3}{3\pi^2 m k_m}\right)^{\frac{m}{m+5}}$$

$$+ \frac{4\sigma B^2 L k_s D_s^{m_s}}{q_s} + E + 4\,k_T \rho g \sigma B L (z_L + H - z_0). \tag{12.60}$$

Dividing Eq. (12.60) by $4\sigma BL$, the system cost per unit discharge F is

$$F = \frac{k_m}{2\sigma B}\left(1 + \frac{m}{5}\right)\left(\frac{320\,k_T \rho f \sigma^3 B^3 L^3}{3\pi^2 m k_m}\right)^{\frac{m}{m+5}}$$

$$+ \frac{k_s D_s^{m_s} B}{q_s} + \frac{E}{4\sigma BL} + k_T \rho g (z_L + H - z_0). \tag{12.61}$$

Following the procedure described for $n = 2$, it is found that for $n = 1$, Eq. (12.55) remained unchanged, whereas Eqs. (12.60) and (12.61) respectively change to

$$F_d = k_m L\left(1 + \frac{m}{5}\right)\left(\frac{320\,k_T \rho f \sigma^3 B^3 L^3}{3\pi^2 m k_m}\right)^{\frac{m}{m+5}}$$

$$+ \frac{2\sigma B^2 L k_s D_s^{m_s}}{q_s} + E + 2\,k_T \rho g \sigma B L (z_L + H - z_0) \tag{12.62}$$

$$F = \frac{k_m}{2\sigma B}\left(1 + \frac{m}{5}\right)\left(\frac{320\,k_T \rho f \sigma^3 B^3 L^3}{3\pi^2 m k_m}\right)^{\frac{m}{m+5}}$$

$$+ \frac{k_s D_s^{m_s} B}{q_s} + \frac{E}{2\sigma BL} + k_T \rho g (z_L + H - z_0). \tag{12.63}$$

Thus for $n \le 2$, Eqs. (12.61) and (12.63) are generalized as

$$F = \frac{k_m}{2\sigma B}\left(1 + \frac{m}{5}\right)\left(\frac{320\,k_T \rho f \sigma^3 B^3 L^3}{3\pi^2 m k_m}\right)^{\frac{m}{m+5}}$$

$$+ \frac{k_s D_s^{m_s} B}{q_s} + \frac{E}{2n\sigma BL} + k_T \rho g (z_L + H - z_0). \tag{12.64}$$

The last term of Eq. (12.64) is constant. Dropping this term, Eq. (12.64) reduces to

$$F_1 = \frac{k_m}{2\sigma B}\left(1 + \frac{m}{5}\right)\left(\frac{320\,k_T \rho f \sigma^3 B^3 L^3}{3\pi^2 m k_m}\right)^{\frac{m}{m+5}}$$

$$+ \frac{k_s D_s^{m_s} B}{q_s} + \frac{E}{2n\sigma BL}. \tag{12.65}$$

Considering B and L as design variables, the minimization of Eq. (12.65) boils down to a geometric programming with zero degree of difficulty (Wilde and Beightler, 1967). The weights w_1, w_2, and w_3 pertaining to Eq. (12.65) were given by

$$w_1 = \frac{k_m}{2\sigma B F_1}\left(1 + \frac{m}{5}\right)\left(\frac{320\, k_T \rho f \sigma^3 B^3 L^3}{3\pi^2 m k_m}\right)^{\frac{m}{m+5}} \tag{12.66}$$

$$w_2 = \frac{k_s D_s^{m_s} B}{q_s F_1} \tag{12.67}$$

$$w_3 = \frac{E}{2n\sigma B L F_1} \tag{12.68}$$

The dual objective function F_2 of Eq. (12.65) is

$$F_2 = \left[\frac{k_m}{2\sigma B w_1}\left(1 + \frac{m}{5}\right)\left(\frac{320\, k_T \rho f \sigma^3 B^3 L^3}{3\pi^2 m k_m}\right)^{\frac{m}{m+5}}\right]^{w_1}\left(\frac{k_s D_s^{m_s} B}{q_s w_2}\right)^{w_2}\left(\frac{E}{2n\sigma B L w_3}\right)^{w_3}. \tag{12.69}$$

The orthogonality conditions for Eq. (12.69) are

$$B: \quad -\frac{5 - 2m}{m+5}w_1^* + w_2^* - w_3^* = 0 \tag{12.70}$$

$$L: \quad \frac{3m}{m+5}w_1^* - w_3^* = 0. \tag{12.71}$$

On the other hand, the normality condition for Eq. (12.69) is

$$w_1^* + w_2^* + w_3^* = 1 \tag{12.72}$$

Solving (12.70)–(12.72), the following optimal weights were obtained:

$$w_1^* = \frac{m+5}{5(m+2)} \tag{12.73}$$

$$w_2^* = \frac{m+5}{5(m+2)} \tag{12.74}$$

$$w_3^* = \frac{3m}{5(m+2)}. \tag{12.75}$$

Equations (12.73) and (12.74) indicate that in a strip zone, the optimal contribution of water distribution network and service connections are equal. Thus, for $m = 1$, the optimal weights are in the proportion 2:2:1. With the increase in m, the optimal weights even out. Thus, for maximum $m = 1.75$, the proportion of weights becomes 1.286:1.286:1. Further, a similar procedure gives the following equation for optimal objective function for a strip zone:

$$F_1^* = (m+2)k_m \left[\frac{5 k_s D_s^{m_s}}{2(m+5)k_m \sigma q_s} \right]^{\frac{m+5}{5(m+2)}} \left(\frac{5000 k_T \rho f E^3}{81 \pi^2 n^3 m^4 k_m^4} \right)^{\frac{m}{5(m+2)}}. \tag{12.76}$$

Using (12.76) for $n = 1$ and 2, the ratio of optimal objective functions is

$$\frac{F_{1,n=1}^*}{F_{1,n=2}^*} = 2^{\frac{3m}{5(m+2)}}. \tag{12.77}$$

Thus, for the practical range $1 \leq m \leq 1.75$, it is 15% to 21% costlier to locate the input point at the end of a strip zone. Using Eqs. (12.67), (12.74), and (12.76), the optimum strip width B^* was found to be

$$B^* = \frac{(m+5)q_s k_m}{5 k_s D_s^{m_s}} \left[\frac{5 k_s D_s^{m_s}}{2(m+5)k_m \sigma q_s} \right]^{\frac{m+5}{5(m+2)}} \left(\frac{5000 k_T \rho f E^3}{81 \pi^2 n^3 m^4 k_m^4} \right)^{\frac{m}{5(m+2)}}. \tag{12.78}$$

According to Eq. (12.78) for $n = 1$ and 2, the optimal strip width ratio is the same as the cost ratio. Thus, the optimal strip width is the 15% to 21% larger if the input point is at one end of the strip. Similarly, using Eqs. (12.68), (12.75), (12.76), and (12.78), the optimum length L^* was obtained as

$$L^* = \frac{25 k_s D_s^{m_s} E}{6mn(m+5)k_m^2 \sigma q_s} \left[\frac{2(m+5)k_m \sigma q_s}{5 k_s D_s^{m_s}} \right]^{\frac{2(m+5)}{5(m+2)}} \left(\frac{81 \pi^2 n^3 m^4 k_m^4}{5000 k_T \rho f E^3} \right)^{\frac{2m}{5(m+2)}}. \tag{12.79}$$

Thus, Eq. (12.79) for $n = 1$ and 2 gives the ratio of optimal strip lengths as

$$\frac{L_1^*}{L_2^*} = 2^{\frac{10-m}{5(m+2)}}. \tag{12.80}$$

For the practical range $1 \leq m \leq 1.75$, the zone length is 23% to 36% longer if the input point is located at the end of the strip zone. Equations (12.78) and (12.80) reveal that both B^* and L^* are inverse functions of σ. On the other hand, use of smoother pipes will reduce the zone width and increase its length.

Example 12.1. Find the optimal circular and strip zone sizes for the following data: $m = 1.2$, $k_T/k_m = 0.05$, $k_s/k_m = 3.0$, $E/k_m = 7000$ (ratios in SI units), $\sigma = 10^{-7}$ m/s, $q_s = 0.001 \, \text{m}^3/\text{s}$, $D_s = 0.025 \, \text{m}$, $m_s = 1.4$, and $f = 0.02$.

Solution. First, for a strip zone using Eqs. (12.73), (12.74), and (12.75) the optimal weights are $w_1^* = w_2^* = 0.3875$, and $w_3^* = 0.2250$. Adopting $n = 1$ for the input point at one end, and using Eq. (12.76), $F_1^* = 27,810\,k_m$. Using Eqs. (12.67) and (12.74), $B^* = 628$ m. Further, using Eqs. (12.68) and (12.75), $L^* = 8900$ m covering an area A^* of 11.19 km^2. Similarly, adopting $n = 2$ for centrally placed input point, the design variables are $B^* = 537$ m, $L^* = 6080$ m, and $A^* = 13.05$ km^2, yielding $F_1^* = 23,749\,k_m$.

For a circular zone with $n = 3$ and using Eq. (12.44), $P = 73.61$. Further, using Eq. (12.46), $w_1^* = 0.1238$; using Eqs. (12.41) and (12.42), $w_2^* = 0.5774$, and $w_3^* = 0.2987$. Using Eq. (12.47), $F_1^* = 31,763\,k_m$; and using Eq. (12.48), $L^* = 1532$ m. The corresponding area $A^* = 7.37$ km^2. Similar calculations for $n > 3$ can be made. The calculations for different n values are depicted in Table 12.8.

A perusal of Table 12.8 shows that for rectangular geometry with $n = 1$ and 2, contribution of the main pipes is about 39% ($w_1^* = 0.3875$) of the total cost. On the other hand, for circular geometry with $n = 3$, the contribution of radial pipes to the total cost is considerably less ($w_1^* = 0.1238$), and this ratio increases slowly with the number of radial lines. Thus, from a consumer point of view, the rectangular zone is superior as the consumer has to bear about 39% of the total cost ($w_2^* = 0.3875$) in comparison with the radial zone, in which his share increases to about 57%. Thus, for a circular zone, the significant part of the cost is shared by the service connections. If this cost has to be passed on to consumers, then the problem reduces considerably. Dropping the service connection cost, for a circular zone, Eq. (12.34) reduces to

$$F_1 = \left(1 + \frac{m}{5}\right)\frac{nk_m}{\pi \sigma L}\left(\frac{64\pi k_T \rho f \sigma^3 L^6}{3mn^3 k_m}\right)^{\frac{m}{m+5}} + \frac{E}{\pi \sigma L^2}. \tag{12.81}$$

Minimization of Eq. (12.81) is a problem of zero degree of difficulty yielding the following optimal weights:

$$w_1^* = \frac{2(m+5)}{7m+5} \tag{12.82}$$

$$w_3^* = \frac{5(m-1)}{7m+5}. \tag{12.83}$$

TABLE 12.8. Variation in Zone Size with Radial Loops

n	w_1^*	w_2^*	w_3^*	F_1^* ($/m^2)	L^* (m)	B^* (m)	A^* (km^2)
1	0.3875	0.3875	0.2250	27,810 k_m	8900	628	11.19
2	0.3875	0.3875	0.2250	23,749 k_m	6080	537	13.05
3	0.1238	0.5774	0.2987	31,763 k_m	1532		7.37
4	0.1626	0.5495	0.2878	27,437 k_m	1680		8.86
5	0.1993	0.5230	0.2473	24,733 k_m	1800		10.19
6	0.2339	0.4981	0.2679	22,897 k_m	1906		11.41

In the current problem for $m = 1.2$, the optimal weight $w_3^* = 0.0746$. That is, the share of establishment cost (in the optimal zone cost per cumec) is about 8%. The corresponding optimal cost and the zone size, respectively, are

$$F_1^* = \frac{(7m+5)nk_m}{10\pi\sigma}\left(\frac{64\pi k_T \rho f \sigma^3}{3mn^3 k_m}\right)^{\frac{2m}{7m+5}}\left[\frac{2E}{(m-1)nk_m}\right]^{\frac{5(m-1)}{7m+5}} \qquad (12.84)$$

$$L^* = \left(\frac{3mn^3 k_m}{64\pi k_T \rho f \sigma^3}\right)^{\frac{m}{7m+5}}\left[\frac{2E}{(m-1)nk_m}\right]^{\frac{m+5}{7m+5}}. \qquad (12.85)$$

By substituting $m = 1$ in Eq. (12.84), a thumb-rule for the optimal cost per cumec is obtained as

$$F_1^* = 1.2 k_m \left(\frac{64\, k_T \rho f n^3}{3\pi^5 k_m \sigma^3}\right)^{\frac{1}{6}}. \qquad (12.86)$$

In the foregoing developments, the friction factor f has been considered as constant. The variation of the friction factor can be considered iteratively by first designing the system with constant f and revising it by using Eq. (2.6a).

In the case of a circular zone, Table 12.8 shows that the zone area A gradually increases with the number of branches. However, the area remains less than that of a strip zone. Thus, a judicious value of A can be selected and the input points in the water distribution network area can be placed at its center. The locations of the input points are similar to optimal well-field configurations (Swamee et al., 1999). Keeping the input points as center and consistent with the pipe network geometry, the zones can be demarcated approximately as circles of diameter $2L$. These zones can be designed as independent entities and nominal connections provided for interzonal water transfer.

EXERCISES

12.1. Write the advantages of decomposing the large multi-input source network to small networks.

12.2. Analyze the network shown in Fig. 12.1 by increasing the population load on each link by a factor of 1.5 (Table 12.1). Use initial pipe diameters equal to 0.20 m. For known pipe discharges, develop Tables 12.4, 12.5, 12.7, and 12.7.

12.3. Write a code for selecting a weak link in the shortest route. Assume suitable parameters for the computation.

12.4. Find the optimal zone size for the following data: $m = 0.935$, $k_T/k_m = 0.07$, $k_s/k_m = 3.5$, $E/k_m = 8500$ (ratios in SI units), $\sigma = 10^{-7}$ m/s, $q_s = 0.001$ m^3/s, $D_s = 0.025$ m, $m_s = 1.2$, and $f = 0.02$.

REFERENCES

Duffin, R.J., Peterson, E.L., and Zener, C. (1967). *Geometric Programming*. John Wiley & Sons, New York.

Swamee, P.K. (1995). Design of sediment-transporting pipeline. *J. Hydr. Eng.* 121(1), 72–76.

Swamee, P.K., and Kumar, V. (2005). Optimal water supply zone size. *Journal of Water Supply: Research and Technology-AQUA 54, IWA* 179–187.

Swamee, P.K., and Sharma, A.K. (1990). Decomposition of a large water distribution system. *J. Env. Eng.* 116(2), 296–283.

Swamee, P.K., Tyagi, A., and Shandilya, V.K. (1999). Optimal configuration of a well-field. *Ground Water* 37(3), 382–386.

Wilde, D.J., and Beightler, C.S. (1967). *Foundations of Optimization*. Prentice Hall, Englewood Cliffs, NJ.

13

REORGANIZATION OF WATER DISTRIBUTION SYSTEMS

Water distribution systems are generally designed for a predecided time span called design period. It varies from 20 to 40 years, whereas the working life of pipelines varies from 60 to 120 years (see Table 5.2). It has been found that the pipelines laid more than 100 years ago are still in operation. For a growing demand scenario, it is always economic to design the system initially for a partial demand of a planning period and then to design an entirely new system or to reorganize the existing system when demand exceeds the capacity of the first system. Because it is costly to replace an existing system after its design period with an entirely new system, for increased demand the networks have to be reorganized using the existing pipelines. Additional

Design of Water Supply Pipe Networks. By Prabhata K. Swamee and Ashok K. Sharma
Copyright © 2008 John Wiley & Sons, Inc.

parallel pipelines are provided to enhance the delivery capacity of the existing system. Moreover, in order to cater to increased discharge and corresponding head loss, a pumping plant of an enhanced capacity would also be required. This process of network upgrading is termed the strengthening process.

The reorganization of a system also deals with the inclusion of additional demand nodes associated with pipe links and additional input source points at predetermined locations (nodes) to meet the increased system demand. Apart from the expansion to new areas, the water distribution network layout is also modified to improve the delivery capacity by adding new pipe links. Generally, 75% to 80% of pipe construction work pertains to reorganization of the existing system and only 20% to 25% constitutes new water supply system.

13.1. PARALLEL NETWORKS

For the increased demand in a parallel network, parallel pipelines along with the corresponding pumping plant are provided. The design of a parallel system is relatively simple.

13.1.1. Parallel Gravity Mains

Figure 13.1 depicts parallel gravity mains. The discharge Q_o flowing in the existing main of diameter D_o can be estimated using Eq. (2.21a), which is modified as

$$Q_o = -0.965D_o^2\left(gD_o\frac{z_0-H-z_L}{L}\right)^{\frac{1}{2}}\ln\left\{\frac{\varepsilon}{3.7D_o}+\frac{1.78v}{D_o}\left[\frac{L}{gD_o(z_0-H-z_L)}\right]^{\frac{1}{2}}\right\}, \quad (13.1)$$

and the discharge Q_n to be shared by the parallel main would be

$$Q_n = Q - Q_o, \quad (13.2)$$

where Q_o is given by Eq. (13.1), and Q = design discharge carried by both the mains jointly.

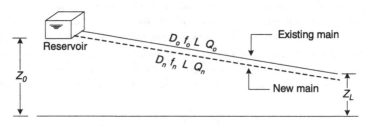

Figure 13.1. Parallel gravity mains.

The diameter for the parallel gravity main can be obtained from Eq. (2.22a), which after modifying and rewriting is

$$D_n = 0.66\left\{\varepsilon^{1.25}\left[\frac{LQ_n^2}{g(z_0 - H - z_L)}\right]^{4.75} + vQ_n^{9.4}\left[\frac{L}{g(z_0 - H - z_L)}\right]^{5.2}\right\}^{0.04}, \quad (13.3)$$

where Q_n is obtained by Eq. (13.2).

13.1.2. Parallel Pumping Mains

Parallel pumping mains are shown in Fig. 13.2. Equation (6.9) gives Q_o the discharge corresponding with the existing pumping main of diameter D_o, that is,

$$Q_o = \left(\frac{\pi^2 m k_m D_o^{m+5}}{40\, k_T \rho f_o}\right)^{\frac{1}{3}}, \quad (13.4)$$

where f_o = friction factor of the existing pumping main. The discharge Q_n to be shared by the parallel main is thus

$$Q_n = Q - \left(\frac{\pi^2 m k_m D_o^{m+5}}{40\, k_T \rho f_o}\right)^{\frac{1}{3}}. \quad (13.5)$$

Equations (6.9) and (13.5) obtain the following equation for the optimal diameter of the parallel pumping main D_n^*:

$$D_n^* = D_o\left[\left(\frac{40\, k_T \rho f_n Q^3}{\pi^2 m k_m D_o^{m+5}}\right)^{\frac{1}{3}} - \left(\frac{f_n}{f_o}\right)^{\frac{1}{3}}\right]^{\frac{3}{m+5}}. \quad (13.6)$$

As both f_o and f_n are unknown functions of D_n^*, Eq. (13.6) will not yield the diameter in a single step. The following iterative method may be used for obtaining D_n^*:

1. Assume f_o and f_n
2. Find Q_o using Eq. (13.4)

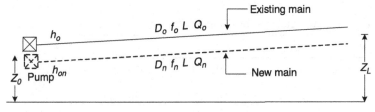

Figure 13.2. Parallel pumping mains.

3. Find Q_n using Eq. (13.5)
4. Find D_n^* using Eq. (13.6)
5. Find f_o and f_n using Eq. (2.6a) or (2.6c)
6. Repeat steps 2–5 until two successive values of D_n^* are close

Knowing D_n^*, the pumping head h_{0n}^* of a new pump can be obtained as

$$h_{0n}^* = \frac{8 f_n L Q_n^2}{\pi^2 g D_n^5} - z_0 + H + z_L. \qquad (13.7)$$

Example 13.1. Design a cast iron, parallel pumping main for a combined discharge of $0.4\,\mathrm{m^3/s}$. The existing main has a diameter of $0.45\,\mathrm{m}$ and is $5\,\mathrm{km}$ long. The pumping station is at an elevation of $235\,\mathrm{m}$, and the elevation of the terminal point is $241\,\mathrm{m}$. The terminal head is prescribed as $15\,\mathrm{m}$. Assume $k_T/k_m = 0.0135$.

Solution. Assuming $f_o = f_n = 0.02$, the various values obtained are tabulated in Table 13.1. A diameter of $0.45\,\mathrm{m}$ may be provided for the parallel pumping main. Using Eq. (13.7), the pumping head for the parallel main is obtained as $36.76\,\mathrm{m}$. Adopt $h_{0n} = 40\,\mathrm{m}$.

13.1.3. Parallel Pumping Distribution Mains

The existing and new parallel distribution mains are shown in Fig. 13.3. The optimal discharges in the existing pipe links can be obtained by modifying Eq. (7.11b) as

$$Q_{oi} = \left(\frac{\pi^2 m k_m D_{oi}^{m+5}}{40 \, k_T \rho f_{oi} Q_{To}} \right)^{\frac{1}{2}}, \qquad (13.8)$$

where Q_{To} = discharge in pipe $i = 1$, which can be estimated as

$$Q_{To} = \left(\frac{\pi^2 m k_m D_{o1}^{m+5}}{40 \, k_T \rho f_{o1}} \right)^{\frac{1}{3}}. \qquad (13.9)$$

Knowing the discharges Q_{oi}, the design discharges Q_{ni} in parallel pipes are obtained as

$$Q_{ni} = Q_i - Q_{oi} \qquad (13.10)$$

TABLE 13.1. Design Iterations for Pumping Main

Iteration No.	$Q_o\,(\mathrm{m^3/s})$	$Q_n\,(\mathrm{m^3/s})$	f_o	f_n	$D_n\,(\mathrm{m})$
1	0.1959	0.2041	0.0180	0.0179	0.4584
2	0.2028	0.1972	0.0180	0.0180	0.4457
3	0.2029	0.1971	0.0180	0.0181	0.4432
4	0.2029	0.1971	0.0180	0.0181	0.4430

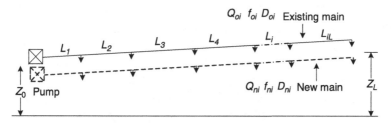

Figure 13.3. Parallel pumping distribution mains.

Using Eq. (7.11b), the optimal diameters D_{ni} are obtained as

$$D_{ni}^* = \left[\frac{40\,k_T\rho f_{ni}(Q_T - Q_{To})Q_{ni}^2}{\pi^2 m k_m} \right]^{\frac{1}{m+5}}, \qquad (13.11)$$

where Q_T = discharge pumped by both the mains (i.e., sum of $Q_{o1} + Q_{n1}$). The friction factors occurring in Eqs. (13.8), (13.9), and (13.11) can be corrected iteratively by using Eq. (2.6c). The pumping head h_{0n} of parallel distribution main is calculated using Eq. (7.12), which is written as

$$h_{0n}^* = z_L + H - z_0 + \frac{8}{\pi^2 g} \left(\frac{\pi^2 m k_m}{40\,k_T\rho Q_{Tn}} \right)^{\frac{5}{m+5}} \sum_{i=1}^{n} L_i \left(f_{ni} Q_{ni}^2 \right)^{\frac{m}{m+5}}. \qquad (13.12)$$

13.1.4. Parallel Pumping Radial System

A parallel radial system can be designed by obtaining the design discharges Q_{oij} flowing in the existing radial system. (See Fig. 8.6 for a radial pumping system.) Q_{oij} can be obtained iteratively using Eq. (8.18), which is rewritten as:

$$Q_{oij} = \left\{ \frac{\pi^2 m k_m D_{oij}^{m+5} \sum_{i=1}^{i_L} \left[\sum_{q=1}^{j_{Li}} L_{iq} \left(f Q_{oiq}^2 \right)^{\frac{m}{m+5}} \right]^{\frac{m+5}{5}}}{40\,k_T\rho f_{oij} Q_{To} \left[\sum_{q=1}^{j_{Li}} L_{iq} \left(f_{oiq} Q_{oiq}^2 \right)^{\frac{m}{m+5}} \right]^{\frac{m+5}{5}}} \right\}^{\frac{1}{2}}. \qquad (13.13)$$

Knowing Q_{oij}, the discharge in the parallel pipe link Q_{nij} is

$$Q_{nij} = Q_{ij} - Q_{oij}. \qquad (13.14)$$

Using Eq. (8.18), the diameters of the parallel pipe links D_n^* are

$$D_{nij}^* = \left\{ \frac{40\,k_T \rho f_{nij} Q_{Tn} Q_{nij} \left[\sum_{q=1}^{j_{Li}} L_{iq} \left(f_{niq} Q_{niq}^2 \right)^{\frac{m}{m+5}} \right]^{\frac{m+5}{5}}}{\pi^2 m k_m \sum_{i=1}^{i_L} \left[\sum_{q=1}^{j_{Li}} L_{iq} \left(f_{niq} Q_{niq}^2 \right)^{\frac{m}{m+5}} \right]^{\frac{m+5}{5}}} \right\}^{\frac{1}{m+5}}. \tag{13.15}$$

Using Eq. (8.17), the pumping head in the parallel pumping station is

$$h_{0n} = z_L + H - z_0 + \frac{8}{\pi^2 g} \left\{ \frac{\pi^2 m k_m}{40\,k_T \rho Q_{Tn}} \sum_{i=1}^{i_L} \left[\sum_{j=1}^{j_{Li}} L_{ij} \left(f_{nij} Q_{nij}^2 \right)^{\frac{m}{m+5}} \right]^{\frac{m+5}{5}} \right\}^{\frac{5}{m+5}}. \tag{13.16}$$

Using Eq. (8.19), the optimal cost of the parallel radial system is

$$F_n^* = \left(1 + \frac{m}{5}\right) k_m \left(\frac{40\,k_T \rho Q_{Tn}}{\pi^2 m k_m} \right)^{\frac{m}{m+5}} \left\{ \sum_{i=1}^{i_L} \left[\sum_{j=1}^{j_{Li}} L_{ij} \left(f_{nij} Q_{nij}^2 \right)^{\frac{m}{m+5}} \right]^{\frac{m+5}{5}} \right\}^{\frac{5}{m+5}}$$

$$+ k_T \rho g Q_{Tn}(z_L + H - z_0). \tag{13.17}$$

13.2. STRENGTHENING OF DISTRIBUTION SYSTEM

In water distribution systems, the provision of a combined pumping plant is desired from the reliability considerations. The pumping head for the parallel mains can be quite different than the existing pumping head; therefore, the existing pumping plant cannot be utilized. Thus in a strengthened network, the entire discharge has to be pumped to the new pumping head h_0.

13.2.1. Strengthening Discharge

If an existing system, originally designed for an input discharge Q_0, has to be improved for an increased discharge Q, the improvement can be accorded in the following ways: (1) increase the pumping capacity and pumping head and (2) strengthen the system by providing a parallel main. If Q is slightly greater than Q_0, then pumping option may be economic. For a large discharge, strengthening will prove to be more economic than by merely increasing the pumping capacity and pumping head. Thus, a rational criterion is required to estimate the minimum discharge Q_s beyond which a distribution main should be strengthened.

Though it is difficult to develop a criterion for Q_s for a water distribution network of an arbitrary geometry, an analytical study can be conducted for a single system like a pumping main. Broadly, the same criterion can be applied to a distribution system.

Thus, considering a horizontal pumping main of length L, the design discharge Q_0 of the existing pipe diameter (optimal) can be estimated using Eq. (6.9) as

$$D_o = \left(\frac{40\,k_T\rho f Q_0^3}{\pi^2 m k_m}\right)^{\frac{1}{m+5}}.$$

(13.18)

At the end of the design period, the same system can be used by enhancing the pumping capacity to cater to an enhanced demand Q_s. Thus, the total system cost in such case is

$$F_1 = kLD_o^m + \frac{8\rho k_T f_o L Q_s^3}{\pi^2 D_o^5}.$$

(13.19)

On the other hand, the same system can be reorganized by strengthening the existing main by providing a parallel additional main of diameter D_n. The head loss in parallel pipes for discharge Q_s:

$$h_f = \frac{8 f_o L Q_s^2}{\pi^2 g}\left[D_o^{2.5} + \left(\frac{f_o}{f_n}\right)^{0.5} D_n^{2.5}\right]^{-2}.$$

(13.20a)

For constant f, Eq. (13.20a) is reduced to

$$h_f = \frac{8 f L Q_s^2}{\pi^2 g}\left(D_o^{2.5} + D_n^{2.5}\right)^{-2}.$$

(13.20b)

Using Eq. (13.20b) the total system cost can be expressed as

$$F_2 = k_m L(D_o^m + D_n^m) + \frac{8\rho k_T f L Q_s^3}{\pi^2}\left(D_o^{2.5} + D_n^{2.5}\right)^{-2}.$$

(13.21)

The optimal diameter D_n^* is obtained by differentiating Eq. (13.21) with respect to D_n, setting $\partial F_2/\partial D_n = 0$ and rearranging terms. Thus,

$$\left(\frac{D_n}{D_o}\right)^{m-2.5}\left[\left(\frac{D_n}{D_o}\right)^{2.5}+1\right]^3 = \left(\frac{Q_s}{Q_o}\right)^3.$$

(13.22)

Equating Eqs. (13.19) and (13.21), one finds the value of Q_s at which both the alternatives are equally economic. This yields

$$\left(\frac{Q_s}{Q_o}\right)^3 = \frac{5}{m} \frac{\left(\frac{D_n}{D_o}\right)^m \left[\left(\frac{D_n}{D_o}\right)^{2.5} + 1\right]^2}{\left[\left(\frac{D_n}{D_o}\right)^{2.5} + 1\right]^2 - 1}. \tag{13.23}$$

Eliminating Q_s/Q_o between Eqs. (13.22) and (13.23) and solving the resulting equation by trial and error, D_n/D_o is obtained as a function of m. Substituting D_n/D_o in Eqs. (13.22) or (13.23), Q_s/Q_o is obtained as a function of m. Swamee and Sharma (1990) approximated such a function to the following linear relationship for the enhanced discharge:

$$Q_s = (2.5 - 0.6\,m)\,Q_o. \tag{13.24}$$

A perusal of Eq. (13.24) reveals that Q_s decreases linearly as m increases.

So long as the increased demand is less than Q_s, no strengthening is required. In such a case, provision of an increased pumping capacity with the existing pipeline will suffice. Equation (13.24) reveals that for the hypothetical case $m = 2.5$, $Q_s = Q_o$. Thus, strengthening is required even for a slight increase in the existing discharge. Although Eq. (13.24) has been developed for a pumping main, by and large, it will hold good for an entire water distribution system.

13.2.2. Strengthening of a Pumping Main

The cost of strengthening of a pumping main is given by

$$F = k_m LD_n^m + k_T \rho g Q h_0. \tag{13.25}$$

The discharge Q_n is obtained by eliminating Q_o between the head-loss equation

$$h_f = \frac{8 f_o L Q_o^2}{\pi^2 g D_o^5} = \frac{8 f_n L Q_n^2}{\pi^2 g D_n^5} \tag{13.26}$$

and the continuity equation

$$Q = Q_o + Q_n \tag{13.27}$$

and solving the resulting equation. Thus

$$Q_n = \frac{Q}{\left(\frac{f_n D_o^5}{f_o D_n^5}\right)^{0.5} + 1}.$$

(13.28)

The constraint to be observed in this case is

$$h_0 = H + z_L - z_0 + \frac{8 f_n L Q_n^2}{\pi^2 g D_n^5}$$

(13.29)

Substituting Q_n from Eq. (13.28), Eq. (13.29) changes to

$$h_0 = H + z_L - z_0 + \frac{8 L Q^2}{\pi^2 g}\left(\frac{D_o^{2.5}}{f_o^{0.5}} + \frac{D_n^{2.5}}{f_n^{0.5}}\right)^{-2}.$$

(13.30)

Substituting h_0 from Eq. (13.30) into Eq. (13.25), the cost function reduces to

$$F = k_m L D_n^m + \frac{8 k_T \rho L Q^3}{\pi^2}\left(\frac{D_o^{2.5}}{f_o^{0.5}} + \frac{D_n^{2.5}}{f_n^{0.5}}\right)^{-2} + k_T \rho g Q (H + z_L - z_0).$$

(13.31)

For optimality, the condition $\partial F / \partial D_n = 0$ reduces Eq. (13.31) to

$$D_n = \left\{\left(\frac{40 \rho k_T f_n Q^3}{\pi^2 m k_m}\right)\left[1 + \left(\frac{f_n D_o^5}{f_o D_n^5}\right)^{0.5}\right]^{-3}\right\}^{\frac{1}{m+5}}.$$

(13.32)

Equation (13.32), being implicit, can be solved by the following iterative procedure:

1. Assume f_o and f_n
2. Assume initially a diameter of new pipe, say 0.2 m, to start the method
3. Find Q_n and Q_o using Eqs. (13.28) and (13.27)
4. Find D_n using Eq. (13.32)
5. Find $\mathbf{R}_o$ and $\mathbf{R}_n$ using Eq. (2.4a) or (2.4c)
6. Find f_o and f_n using Eq. (2.6a) or (2.6b)
7. Repeat steps 3–5 until the two successive values of D_n are close
8. Round off D_n to the nearest commercially available size
9. Calculate the pumping head h_0 using Eq. (13.29)

13.2.3. Strengthening of a Distribution Main

Figure 13.4 shows a distribution main having i_L number of withdrawals at intervals separated by pipe sections of length $L_1, L_2, L_3, \ldots, L_{i_L}$ and the existing pipe diameters $D_{o1}, D_{o2}, D_{o3}, \ldots, D_{oi_L}$. Designating the sum of the withdrawals as Q_T, the system cost of new links and pumping is given by

$$F = \sum_i^{i_L} k_m L_i D_{ni}^m + \rho g k_T Q_T h_0, \qquad (13.33)$$

where $h_0 =$ the pumping head is expressed as

$$h_0 = H + z_L - z_0 + \frac{8}{\pi^2 g} \sum_{i=1}^{i_L} \left(\frac{D_{oi}^{2.5}}{f_{oi}^{0.5}} + \frac{D_{ni}^{2.5}}{f_{ni}^{0.5}} \right)^{-2} L_i Q_i^2. \qquad (13.34)$$

Eliminating h_0 between Eqs. (13.33) and (13.34) and then equating the partial differential coefficient with respect to D_{ni} to zero and simplifying,

$$D_{ni} = \left\{ \left(\frac{40 \rho k_T f_{ni} Q_T Q_i^2}{\pi^2 m k_m} \right) \left[1 + \left(\frac{f_{ni} D_{oi}^5}{f_{oi} D_{ni}^5} \right)^{0.5} \right]^{-3} \right\}^{\frac{1}{m+5}}. \qquad (13.35)$$

Equation (13.35) can be solved iteratively by the procedure similar to that described for strengthening of a pumping main. However, an approximate solution can be obtained using the method described below.

An approximate solution of strengthening of distribution mains can also be obtained by considering constant f for all the pipes and simplifying Eq. (13.34),

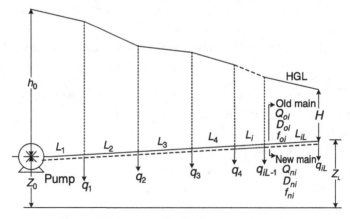

Figure 13.4. Water distribution pumping main.

which is written as

$$h_0 = H + z_L - z_0 \frac{8f}{\pi^2 g} \sum_i^{i_L} L_i Q_i^2 \left(D_{oi}^{2.5} + D_{ni}^{2.5} \right)^{-2}. \tag{13.36}$$

Substituting h_0 from Eq. (13.36) in Eq. (13.33) and differentiating the resulting equation with respect to D_{ni} and setting $\partial F/\partial D_{ni} = 0$ yields

$$D_{ni}^{m-2.5} = \frac{40 \rho k_T f Q_T Q_i^2}{\pi^2 km} \left(D_{oi}^{2.5} + D_{ni}^{2.5} \right)^{-3}. \tag{13.37}$$

Designating

$$D_{*i} = \left(\frac{40 \rho k_T f Q_T Q_i^2}{\pi^2 km} \right)^{\frac{1}{m+5}}, \tag{13.38}$$

where D_{*i} is the diameter of the ith pipe link without strengthening (Eq. 7.11b). Combining Eqs. (13.37) and (13.38) yields

$$\left(\frac{D_{ni}}{D_{*i}} \right)^{2.5-m} = \left[\left(\frac{D_{oi}}{D_{*i}} \right)^{2.5} + \left(\frac{D_{ni}}{D_{*i}} \right)^{2.5} \right]^3. \tag{13.39}$$

Figure 13.5 shows the plot of Eq. (13.39) for $m = 1.4$. A perusal of Fig. 13.5 reveals that for each value of $D_{oi}/D_{*i} < 0.82$, there are two values of D_{ni}/D_{*i}. For $D_{oi}/D_{*i} > 0.82$, only the pumping head has to be increased and no strengthening is required. The upper limb of Fig. 13.5 represents a lower stationary point, whereas the lower limb represents a higher stationary point in cost function curve. Thus, the upper limb represents the optimal solution. Unfortunately, Eq. (13.39) is implicit in D_{ni} and as such it cannot be used easily for design purposes. Using the plotted coordinates of Eq. (13.39) and adopting a method of curve fitting, the following explicit equation has been obtained:

$$\frac{D_{ni}}{D_{*i}} = \left[1 + 0.05 \left(\frac{D_{oi}}{D_{*i}} \right)^{3.25} \right]^{-17.5}. \tag{13.40}$$

Equation (13.40) can provide a good trial solution for strengthening a network of arbitrary geometry. Similarly, Eq. (13.40) can also be used for starting a solution for strengthening a pipe network using LP technique. The aim of developing Eq. (13.40) is to provide a starting solution, thus it does not require high accuracy. As an approximate solution, Eq. (13.40) holds good for all values of m.

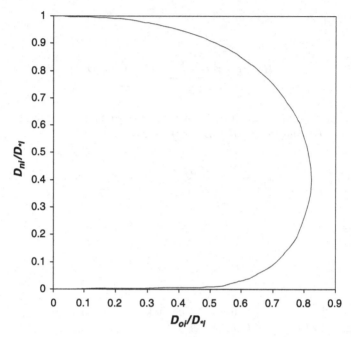

Figure 13.5. Variation of D_{ni}/D_{*i} with D_{oi}/D_{*i}.

13.2.4. Strengthening of Water Distribution Network

A water distribution system having i_L number of pipes, k_L number of loops, n_L number of input points with existing pipe diameters D_{oi} has to be restrengthened for increased water demand due to increase in population. Swamee and Sharma (1990) developed a method for the reorganization/restrengthening of existing water supply systems, which is described in the following section.

The method is presented by taking an example of an existing network as shown in Fig. 13.6. It contains 55 pipes, 33 nodes, 23 loops, and 3 input source points at nodes 11, 22, and 28. The pipe network geometry data including existing population load and existing pipe sizes are given in Table 13.2.

The existing population of 20,440 is increased to 51,100 for restrengthening the network. The rate of water supply as 175 L per person per day, a peak factor of 2.5, and terminal head of 10 m are considered for the design. For the purpose of preliminary analysis of the network, it is assumed that all the existing pipe links are to be strengthened by providing parallel pipe links of diameter 0.2 m. The network is analyzed using the algorithm described in Chapter 3, and Eq. (13.20) is used for head-loss computation in parallel pipes. The analysis results in the pipe discharges in new and old pipes and nodal heads. In addition to this, the discharges supplied by the input points are also obtained for arbitrary assumed parallel pipe link diameter and input heads.

Once the pipe link discharges are obtained, it is required to find a good starting solution so that the system can be restrengthened with a reasonable computational effort. A method for estimating approximate diameter of parallel pipe links is presented

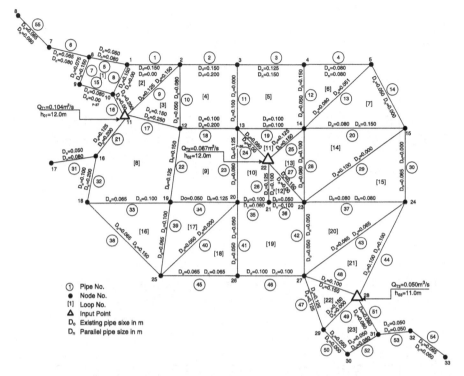

Figure 13.6. Strengthening of a water distribution system.

in Section 13.2.3. (Eq. 13.40). The pipe discharges obtained from initial analysis of the network are used in Eq. (13.38) to calculate $D_{\cdot i}$ for each pipe link. Equation (13.40) provides the starting solution of the parallel pipes.

As the starting solution obtained by Eq. (13.40) is continuous in nature, for LP application two discrete diameters D_{ni1} and D_{ni2} are selected out of commercially available sizes such that D_{ni} in the parallel pipe link i is $D_{ni1} < D_{ni} < D_{ni2}$. The LP problem for the system to be reorganized is

$$\min \ F = \sum_{i=1}^{i_L} (c_{i1} x_{i1} + c_{i2} x_{i2}) + \rho g \, k_T \sum_{n}^{n_L} Q_{Tn} h_{0n}, \tag{13.41}$$

subject to

$$x_{i1} + x_{i2} = L_i; \ i = 1, 2, 3, \ldots, i_L, \tag{13.42}$$

$$\sum_{p=I_t(i,\ell)} \left(\frac{8 f_{np1} Q_{np}^2}{\pi^2 g D_{np1}^5} x_{p1} + \frac{8 f_{np2} Q_{np}^2}{\pi^2 g D_{np2}^5} x_{p2} \right) \le z_{Js(i)} + h_{0Js(i)} - z_{J_t(i)} - H - \sum_{p=I_t(i,\ell)} \frac{8 k_{fp} Q_p^2}{\pi^2 g D_{p2}^4}$$

$$l = 1, 2, 3 \ N_t(i) \qquad \text{For } i = 1, 2, 3 \ldots i_L \tag{13.43}$$

TABLE 13.2. Existing Pipe Network Geometry

Pipe/Node i/j	First Node $J_1(i)$	Second Node $J_2(i)$	Loop 1 $K_1(i)$	Loop 2 $K_2(i)$	Pipe Length $L(i)$	Form-Loss Coefficient $k_f(i)$	Population Load $P(i)$	Existing Pipe D_{oi} (m)	Nodal Elevation z_j (m)
1	1	2	2	0	380	0.0	500	0.150	101.85
2	2	3	4	0	310	0.0	385	0.150	101.90
3	3	4	5	0	430	0.2	540	0.125	101.95
4	4	5	6	0	270	0.0	240	0.080	101.60
5	1	6	1	0	150	0.0	190	0.050	101.75
6	6	7	0	0	250	0.0	250	0.065	101.80
7	6	9	1	0	150	0.0	190	0.065	101.00
8	1	10	1	2	150	0.0	190	0.150	100.40
9	2	11	2	3	390	0.0	490	0.125	101.85
10	2	12	3	4	320	0.0	400	0.050	101.90
11	3	13	4	5	320	0.0	400	0.100	102.00
12	4	14	5	6	330	0.0	415	0.050	101.80
13	5	14	6	7	420	0.0	525	0.080	101.80
14	5	15	7	0	320	0.0	400	0.050	101.90
15	9	10	1	0	160	0.0	200	0.080	100.50
16	10	11	2	0	120	0.0	150	0.200	100.80
17	11	12	3	8	280	0.0	350	0.150	100.70
18	12	13	4	9	330	0.0	415	0.100	101.40
19	13	14	5	11	450	0.2	560	0.100	101.60
20	14	15	7	14	360	0.2	450	0.080	101.80
21	11	16	8	0	230	0.0	280	0.125	101.85
22	12	19	8	9	350	0.0	440	0.125	101.95
23	13	20	9	10	360	0.0	450	0.080	101.80
24	13	22	10	11	260	0.0	325	0.080	101.10
25	14	22	11	13	320	0.0	400	0.125	101.40
26	21	22	10	12	160	0.0	200	0.125	101.20

27	22	23	12	13	290	0.0	365	0.150	101.70
28	14	23	13	14	320	0.0	400	0.080	101.90
29	15	23	14	15	500	0.0	625	0.100	101.70
30	15	24	15	0	330	0.0	410	0.065	101.80
31	16	17	0	0	230	0.0	290	0.050	101.80
32	16	18	8	0	220	0.0	275	0.100	100.50
33	18	19	8	18	350	0.0	440	0.065	99.50
34	19	20	9	17	330	0.0	410	0.050	
35	20	21	10	19	220	0.0	475	0.100	
36	21	23	12	19	250	0.0	310	0.050	
37	23	24	15	20	370	0.0	460	0.080	
38	18	25	16	0	470	0.0	590	0.080	
39	19	25	16	17	320	0.0	400	0.065	
40	20	25	17	18	460	0.0	575	0.050	
41	20	26	18	19	310	0.0	390	0.050	
42	23	27	19	20	330	0.0	410	0.050	
43	24	27	20	21	510	0.0	640	0.065	
44	24	28	21	0	470	0.0	590	0.100	
45	25	26	18	0	300	0.0	375	0.065	
46	26	27	19	0	490	0.0	610	0.100	
47	27	29	22	0	230	0.0	290	0.125	
48	27	28	21	22	290	0.0	350	0.100	
49	28	29	22	23	190	0.0	240	0.150	
50	29	30	23	0	200	0.0	250	0.050	
51	28	31	23	0	160	0.0	200	0.080	
52	30	31	23	0	140	0.0	175	0.050	
53	31	32	0	0	200	0.0	110	0.050	
54	32	33	0	0	200	0.0	200	0.065	
55	7	8	0	0	200	0.0	250	0.065	

where c_{i1} and c_{i2} are per meter cost of pipe sizes D_{ni1} and D_{ni2}, and f_{np1} and f_{np2} are the friction factors in parallel pipes of diameters D_{np1} and D_{np2}, respectively. The LP problem can be solved using the algorithm described in Appendix 1. The starting solution can be obtained using Eq. (13.40). Once the new pipe diameters and pumping heads are obtained, the analysis process is repeated to get a new set of pipe discharges and input point discharges. The starting solution is recomputed for new LP formulation. The process of analysis and synthesis by LP is repeated until two successive designs are close. The obtained parallel pipe sizes are depicted in Fig. 13.6 along with input point discharges and pumping heads.

EXERCISES

13.1. Assuming suitable parameters for the gravity system shown in Fig. 13.1, design a parallel pipe system for assumed increased flows.

13.2. For the pumping system shown in Fig. 13.2, obtain the parallel main for $L = 1500$ m, $D_o = 0.30$ m, and design $Q = 0.3 \, \text{m}^3/\text{s}$. The elevation difference between z_0 and z_L is 20 m. The prescribed terminal head is 15 m, and $k_T/k_m = 0.014$ SI units.

13.3. Consider a distribution main similar to Fig. 13.4 for five pipe links, terminal head $= 20$ m, and the topography is flat. The existing nodal withdrawals were increased from $0.05 \, \text{m}^3/\text{s}$ to $0.08 \, \text{m}^3/\text{s}$. Design the system for existing and increased discharges. Assume suitable data for the design.

13.4. Reorganize a pipe network of 30 pipes, 10 loops, and a single source if the existing population has doubled. Assume flat topography and apply local data for pipe network design.

REFERENCE

Swamee, P.K., and Sharma, A.K. (1990). Reorganization of water distribution system. *J. Envir. Eng.* 116(3), 588–600.

14

TRANSPORTATION OF SOLIDS THROUGH PIPELINES

Solids through a pipeline can be transported as a slurry or containerized in capsules, and the capsules can be transported along with a carrier fluid. Slurry transport through pipelines includes transport of coal and metallic ores, carried in water suspension; and pneumatic conveyance of grains and solid wastes. Compared with slurry transport, the attractive features of capsule transport are that the cargo is not wetted or contaminated by the carrier fluid; no mechanism is required to separate the transported material from the carrier fluid; and it requires less power to maintain flow. Bulk transport through a pipeline can be economic in comparison with other modes of transport.

Design of Water Supply Pipe Networks. By Prabhata K. Swamee and Ashok K. Sharma
Copyright © 2008 John Wiley & Sons, Inc.

259

14.1. SLURRY-TRANSPORTING PIPELINES

The continuity equation for slurry flow is written as

$$V = \frac{4(Q + Q_s)}{\pi D^2},$$

(14.1)

where Q_s = sediment discharge expressed as volume per unit time. Assuming the average velocity of sediment and the fluid to be the same, the sediment concentration can be expressed as

$$C_v = \frac{Q_s}{Q + Q_s}.$$

(14.2)

Using Eqs. (14.1) and (14.2), the resistance equation (2.31) is reduced to

$$h_f = \frac{8fL(Q + Q_s)^2}{\pi^2 g D^5} + \frac{81\pi(s - 1)fLQ_s}{8(Q + Q_s)^2 C_D^{0.75}} D^2 \sqrt{(s - 1)gD}.$$

(14.3)

In the design of a sediment-transporting pipeline, Q is a design variable. If the selected Q is too low, there will be flow with bed load; that is, the sediment will be dragging on the pipe bed. Such a movement creates maintenance problems at pipe bends and inclines and thus is not preferred. On the other hand, if it is too high, there is a significant head loss amounting to high cost of pumping. Durand (Stepanoff, 1969) found that the velocity at the lower limit of the transition between heterogeneous flow and with moving bed corresponds fairly accurately to minimum head loss. This velocity has been named *limit deposit velocity*. The discharge corresponding with the limit deposit velocity can be obtained by differentiating h_f in Eq. (14.3) with respect to Q and equating the resulting expression to zero. Thus,

$$Q = 2.5Q_s^{0.25}\left[\frac{D^2\sqrt{(s - 1)gD}}{C_D^{0.25}}\right]^{0.75} - Q_s.$$

(14.4)

Combining Eqs. (14.3) and (14.4), one gets

$$h_f = 10.16(s - 1)fL\left[\frac{Q_s}{C_D^{0.75}D^2\sqrt{(s - 1)gD}}\right]^{0.5}.$$

(14.5)

14.1.1. Gravity-Sustained, Slurry-Transporting Mains

In a situation where material has to be transported from a higher elevation to a lower elevation, it may be transported through a gravity main without any expenditure on maintaining the flow. As the water enters from the intake chamber to the gravity main, the granular material is added to it. The grains remain in suspension on account of vertical turbulent velocity fluctuations. At the pipe exit, the material is separated from water and dried. A gravity-sustained system is shown in Fig. 14.1.

Eliminating the head loss between Eqs. (6.1) and (14.5) and simplifying, the pipe diameter is obtained as

$$D = 6.39 \left[\frac{(s-1)^3}{C_D^{1.5}} \left(\frac{fL}{z_0 - z_L - H} \right)^4 \frac{Q_s^2}{g} \right]^{0.2}. \tag{14.6}$$

Eliminating D between Eqs. (14.4) and (14.6), the carrier fluid discharge is obtained as

$$Q = Q_s \left\{ \left[\frac{18.714(s-1)fL}{C_D^{0.5}(z_0 - z_L - H)} \right]^{1.5} - 1 \right\}. \tag{14.7}$$

Using Eqs. (14.1), (14.6), and (14.7), the average velocity is found to be

$$V = 2.524 \left[\frac{(s-1)^3 (z_0 - z_L - H)}{C_D^{1.5} fL} \right]^{0.1}. \tag{14.8}$$

Combining Eqs. (14.2) and (14.7), the sediment concentration is expressed as

$$C_v = \left[\frac{C_D^{0.5}(z_0 - z_L - H)}{18.714(s-1)fL} \right]^{1.5}. \tag{14.9}$$

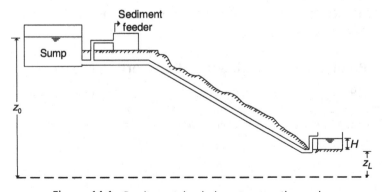

Figure 14.1. Gravity-sustained, slurry-transporting main.

TABLE 14.1. Design Iterations

Iteration No.	f	D (m)	Q (m³/s)	C_v	V (m/s)	R
1	0.010	0.5533	0.3286	0.0707	1.4707	8,137,610
2	0.0136	0.7082	0.5367	0.0445	1.4261	10,098,900
3	0.0130	0.6838	0.5009	0.0475	1.4323	9,793,710
4	0.0131	0.6871	0.5058	0.0471	1.4314	9,830,090

Equations (14.8) and (14.9) reveal that V and C_v are independent of sediment discharge. Using Eqs. (4.4) and (14.6), the corresponding cost is

$$F = 6.39^m k_m L \left[\frac{(s-1)^3}{C_D^{1.5}} \left(\frac{fL}{z_0 - z_L - H} \right)^4 \frac{Q_s^2}{g} \right]^{\frac{m}{5}} \qquad (14.10)$$

The friction factor f occurring in Eq. (14.6) is unknown. Assume a suitable value of f to start the design procedure. Knowing D and V, the Reynolds number R can be obtained by Eq. (2.4a) and subsequently f can be obtained by Eq. (2.6a) or Eq. (2.6b). Substituting revised values in Eq. (14.6), the pipe diameter is calculated again. The process can be repeated until two consecutive diameters are close.

Example 14.1. Design a steel pipeline for transporting coal at the rate of $0.25\,m^3/s$. The coal has a grain size of 0.2 mm and $s = 1.5$. The transportation has to be carried out to a place that is 200 m below the entry point and at a distance of 50 km. The pipeline has $\varepsilon = 0.5$ mm. The terminal head $H = 5$ m.

Solution. Taking $v = 1 \times 10^{-6}$ m²/s and using Eq. (2.34), $w = 0.0090$ m/s; on using Eq. (2.33), $R_s = 4.981$; and using Eq. (2.32), $C_D = 15.996$. For starting the algorithm, $f = 0.01$ is assumed and the iterations are carried out. These iterations are shown in Table 14.1. Thus, a diameter of 0.7 m is provided. For this diameter, Eq. (14.4) yields $Q = 0.525\,m^3/s$; and using Eq. (14.2), this discharge gives $C_v = 0.045$.

14.1.2. Pumping-Sustained, Slurry-Transporting Mains

Swamee (1995) developed a method for the design of pumping-sustained, slurry-transporting pipelines, which is described in this section.

The pumping head h_0 can be expressed as

$$h_0 = z_L - z_0 + H + h_f. \qquad (14.11)$$

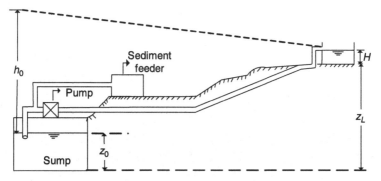

Figure 14.2. Pumping-sustained, slurry-transporting main.

Eliminating h_f between Eqs. (14.5) and (14.11), one gets

$$h_0 = 10.16(s-1)fL\left[\frac{Q_s}{C_D^{0.75}D^2\sqrt{(s-1)gD}}\right]^{0.5} + z_L - z_0 + H. \qquad (14.12)$$

The pumping-sustained, slurry-transporting main shown in Fig. 2.20 is included in this section again as Fig. 14.2.

14.1.2.1. *Optimization.* In this case, Eq. (6.4) is modified to

$$F = k_m L D^m + k_T \rho g (Q + sQ_s)h_0. \qquad (14.13)$$

Eliminating Q and h_0 in Eqs. (14.4), (14.12), and (14.13), one gets

$$F = k_m L D^m + 25.4\frac{\rho k_T[(s-1)g]^{1.125}fLQ_s^{0.75}D^{0.625}}{C_D^{0.5625}}$$

$$+ 10.16\frac{\rho k_T(s-1)^{1.75}g^{0.75}fLQ_s^{1.5}}{C_D^{0.375}D^{1.25}}$$

$$+ 2.5\frac{\rho k_T(s-1)^{0.375}g^{1.375}(z_L - z_0 + H)Q_s^{0.25}D^{1.875}}{C_D^{0.1875}}$$

$$+ \rho k_T(s-1)g(z_L - z_0 + H)Q_s. \qquad (14.14)$$

Considering $z_L - z_0$ and H to be small in comparison with h_f, the fourth term on the right-hand side of Eq. (14.14) can be neglected. Furthermore, the last term on the right-hand side of Eq. (14.14), being constant, can be dropped. Thus, Eq. (14.14) reduces to

$$F_1 = \phi^m + \phi^{0.625} + 0.4G_s^{0.75}\phi^{-1.25} \qquad (14.15)$$

where

$$F_1 = \frac{F}{k_m L D_s^m} \tag{14.16a}$$

$$\phi = \frac{D}{D_s} \tag{14.16b}$$

$$G_s = \frac{(s-1)^{5/6} C_D^{0.25} Q_s}{D_s^2 \sqrt{g D_s}} \tag{14.16c}$$

$$D_s = \left\{ \frac{25.4 \rho k_T [(s-1)g]^{1.125} f Q_s^{0.75}}{k_m C_D^{0.5625}} \right\}^{\frac{1.6}{1.6m-1}}. \tag{14.16d}$$

Equation (14.15) is in the form of a positive posynomial in ϕ. Thus, the minimization of Eq. (14.15) gives rise to a geometric programming problem having a single degree of difficulty. The following weights w_1, w_2, and w_3 define contributions of various terms of Eq. (14.15):

$$w_1 = \frac{\phi^m}{F_1} \tag{14.17a}$$

$$w_2 = \frac{\phi^{0.625}}{F_1} \tag{14.17b}$$

$$w_3 = \frac{0.4 G_s^{0.75}}{\phi^{1.25} F_1}. \tag{14.17c}$$

The dual objective function F_2 of Eq. (14.15) is written as

$$F_2 = \left(\frac{\phi^m}{w_1} \right)^{w_1} \left(\frac{\phi^{0.625}}{w_2} \right)^{w_2} \left(\frac{0.4 G_s^{0.75}}{\phi^{1.25} w_3} \right)^{w_3}. \tag{14.18}$$

The orthogonality condition of Eq. (14.18) for ϕ can be written as in terms of optimal weights w_1^*, w_2^*, and w_3^* ($*$ corresponds with optimality):

$$mw_1^* + 0.625 w_2^* - 1.25 w_3^* = 0, \tag{14.19a}$$

and the corresponding normality condition is

$$w_1^* + w_2^* + w_3^* = 1. \tag{14.19b}$$

Solving Eq. (14.19a, b), one gets

$$w_1^* = -\frac{1}{1.6\,m - 1} + \frac{3}{1.6\,m - 1} w_3^*. \tag{14.20a}$$

$$w_2^* = \frac{1.6\,m}{1.6\,m - 1} - \frac{1.6\,m + 2}{1.6\,m - 1} w_3^*. \tag{14.20b}$$

Substituting Eq. (14.20a,b) in Eq. (14.18), one obtains

$$F_2^* = \frac{1.6\,m - 1}{3w_3^* - 1} \left[\frac{3w_3^* - 1}{1.6\,m - (1.6\,m + 2)w_3^*} \right]^{\frac{1.6\,m}{1.6\,m-1}}$$

$$\times \left\{ \frac{1.6\,m - (1.6\,m + 2)w_3^*}{(4\,m - 2.5)w_3^*} \left[\frac{1.6\,m - (1.6\,m + 2)w_3^*}{3w_3^* - 1} \right]^{\frac{3}{1.6\,m-1}} G_s^{0.75} \right\}^{w_3^*}. \tag{14.21}$$

Equating the factor having the exponent w_3^* on the right-hand side of Eq. (14.21) to unity, the following optimality condition of Eq. (14.21) is obtained (Swamee, 1995):

$$G_s = \left[\frac{(4\,m - 2.5)w_3^*}{1.6\,m - (1.6\,m + 2)w_3^*} \right]^{\frac{4}{3}} \left[\frac{3w_3^* - 1}{1.6\,m - (1.6\,m + 2)w_3^*} \right]^{\frac{4}{1.6\,m-1}}. \tag{14.22}$$

Equation (14.22) is an implicit equation in w_3^*. For the practical range $0.9 \le m \le 1.7$, Eq. (14.22) is fitted to the following explicit form in w_3^*:

$$w_3^* = \frac{m + 1.375 m G_s^p}{3\,m + 1.375(m + 1.25)G_s^p}, \tag{14.23a}$$

where

$$p = 0.15 m^{1.5}. \tag{14.23b}$$

The maximum error involved in the use of Eq. (14.23a) is about 1%. Using Eqs. (14.16a–d), (14.21), and (14.22) with the condition at optimality $F_1^* = F_2^*$, one gets

$$F^* = \frac{(1.6\,m - 1)k_m L}{3w_3^* - 1} \left[\frac{3w_3^* - 1}{1.6\,m - (1.6\,m + 2)w_3^*} \right.$$

$$\times \left. \frac{25.4\,k_T \rho\,[(s - 1)g]^{1.125} f Q_s^{0.75}}{k_m C_D^{0.5625}} \right]^{\frac{1.6\,m}{1.6\,m-1}}, \tag{14.24}$$

TABLE 14.2. Design Iterations

Iteration No.	f	D (m)	Q (m³/s)	w_3	V (m/s)	$\mathbf{R}$
1	0.0100	0.100	0.00567	0.4003	1.98	198,780
2	0.0205	0.120	0.01188	0.3784	1.94	232,310
3	0.0203	0.120	0.01178	0.3787	1.94	231,816
4	0.0203	0.120	0.01178	0.3787	1.94	231,816

where w_3^* is given by Eqs. (14.23a, b). Using Eqs. (14.16a–d), (14.17a), and (14.24), the optimal diameter D^* is

$$D^* = \left[\frac{3w_3^* - 1}{1.6\,m - (1.6\,m + 2)w_3^*} \frac{25.4\rho k_T [(s-1)g]^{1.125} f Q_s^{0.75}}{k_m C_D^{0.5625}} \right]^{\frac{1.6}{1.6\,m-1}}. \tag{14.25}$$

For a given data, G_s is obtained by Eqs. (14.16c,d). Using Eqs. (14.23a,b), the optimal weight w_3^* is obtained. As the friction factor f is unknown, a suitable value of f is assumed and D is obtained by Eq. (14.25). Further, Q is found by using Eq. (14.4). Thus, the average velocity V is obtained by the continuity equation (2.1). Equation (2.4a) then obtains the Reynolds number $\mathbf{R}$ and subsequently Eq. (2.6a) or Eq. (2.6b) finds f. The process is repeated until two successive diameters are close. The diameter is then reduced to the nearest commercially available size, or using Eq. (14.13), two values of F are calculated for lower and upper values of commercially available pipe diameters and the lower-cost diameter is adopted. Knowing the design diameter, the pumping head h_0 is found by using Eq. (14.3).

Example 14.2. Design a cast iron pipeline for carrying a sediment discharge of $0.01\,\mathrm{m}^3/\mathrm{s}$ having $s = 2.65$ and $d = 0.1\,\mathrm{mm}$ from a place at an elevation of 200 m to a location at an elevation of 225 m and situated at a distance of 10 km. The terminal head $H = 5\,\mathrm{m}$. The ratio $k_T/k_m = 0.018$ units.

Solution. For cast iron pipes, Table 2.1 gives $\varepsilon = 0.25\,\mathrm{mm}$. Taking $m = 1.2$, $\rho = 1000$ kg/m³, and $v = 1.0 \times 10^{-6}$ m²/s for fluid, $g = 9.80\,\mathrm{m/s}^2$, and using Eq. (2.34), w $= 0.00808\,\mathrm{m/s}$; on using Eq. (2.33), $\mathbf{R_s} = 0.8084$; and using Eq. (2.32), $C_D = 32.789$. Assuming $f = 0.01$, the iterations are carried out. These iterations are shown in Table 14.2. Thus, a diameter of 0.15 m can be provided. For this diameter, $Q = 0.023\,\mathrm{m}^3/\mathrm{s}$; $C_v = 0.30$; and $h_0 = 556\,\mathrm{m}$. Using Eq. (14.20a), $w_1 = 0.137$, indicating that the pipe cost is less than 14% of the overall cost.

14.2. CAPSULE-TRANSPORTING PIPELINES

The carrier fluid discharge Q is a design variable. It can be obtained by dividing the fluid volume in one characteristic length by the characteristic time. That is,

$$Q = \frac{\pi}{4t_c} a(1 - k^2 + \beta)D^3. \tag{14.26a}$$

Eliminating t_c between Eqs. (2.37) and (14.26a), Q is obtained as

$$Q = \frac{a(1 - k^2 + \beta)s_s Q_s}{k^2 a - 2s_c \theta[k(k + 2a) - 2\theta(2k + a - 2\theta)]}. \qquad (14.26b)$$

14.2.1. Gravity-Sustained, Capsule-Transporting Mains

A typical gravity-sustained, capsule-transporting system is shown in Fig. 14.3.

Eliminating the head loss between Eqs. (6.1), (2.39), and (2.40) and simplifying, the pipe diameter is obtained as

$$D = \left(\frac{8LQ_s^2}{\pi^2 g(z_0 - z_L - H)} \right.$$

$$\left. \frac{a(1 + \beta)s_s^2 \left[f_p a + f_b \beta a \left(1 + k^2 \sqrt{k\lambda} \right)^2 + k^5 \lambda \right]}{\left(1 + \sqrt{k\lambda} \right)^2 \{k^2 a - 2s_c \theta[k(k + 2a) - 2\theta(2k + a - 2\theta)]\}^2} \right)^{0.2}, \qquad (14.27)$$

where $\lambda = f_p/f_c$. To use Eq. (14.27), several provisions have to be made. As indicated in Chapter 2, the capsule diameter coefficient k may be selected between 0.85 and 0.95. Thickness has to be decided by handling and strength viewpoint. Adopting $D = 0.3\,\mathrm{m}$ initially, the capsule thickness coefficient may be worked out. A very large value of a will have problems in negotiating the capsules at bends in the pipeline. Thus, a can be selected between 1 and 2. The ideal value of β is zero. However, β may be assumed between 1 and 2 leaving the scope of increasing the cargo transport rate in the future. The capsule material selected should satisfy the following conditions:

$$s_c > s_s - \frac{(s_s - 1)k^2 a}{2\theta[k(k + 2a) - 2\theta(2k + a - 2\theta)]} \qquad (14.28a)$$

$$s_c < \frac{k^2 a}{2\theta[k(k + 2a) - 2\theta(2k + a - 2\theta)]}. \qquad (14.28b)$$

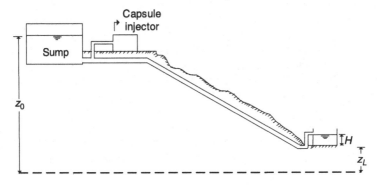

Figure 14.3. Gravity-sustained, capsule-transporting main.

Thus, the capsule material may be selected by knowing the lower and upper bounds of s_c given by Eqs. (14.28a, b), respectively. Initially, $f_b = f_c = f_p = 0.01$ may be assumed. This gives $\lambda = f_p/f_c = 1$. With these assumptions and initializations, a preliminary value of D is obtained by using Eq. (14.27). Using Eq. (2.38), the capsule velocity V_c can be obtained. Further, using Eqs. (2.41) and (2.42), V_a and V_b, respectively are obtained. This enables computation of corresponding Reynolds numbers $\mathbf{R} = V_b D/v$, $(1 - k)(V_c - V_a)D/v$ and $(1 - k)V_a D/v$ to be used in Eq. (2.6a) for obtaining the friction factors $f_b, f_c,$ and f_p, respectively. Using these friction factors, an improved diameter is obtained by using Eq. (14.27). The process is repeated until two consecutive diameters are close. The diameter is then reduced to the nearest available size.

14.2.2. Pumping-Sustained, Capsule-Transporting Mains

Swamee (1998) presented a method for the pumping capsule-transporting mains. As per the method, the number of capsules n is given by

$$n = \frac{(1 + s_a)L}{(1 + \beta)aD}, \tag{14.29}$$

where s_a = part of capsules engaged in filling and emptying the cargo. The cost of capsules C_c is given by

$$C_c = k_c LD^2, \tag{14.30}$$

where k_c = cost coefficient given by

$$k_c = \frac{\pi c_c(1 + s_a)\theta}{2(1 + \beta)a}[k(k + 2a) - 2\theta(2k + a - 2\theta)], \tag{14.31}$$

where c_c = volumetric cost of capsule material. Augmenting the cost function of pumping main (Eq. 6.4) by the capsule cost (Eq. 14.30), the cost function of

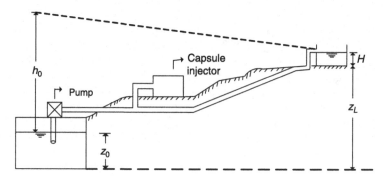

Figure 14.4. Pumping-sustained, capsule-transporting main.

capsule-transporting main is obtained as

$$F = k_m L D^m + k_c L D^2 + k_T \rho g Q_e h_0. \tag{14.32}$$

A typical pumping-sustained, capsule-transporting main shown in Fig. 2.22 is depicted again in this section as Fig. 14.4.

14.2.2.1. Optimization. Using Eqs. (14.11) and (2.39), the pumping head is expressed as

$$h_0 = \frac{8 f_e L Q_s^2}{\pi^2 g D^5} + z_L - z_0 + H. \tag{14.33}$$

Elimination of h_0 between Eqs. (14.32) and (14.33) gives

$$F = k_m L D^m + k_c L D^2 + \frac{8 \, k_T \rho g f_e L Q_e Q_s^2}{\pi^2 D^5} + k_T \rho g Q_e (z_L - z_0 + H). \tag{14.34}$$

The last term of Eq. (14.34) is constant. Dropping this term and simplifying, Eq. (14.34) reduces to

$$F_1 = \phi^m + G_c \phi^2 + \phi^{-5}, \tag{14.35}$$

where

$$F_1 = \frac{F}{k_m L D_0^m} \tag{14.36a}$$

$$\phi = \frac{D}{D_0} \tag{14.36b}$$

$$G_c = \frac{k_c D_0^{2-m}}{k_m} \tag{14.36c}$$

$$D_0 = \left(\frac{8 \, k_T \rho f_e Q_e Q_s^2}{\pi^2 \, k_m} \right)^{\frac{1}{m+5}}. \tag{14.36d}$$

Equation (14.35) is a positive polynomial in ϕ. Thus, the minimization of Eq. (14.35) is a geometric programming problem having a single degree of difficulty. Defining the weights w_1, w_2, and w_3 as

$$w_1 = \frac{\phi^m}{F_1} \tag{14.37a}$$

$$w_2 = \frac{G_c \phi^2}{F_1} \tag{14.37b}$$

$$w_3 = \frac{1}{\phi^5 F_1} \tag{14.37c}$$

and assuming constant friction factors, the dual of Eq. (14.35) is written as

$$F_2 = \left(\frac{\phi^m}{w_1}\right)^{w_1} \left(\frac{G_c \phi^2}{w_2}\right)^{w_2} \left(\frac{1}{\phi^5 w_3}\right)^{w_3}. \tag{14.38}$$

The orthogonality and normality conditions of Eq. (14.38) for ϕ can be written as in terms of optimal weights w_1^*, w_2^*, and w_3^* as

$$mw_1^* + 2w_2^* - 5w_3^* = 0 \tag{14.39a}$$

$$w_1^* + w_2^* + w_3^* = 1. \tag{14.39b}$$

Solving Eq. (14.39a, b) in terms of w_2^*, one gets

$$w_1^* = \frac{5}{m+5} - \frac{7}{m+5} w_2^* \tag{14.40a}$$

$$w_3^* = \frac{m}{m+5} + \frac{2-m}{m+5} w_2^*. \tag{14.40b}$$

Substituting Eq. (14.40a, b) in Eq. (14.38), the optimal dual is

$$F_2^* = \frac{m+5}{5-7w_2^*} \left[\frac{5-7w_2^*}{m+(2-m)w_2^*}\right]^{\frac{m}{m+5}} \left\{ \left[\frac{5-7w_2^*}{m+(2-m)w_2^*}\right]^{\frac{7}{m+5}} \right.$$

$$\left. \times \left[\frac{m+(2-m)w_2^*}{(m+5)w_2^*} G_c\right]\right\}^{w_2^*}. \tag{14.41}$$

Equating the factor having the exponent w_2^* on the right-hand side of Eq. (14.41) to unity (Swamee, 1995), the optimality condition of Eq. (14.41) is

$$G_c = \frac{(m+5)w_2^*}{m+(2-m)w_2^*} \left[\frac{m+(2-m)w_2^*}{5-7w_2^*}\right]^{\frac{7}{m+5}}. \tag{14.42}$$

The implicit equation Eq. (14.42) is fitted to the following explicit form:

$$w_2^* = \frac{5}{7} \left\{ \left[\frac{m+5}{7G_c} \left(\frac{m}{5} \right)^{\frac{2-m}{m+5}} \right]^{\frac{9}{11-m}} + 1 \right\}^{-\frac{11-m}{9}}. \tag{14.43}$$

For $m = 2$, Eq. (14.43) is exact. The maximum error involved in the use of Eq. (14.43) is about 1.5%. Using Eqs. (14.41) and (14.42) with the condition at optimality $F_1^* = F_2^*$, the following equation is obtained:

$$F_1^* = \frac{m+5}{5 - 7w_2^*} \left[\frac{5 - 7w_2^*}{m + (2-m)w_2^*} \right]^{\frac{m}{m+5}}, \tag{14.44}$$

where w_2^* is given by Eqs. (14.43). Using Eqs. (14.34), (14.36a), and (14.44), the optimal cost is found to be

$$F^* = \frac{(m+5)k_m L}{5 - 7w_2^*} \left[\frac{5 - 7w_2^*}{m + (2-m)w_2^*} \frac{8\, k_T \rho f_e Q_e Q_s^2}{\pi^2 k_m} \right]^{\frac{m}{m+5}}$$
$$+ k_T \rho g Q_e (z_L + H - z_0). \tag{14.45}$$

Using Eqs. (14.37a), (14.40a), and (14.44), the optimal diameter D^* is

$$D^* = \left[\frac{5 - 7w_2^*}{m + (2-m)w_2^*} \frac{8\, k_T \rho f_e Q_e Q_s^2}{\pi^2 k_m} \right]^{\frac{1}{m+5}}. \tag{14.46}$$

The above methodology is summarized in the following steps:
For starting the calculations, initially assume $\lambda = 1$.

1. Find k_c using Eq. (14.31).
2. Find V_c using Eq. (2.38).
3. Find V_a and V_b using Eqs. (2.41) and (2.42).
4. Find t_c using Eq. (2.36).
5. Find f_b, f_c, and f_p using Eq. (2.6a) and λ. Use corresponding Reynolds numbers $R = V_b D/v$, $(1-k)(V_c - V_a)D/v$, and $(1-k)V_a D/v$ for f_b, f_c, and f_p, respectively.
6. Find f_e using Eq. (2.40).
7. Find Q_e using Eq. (2.43).
8. Find D_0 using Eq. (14.36d).
9. Find G_c using Eq. (14.36c).

10. Find w_2^* using Eq. (14.43).
11. Find D using Eq. (14.46).
12. Knowing the capsule thickness and D, revise θ.
13. If s_c violates the range, use Eqs. (14.28a) and (14.28b), revise the capsule thickness to satisfy the range, and obtain θ.
14. Repeat steps 1–12 until two consecutive values of D are close.
15. Reduce D to the nearest commercially available size; or use Eq. (14.32) to calculate F for lower and upper values of commercially available D and adopt the lowest cost pipe size.
16. Find n using Eq. (14.29) and round off to the nearest integer.
17. Find Q using Eq. (14.26a).
18. Find V_s using Eq. (2.35).
19. Find h_0 using Eq. (14.33).
20. Find F using Eq. (14.32).

Example 14.3. Design a pipeline for a cargo transport rate of $0.01\,\text{m}^3/\text{s}$, with $s_s = 1.75$, $z_L - z_0 = 12\,\text{m}$, $H = 5\,\text{m}$, and $L = 5\,\text{km}$. The ratio $k_T/k_m = 0.018$ units, and $c_c/k_m = 4$ units. Consider pipe cost exponent $m = 1.2$ and $\varepsilon = 0.25\,\text{mm}$. Use $g = 9.80\,\text{m/s}^2$, $\rho = 1000\,\text{kg/m}^3$, and $v = 1 \times 10^{-6}\,\text{m}^2/\text{s}$.

Solution. For the design proportions $k = 0.9$, $a = 1.5$, $\beta = 1.5$ are assumed. Aluminum ($s_c = 2.7$) capsules having wall thickness of 10 mm are used in this design. Further, assuming $D = 0.3\,\text{m}$, iterations were carried out. The iterations are listed in Table 14.3.

Considering any unforeseen increase in cargo transport rate, a pipe diameter of 0.40 m is provided. Thus, $\theta = 0.01/0.45 = 0.025$. Using Eqs. (14.28a, b), the range of specific gravity of capsule material is $-3.694 < s_c < 7.259$. Thus, there is no necessity to revise capsule thickness. Capsule diameter $= kD = 0.36\,\text{m}$, the capsule length $= aD = 0.60\,\text{m}$, and the intercapsule distance $= \beta aD = 0.9\,\text{m}$. Adopting $s_a = 1$, in Eq. (14.29), the number of capsules obtained is 6666. Thus 6670 capsules are provided. Using Eq. (2.36), $t_c = 2.12\,\text{s}$. Cargo volume in capsule $V_s = Q_s t_c = 0.0212\,\text{m}^3$ (21.2L). Furthermore, using Eq. (14.26a), $Q = 0.058\,\text{m}^3/\text{s}$. Using Eqs. (2.40) and (2.43), respectively, $f_e = 3.29$ and $Q_e = 0.085\,\text{m}^3/\text{s}$. Using Eq. (14.33), $h_f = 16.06\,\text{m}$ yielding

TABLE 14.3. Design Iterations

Iteration No.	f_b	f_c	f_p	θ	w_2^*	D (m)
1	0.01975	0.02734	0.02706	0.03333	0.11546	0.3918
2	0.01928	0.02976	0.02936	0.02558	0.08811	0.3606
3	0.01939	0.02898	0.02860	0.02778	0.09576	0.3687
4	0.01936	0.02918	0.02880	0.02717	0.09365	0.3664

$h_0 = h_f + H + z_L - z_0 = 33.06$ m. Adopting $\eta = 0.75$, the power consumed $= \rho g Q_e h_0 / \eta = 37.18$ kW. Considering $s_b = 0.5$, three pumps of 20 kW are provided.

EXERCISES

14.1. Design a steel pipeline for transporting coal at the rate of $0.3 \, \text{m}^3/\text{s}$. The coal has a grain size of 0.25 mm and $s = 1.6$. The transportation has to be carried out to a place that is 100 m below the entry point and at a distance of 25 km. The pipeline has $\varepsilon = 0.5$ mm. The terminal head $H = 5$ m.

14.2. Design a cast iron pipeline for carrying a sediment discharge of $0.015 \, \text{m}^3/\text{s}$ having $s = 2.65$ and $d = 0.12$ mm from a place at an elevation of 250 m to a location at an elevation of 285 m and situated at a distance of 20 km. The terminal head $H = 5$ m. The ratio $k_T/k_m = 0.02$ units.

14.3. Design a pipeline for a cargo transport rate of $0.015 \, \text{m}^3/\text{s}$, with $s_s = 1.70$, $z_L - z_0 = 15$ m, $H = 2$ m, and $L = 5$ km. The ratio $k_T/k_m = 0.017$ units, and $c_c/k_m = 4.5$ units. Consider pipe cost exponent $m = 1.4$ and $\varepsilon = 0.25$ mm. Use $g = 9.80 \, \text{m/s}^2$, $\rho = 1000 \, \text{kg/m}^3$, and $v = 1 \times 10^{-6} \, \text{m}^2/\text{s}$.

REFERENCES

Stepanoff, A.J. (1969). *Gravity Flow of Solids and Transportation of Solids in Suspension*. John Wiley & Sons, New York.

Swamee, P.K. (1995). Design of sediment-transporting pipeline. *J. Hydraul. Eng.* 121(1), 72–76.

Swamee, P.K. (1998). Design of pipelines to transport neutrally buoyant capsules. *J. Hydraul. Eng.* 124(11), 1155–1160.

Appendix 1

LINEAR PROGRAMMING

The application of linear programming (LP) for the optimal design of water distribution is demonstrated in this section. In an LP problem, both the objective function and the constraints are linear functions of the decision variables.

PROBLEM FORMULATION

As an example, optimal design problem for a branched gravity water distribution system is formulated. In order to make LP application possible, it is considered that each pipe link L_i consists of two commercially available discrete sizes of diameters D_{i1} and D_{i2} having lengths x_{i1} and x_{i2}, respectively. Thus, the cost function F is written as

$$F = \sum_{i=1}^{i_L} (c_{i1}x_{i1} + c_{i2}x_{i2}), \qquad (A1.1)$$

Design of Water Supply Pipe Networks. By Prabhata K. Swamee and Ashok K. Sharma
Copyright © 2008 John Wiley & Sons, Inc.

where c_{i1} and c_{i2} are the cost of 1 m of pipe of diameters D_{i1} and D_{i2}, respectively, and i_L is the number of pipe links in the network. The network is subject to the following constraints:

- The pressure head at each node should be equal to or greater than the prescribed minimum head H; that is,

$$\sum_{i \varepsilon T_j} \left[\left(\frac{8f_{i1}Q_i^2}{\pi^2 g D_{i1}^5} \right) x_{i1} + \left(\frac{8f_{i2}Q_i^2}{\pi^2 g D_{i2}^5} \right) x_{i2} \right] \le z_0 - z_j - H, \qquad (A1.2)$$

where $Q_i =$ discharge in the ith link, f_{i1} and f_{i2} are the friction factors for the two pipe sections of the link i, $z_0 =$ elevation at input source, $z_j =$ ground level of node j, and $T_j =$ a set of pipes connecting input point to the node j.
- The sum of lengths x_{i1} and x_{i2} is equal to the pipe link length L_i; that is,

$$x_{i1} + x_{i2} = L_i \qquad \text{for } i = 1, 2, 3, \dots i_L \qquad (A1.3)$$

Considering x_{i1} and x_{i2} as the decision variables, Eqs. (A1.1), (A1.2), and (A1.3) constitute a LP problem. Taking lower and upper sizes in the range of the commercially available pipe diameters as D_{i1} and D_{i2} and solving the LP problem, the solution gives either $x_{i1} = L_i$ or $x_{i2} = L_i$, thus indicating the preference for either the lower diameter or the upper diameter for each link. Retaining the preferred diameter and altering the other diameter, the range of pipe diameters D_{i1} and D_{i2} is reduced in the entire network, and the new LP problem is solved again. The process is repeated until a final solution is obtained for all pipe links of the entire network.

The solution methodology for LP problem is called simplex algorithm. For the current formulation, the simplex algorithm is described below.

SIMPLEX ALGORITHM

For illustration purposes, the following problem involving only two-decision variables x_1 and x_2 is considered:
Minimize

$$x_0 = 10x_1 + 20x_2 ; \qquad \text{Row 0}$$

subject to the constraints

$$0.2x_1 + 0.1x_2 \le 15 \qquad \text{Row 1}$$
$$1x_1 + 1x_2 = 100 \qquad \text{Row 2}$$
$$\text{and} \quad x_1, x_2 \ge 0.$$

In the example, Row 0 represents the cost function, and Row 1 and Row 2 are the constraints similar to Eqs. (A1.2) and (A1.3), respectively. Adding a nonnegative variable x_3 (called *slack variable*) in the left-hand side of Row 1, the inequation is converted to the following equation:

$$0.2x_1 + 0.1x_2 + 1x_3 = 15.$$

Row 2 is an equality constraint. In this case, Row 2 is augmented by adding an artificial variable x_4:

$$1x_1 + 1x_2 + 1x_4 = 100.$$

The artificial variable x_4 has no physical meaning. The procedure is valid if x_4 is forced to zero in the final solution. This can be achieved if the effect of x_4 is to increase the cost function x_0 in a big way. This can be achieved by multiplying it by a large coefficient, say 200, and adding to the cost function. Thus, Row 0 is modified to the following form:

$$x_0 - 10x_1 - 20x_2 - 200x_4 = 0.$$

Thus, the revised formulation takes the following form:

$$x_0 - 10x_1 - 20x_2 - 0x_3 - 200x_4 = 0 \qquad \text{Row 0}$$
$$0.2x_1 + 0.1x_2 + 1x_3 + 0x_4 = 15 \qquad \text{Row 1}$$
$$1x_1 + 1x_2 + 0x_3 + 1x_4 = 100 \qquad \text{Row 2}$$

Row 1 and Row 2 constituting two equations contain four variables. Assuming x_1 and x_2 as zero, the solution for the other two x_3, x_4 can be obtained. These nonzero variables are called *basic variables*. The coefficient of x_3 in Row 0 is already zero; and by multiplying Row 2 by 200 and adding it to Row 0, the coefficient of x_4 in Row 0 is made to zero. Thus, the following result is obtained:

$$x_0 + 190x_1 + 180x_2 + 0x_3 + 0x_4 = 20,000 \qquad \text{Row 0}$$
$$0.2x_1 + 0.1x_2 + 1x_3 + 0x_4 = 15 \qquad \text{Row 1}$$
$$1x_1 + 1x_2 + 0x_3 + 1x_4 = 100 \qquad \text{Row 2}$$

Discarding the columns containing the variables x_1 and x_2 (which are zero), the above set of equations is written as

$$x_0 = 20,000$$
$$x_3 = 15$$
$$x_4 = 100,$$

which is the initial solution of the problem. Now according to Row 0, if x_1 is increased from zero to one, the corresponding decrease in the cost function is 190. A similar increase in x_2 produces a decrease of 180 in x_0. Thus, to have maximum decrease in x_0, the variable x_1 should be nonzero. We can get only two variables by solving two equations (of Row 1 and Row 2) out of them; as discussed, one variable is x_1, and the other variable has to be decided from the condition that all variable are nonnegative. The equation of Row 1 can be written as

$$x_3 = 15 - 0.2x_1.$$

Thus for x_3 to become zero, $x_1 = 15/0.2 = 75$. On the other hand, the equation of Row 2 is written as

$$x_4 = 100 - 1x_1.$$

Now for x_4 to become zero, $x_1 = 100/1 = 100$. Taking the lower value, thus for $x_1 = 75$, $x_3 = 0$ and $x_4 = 25$. In the linear programming terminology, x_1 will enter the basis and, as a consequence, x_3 will leave the basis. Dividing the Row 1 by 0.2, the coefficient of x_1 becomes unity; that is,

$$1x_1 + 0.5x_2 + 5x_3 + 0x_4 = 75.$$

Multiplying Row 1 by 190 and subtracting from Row 0, the coefficient of x_1 becomes zero. Similarly, multiplying Row 1 by 1 and subtracting from Row 2, the coefficient of x_1 becomes zero. The procedure of making all but one coefficients of column 1 is called pivoting. Thus, the resultant system of equations is

$$x_0 + 0x_1 + 85x_2 - 950x_3 + 0x_4 = 5750 \qquad \text{Row 0}$$
$$1x_1 + 0.5x_2 + 5x_3 + 0x_4 = 75 \qquad \text{Row 1}$$
$$0x_1 + 0.5x_2 - 5x_3 + 1x_4 = 25 \qquad \text{Row 2}$$

Discarding the columns containing the variables x_2 and x_3 (which are zero, thus out of the basis), the above set of equation is written as the following solution form:

$$x_0 = 5750$$
$$x_1 = 75$$
$$x_4 = 25.$$

Further, in Row 0, if x_2 is increased from zero to one, the corresponding decrease in the cost function is 85. A similar increase in x_3 produces an increase of 950 in x_0. Thus, to have decrease in x_0, the variable x_2 should be nonzero (i.e., it should enter in the basis). Now the variable leaving the basis has to be decided. The equation of Row 1 can be written as

$$x_1 = 75 - 0.5x_2.$$

For $x_1 = 0$ (i.e., x_1 leaving the basis), $x_2 = 75/0.5 = 150$. On the other hand, for x_4 leaving the basis, the equation of Row 2 is written as

$$x_4 = 25 - 0.5x_2.$$

For $x_4 = 0$, $x_2 = 25/0.5 = 50$. Of the two values of x_2 obtained, the lower value will not violate nonnegativity constraints. Thus, x_2 will enter the basis, x_4 will leave the basis.

Performing pivoting operation so that x_2 has a coefficient of 1 in Row 2 and 0 in the other rows, the following system of equations is obtained:

$$x_0 + 0x_1 + 0x_2 - 100x_3 - 170x_4 = 1500 \qquad \text{Row 0}$$
$$1x_1 + 0x_2 + 5x_3 - 1x_4 = 50 \qquad \text{Row 1}$$
$$0x_1 + 1x_2 - 10x_3 + 2x_4 = 50 \qquad \text{Row 2}$$

The system of equations yields the solution

$$x_0 = 1500$$
$$x_1 = 50$$
$$x_2 = 50.$$

Now in Row 0, one can see that, as the coefficients of x_3 and x_4 are negative, increasing their value from zero increases the cost function x_0. Thus, the cost function has been minimized at $x_1 = 50$, $x_2 = 50$ giving $x_0 = 1500$.

It can be concluded from the above solution that the pipe link L_i can also have two discrete sizes of diameters D_{i1} and D_{i2} having lengths x_{i1} and x_{i2}, respectively, in the final solution such that the lengths $x_{i1} + x_{i2} = L_i$. A similar condition can be seen in Table 9.8 for pipe $i = 1$ of length $L_1 = 1400$ m having 975 m length of 0.3 m pipe size and 425 m length of 0.250 m pipe size in the solution. Such a condition is generally seen in pipe links of significant lengths (1400 m in this case) in the pipe network.

Appendix 2

GEOMETRIC PROGRAMMING

Geometric programming (GP) is another optimization technique used commonly for the optimal design of water supply systems. The application of GP is demonstrated in this section. In a GP problem, both the objective function and the constraints are in the form of *posynomials*, which are polynomials having positive coefficients and variables and also real exponents. In this technique, the emphasis is placed on the relative magnitude of the terms of the objective function rather than on the variables. In this technique, the value of the objective function is calculated first and then the optimal values of the variables are obtained.

The objective function is the following general form of the posynomial:

$$F = \sum_{t=1}^{T} c_t \prod_{n=1}^{N} x_n^{a_{tn}}, \tag{A2.1}$$

where c_t's are the positive cost coefficients of term t, the x_n's are the independent variables, and a_{tn}'s are the exponents of the independent variables. T is the total number of terms, and N is the total number of independent variables in the cost function. The

Design of Water Supply Pipe Networks. By Prabhata K. Swamee and Ashok K. Sharma
Copyright © 2008 John Wiley & Sons, Inc.

contribution of various terms in Eq. (A2.1) is given by the weights w_t defined as

$$w_t = \frac{c_t}{F} \prod_{n=1}^{N} x_n^{a_{tn}} \quad \text{for } t = 1, 2, 3 \ldots T. \tag{A2.2}$$

The weights should sum up to unity. That is the normality condition:

$$\sum_{t=1}^{T} w_t = 1.$$

The optimum of Eq. (A2.1) is given by

$$F^* = \prod_{t=1}^{T} \left(\frac{c_t}{w_t^*} \right)^{w_t^*}, \tag{A2.3}$$

where the optimal w_t^* are weights given by solution. The following N equations constitute the orthogonality conditions

$$\sum_{t=1}^{T} a_{tn} w_t^* = 0; \quad \text{for } n = 1, 2, 3, \ldots N \tag{A2.4}$$

and of the normality condition for optimum weights

$$\sum_{t=1}^{T} w_t^* = 1. \tag{A2.5}$$

Equations (A2.4) and (A2.5) provide unique solution for $T = N + 1$. Thus, the geometric programming is attractive when the degree of difficulty D defined as $D = T - (N + 1)$ is zero. Knowing the optimal weights and the objective function, the corresponding variables are obtained by solving Eq. (A2.2).

Example 1 (with zero degree of difficulty). In a water supply reservoir–pump installation, the cost of the pipe is given by $5000D^{1.5}$, where D is the diameter of the pipe in meters. The cost of the reservoir is the function of discharge Q as $1500/Q$, where Q is the rate of pumping in m^3/s and the pumping cost is given by $5000Q^2/D^5$.

Solution. The cost function is expressed as

$$F = 5000D^{1.5}Q^0 + 1500D^0Q^{-1} + 5000D^{-5}Q^2. \tag{A2.6}$$

Thus, the coefficients and exponents involved in this equation are $c_1 = 5000$; $c_2 = 1500$; $c_3 = 5000$; $a_{11} = 1.5$; $a_{12} = 0$; $a_{21} = 0$; $a_{22} = -1$; $a_{31} = -5$; and $a_{32} = 2$. Thus, the orthogonality conditions corresponding with Eq. (A2.4) and normality condition of Eq. (A2.5) are

$$1.5w_1^* - 5w_3^* = 0 \tag{A2.7a}$$
$$-w_2^* + 2w_3^* = 0 \tag{A2.7b}$$
$$w_1^* + w_2^* + w_3^* = 1. \tag{A2.7c}$$

Solving the Eqs. (A2.7a–c), the optimal weights are $w_1^* = 0.5263$, $w_2^* = 0.3158$, and $w_3^* = 0.1579$. Substituting the optimal weights in Eq. (A2.3), the minimum cost is obtained as

$$F^* = \left(\frac{5000}{0.5263}\right)^{0.5263} \left(\frac{1500}{0.3158}\right)^{0.3158} \left(\frac{5000}{0.1579}\right)^{0.1579} = 9230.$$

Using the definition of weights as given by Eq. (A2.2), the definition of w_1^* gives

$$0.5263 = \frac{5000D^{1.5}}{9230}$$

yielding $D = 0.980$ m. Similarly, the definition of w_2^* as given by Eq. (A2.2) is

$$0.3158 = \frac{1500}{9230}Q^{-1}$$

and gives $Q = 0.5102$ m^3/s. On the other hand, the definition of w_3^* leads to

$$w_3^* = \frac{5000}{9230}D^{-5}Q^2.$$

Substituting D and Q, the optimal weight w_3^* is obtained as 0.156, which is $\cong 0.1579$ as obtained earlier. Similarly, substituting values of Q and D in Eq. (A2.6),

$$F = 5000(0.98)^{1.5} + 1500(0.5102)^{-1} + 5000(0.98)^{-5}(0.5102)^2 = 9230.63$$

verifies the earlier obtained result.

Example 2 (with 1 degree of difficulty). In a water supply reservoir–pump installation, the cost of the pipe is given by $5000D^2$, where D is the diameter of the pipe in meters. The cost of the reservoir is the function of discharge Q as $1500/Q$, where Q

is the rate of pumping in m^3/s, the pumping cost is $5000Q^2/D^5$, and the cost of pumping station is given by $300QD$.

Solution. The cost function is expressed as

$$F = 5000D^2Q^0 + 1500D^0Q^{-1} + 5000D^{-5}Q^2 + 300DQ. \qquad \text{(A2.8)}$$

Thus, the orthogonality conditions corresponding with Eq. (A2.4) and normality condition of Eq. (A2.5) are

$$2w_1^* - 5w_3^* + w_4^* = 0 \qquad \text{(A2.9a)}$$

$$-w_2^* + 2w_3^* + w_4^* = 0 \qquad \text{(A2.9b)}$$

$$w_1^* + w_2^* + w_3^* + w_4^* = 1. \qquad \text{(A2.9c)}$$

In this geometric programming example, the total number of terms is $T = 4$ and independent variables $N = 2$, thus the degree of difficulty $= T - (N + 1)$ is 1. Such a problem can be solved by first obtaining w_1^*, w_2^* and w_3^* in terms of w_4^* from Eqs. (A2.9a–c). Thus

$$w_1^* = \frac{5}{11} - \frac{13}{11}w_4^* \qquad \text{(A2.10a)}$$

$$w_2^* = \frac{4}{11} + \frac{5}{11}w_4^* \qquad \text{(A2.10b)}$$

$$w_3^* = \frac{2}{11} - \frac{3}{11}w_4^*. \qquad \text{(A2.10c)}$$

The optimal cost function F^* for Eq. (A2.8) is

$$F^* = \left(\frac{5000}{w_1^*}\right)^{w_1^*} \left(\frac{1500}{w_2^*}\right)^{w_2^*} \left(\frac{5000}{w_3^*}\right)^{w_3^*} \left(\frac{300}{w_4^*}\right)^{w_4^*}. \qquad \text{(A2.11)}$$

Substituting w_1, w_2, and w_3 in terms of w_4, the above equation can be written as

$$F^* = \left(\frac{5000}{\dfrac{5}{11} - \dfrac{13}{11}w_4^*}\right)^{\frac{5}{11} - \frac{13}{11}w_4^*} \left(\frac{1500}{\dfrac{4}{11} + \dfrac{5}{11}w_4^*}\right)^{\frac{4}{11} + \frac{5}{11}w_4^*}$$

$$\times \left(\frac{5000}{\dfrac{2}{11} - \dfrac{3}{11}w_4^*}\right)^{\frac{2}{11} - \frac{3}{11}w_4^*} \left(\frac{300}{w_4^*}\right)^{w_4^*},$$

which further simplifies to

$$F^* = \left[\left(\frac{55000}{5-13w_4^*}\right)^{\frac{5}{11}}\left(\frac{16500}{4+5w_4^*}\right)^{\frac{4}{11}}\left(\frac{55000}{2-3w_4^*}\right)^{\frac{2}{11}}\right]$$

$$\times \left[\left(\frac{5-13w_4^*}{55000}\right)^{\frac{13}{11}}\left(\frac{16500}{4+5w_4^*}\right)^{\frac{5}{11}}\left(\frac{2-3w_4^*}{55000}\right)^{\frac{3}{11}}\left(\frac{300}{w_4^*}\right)\right]^{w_4^*}.$$

Traditionally w_4^* is obtained by differentiating this equation with respect to w_4^*, equating it to zero. This method would be very cumbersome. Swamee (1995)[1] found a short cut to this method by equating the factor having the exponent w_4^* on the right-hand side of the above equation to unity. The solution of the resulting equation gives w_4^*. Thus the optimality condition is written as

$$\left(\frac{5-13w_4^*}{55000}\right)^{\frac{13}{11}}\left(\frac{16500}{4+5w_4^*}\right)^{\frac{5}{11}}\left(\frac{2-3w_4^*}{55000}\right)^{\frac{3}{11}}\left(\frac{300}{w_4^*}\right) = 1.$$

This equation is rewritten as

$$\frac{\left(5-13w_4^*\right)^{13/11}\left(2-3w_4^*\right)^{3/11}}{\left(4+5w_4^*\right)^{5/11}w_4^*} = 316.9.$$

Solving this equation by trial and error, w_4^* is obtained as 0.0129. Thus, $w_1^* = 0.4393$, $w_2^* = 0.3695$, and $w_3^* = 0.1783$ are obtained from Eqs. (A2.10a–c). Using Eq. (A2.11), the optimal cost

$$F^* = \left(\frac{5000}{0.4393}\right)^{0.4393}\left(\frac{1500}{0.3695}\right)^{0.3695}\left(\frac{5000}{0.1783}\right)^{0.1783}\left(\frac{300}{0.0129}\right)^{0.0129} = 9217.$$

Using the definition of weights as given by Eq. (A2.2), the definition of w_1^* gives

$$0.4393 = \frac{5000D^2}{9217}$$

[1]Swamee, P.K. (1995). Design of sediment-transporting pipeline. *Journal of Hydraulics Engineering*, 121(1), 72–76.

yielding $D = 0.899$ m. Similarly, the definition of w_2^* gives

$$0.3695 = \frac{1500}{9217} Q^{-1}$$

which gives $Q = 0.44$ m^3/s. On the other hand, the definition of w_3^* leads to

$$w_3^* = \frac{5000}{9217} D^{-5} Q^2.$$

Substituting D and Q the optimal weight w_3^* is obtained as 0.1783, which is same as obtained earlier. Substituting values of $Q = 0.44$ m^3/s and $D = 0.9$ m in Eq. (A2.8)

$$F^* = 5000(0.90)^2 + 1500(0.44)^{-1} + 5000(0.90)^{-5}(0.44)^2 + 300 \times 0.9 \times 0.44$$
$$= 9217$$

verifying the result obtained earlier.

Appendix 3

WATER DISTRIBUTION NETWORK ANALYSIS PROGRAM

Computer programs for water distribution network analysis having single-input and multi-input water sources are provided in this section. The explanation of the algorithm is also described line by line to help readers understand the code. The aim of this section is to help engineering students and water professionals to develop skills in writing water distribution network analysis algorithms and associated computer programs, although numerous water distribution network analysis computer programs are available now and some of them even can be downloaded free from their Web sites. EPANET developed by the United States Environmental Protection Agency is one such popular program, which is widely used and can be downloaded free.

The computer programs included in this section were initially written in FORTRAN 77 but were upgraded to run on FORTRAN 90 compilers. The program can be written in various ways to code an algorithm, which depends upon the language used and the skills of the programmer. Readers are advised to follow the algorithm and rewrite a program in their preferred language using a different method of analysis.

SINGLE-INPUT WATER DISTRIBUTION NETWORK ANALYSIS PROGRAM

In this section, the algorithm and the software for a water distribution network having single-input source is described. Information about data collection, data input, and the

Design of Water Supply Pipe Networks. By Prabhata K. Swamee and Ashok K. Sharma
Copyright © 2008 John Wiley & Sons, Inc.

output and their format is discussed first. Nodal continuity equations application and Hardy Cross method for loop pipes discharge balances are then discussed. Readers can modify the algorithm to their preferred analysis method as described in Chapter 3.

As discussed in Chapter 3, water distribution networks are analyzed for the determination of pipe link discharges and pressure heads. The other important reasons for analysis are to find deficiencies in the pipe network in terms of flow and nodal pressure head requirements and also to understand the implications of closure of some of the pipes in the network. The pipe network analysis is also an integral part of the pipe network design or synthesis irrespective of design technique applied.

A single-input source water distribution network as shown in Fig. A3.1 is referred in describing the algorithm for analysis. Figure A3.1 depicts the pipe numbers, nodes, loops, input point, and existing pipe diameter as listed in Tables A3.1, A3.2, A3.3, and A3.4.

Data Set

The water distribution network has a total of 55 pipes (i_L), 33 nodes (j_L), 23 loops (k_L), and a single-input source (m_L). In the book text, the input points are designated as n_L.

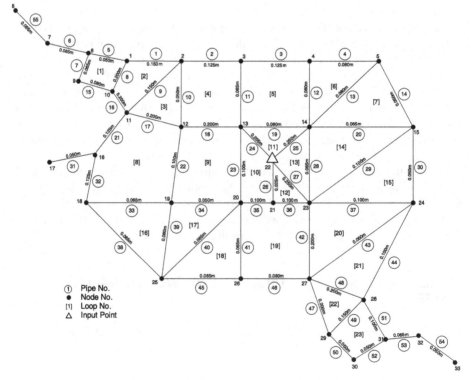

Figure A3.1. Single-input source water distribution system.

TABLE A3.1. Pipe Network Size

i_L	j_L	k_L	m_L
IL	JL	KL	ML
55	33	23	1

TABLE A3.2. Data on Pipes in the Network

i	$J_1(i)$	$J_2(i)$	$K_1(i)$	$K_2(i)$	$L(i)$ (m)	$k_f(i)$	$P(i)$ (no.)	$D(i)$ (m)
I	JLP(I,1)	JLP(I,2)	IKL(I,1)	IKL(I,2)	AL(I)	FK(I)	PP(I)	D(I)
1	1	2	2	0	380	0	500	0.150
2	2	3	4	0	310	0	385	0.125
3	3	4	5	0	430	0.2	540	0.125
4	4	5	6	0	270	0	240	0.080
5	1	6	1	0	150	0	190	0.050
6	6	7	0	0	200	0	500	0.065
7	6	9	1	0	150	0	190	0.065
8	1	10	1	2	150	0	190	0.200
9	2	11	2	3	390	0	490	0.150
10	2	12	3	4	320	0	400	0.050
11	3	13	4	5	320	0	400	0.065
12	4	14	5	6	330	0	415	0.080
13	5	14	6	7	420	0	525	0.080
14	5	15	7	0	320	0	400	0.050
15	9	10	1	0	160	0	200	0.080
16	10	11	2	0	120	0	150	0.200
17	11	12	3	8	280	0	350	0.200
18	12	13	4	9	330	0	415	0.200
19	13	14	5	11	450	0.2	560	0.080
20	14	15	7	14	360	0.2	450	0.065
21	11	16	8	0	230	0	280	0.125
22	12	19	8	9	350	0	440	0.100
23	13	20	9	10	360	0	450	0.100
24	13	22	10	11	260	0	325	0.250
25	14	22	11	13	320	0	400	0.250
26	21	22	10	12	160	0	200	0.250
27	22	23	12	13	290	0	365	0.250
28	14	23	13	14	320	0	400	0.065
29	15	23	14	15	500	0	625	0.100
30	15	24	15	0	330	0	410	0.050
31	16	17	0	0	230	0	290	0.050
32	16	18	8	0	220	0	275	0.125
33	18	19	8	16	350	0	440	0.065
34	19	20	9	17	330	0	410	0.050
35	20	21	10	19	220	0	475	0.100
36	21	23	12	19	250	0	310	0.100

(*Continued*)

TABLE A3.2. *Continued*

i	$J_1(i)$	$J_2(i)$	$K_1(i)$	$K_2(i)$	$L(i)$ (m)	$k_f(i)$	$P(i)$ (no.)	$D(i)$ (m)
37	23	24	15	20	370	0	460	0.100
38	18	25	16	0	470	0	590	0.065
39	19	25	16	17	320	0	400	0.080
40	20	25	17	18	460	0	575	0.065
41	20	26	18	19	310	0	390	0.065
42	23	27	19	20	330	0	410	0.200
43	24	27	20	21	510	0	640	0.050
44	24	28	21	0	470	0	590	0.100
45	25	26	18	0	300	0	375	0.065
46	26	27	19	0	490	0	610	0.080
47	27	29	22	0	230	0	290	0.200
48	27	28	21	22	290	0	350	0.200
49	28	29	22	23	190	0	240	0.150
50	29	30	23	0	200	0	250	0.050
51	28	31	23	0	160	0	200	0.100
52	30	31	23	0	140	0	175	0.050
53	31	32	0	0	250	0	310	0.065
54	32	33	0	0	200	0	250	0.050
55	7	8	0	0	200	0	250	0.065

TABLE A3.3. Nodal Elevation Data

j	$Z(j)$	j	$Z(j)$	j	$Z(j)$	j	$Z(j)$	j	$Z(j)$	j	$Z(j)$
J	Z(J)	J	Z(J)	J	Z(J)	J	Z(J)	J	Z(J)	J	Z(J)
1	101.85	7	101.80	13	101.80	19	101.60	25	101.40	31	101.80
2	101.90	8	101.40	14	101.90	20	101.80	26	101.20	32	101.80
3	101.95	9	101.85	15	100.50	21	101.85	27	101.70	33	100.40
4	101.60	10	101.90	16	100.80	22	101.95	28	101.90		
5	101.75	11	102.00	17	100.70	23	101.80	29	101.70		
6	101.80	12	101.80	18	101.40	24	101.10	30	101.80		

TABLE A3.4. Input Source Data

m	$S(m)$	$h_0(m)$
M	INP(M)	HA(M)
1	22	20

The data set is shown in Table A3.1. The notations used in the computer program are also included in this table for understanding the code.

Another data set listed in Table A3.2 is for pipe number (i), both nodes $J_1(i)$ and $J_1(i)$ of pipe i, loop numbers $K_1(i)$ and $K_2(i)$, pipe length $L(i)$, form-loss coefficient due to pipe fittings and valves $k_f(i)$, population load on pipe $P(i)$, and the pipe diameter $D(i)$. The notations used in developing the code are also provided in this table. It is important to note here that the pipe node $J_1(i)$ is the lower-magnitude node of the two. The data set can be generated without such limitation, and the program can modify these node numbers accordingly. Readers are advised to make necessary changes in the code as an exercise. Hint: Check after read statement (Lines 128 and 129 of code) if $J_1(i)$ is greater than $J_2(i)$, then redefine $J_1(i) = J_2(i)$ and $J_2(i) = J_1(i)$.

The next set of data is for nodal number and nodal elevations, which are provided in Table A3.3.

The final set of data is for input source node $S(m)$ and input head $h_0(m)$. In case of a single-input source network, $m_L = 1$. The notations used for input source node and input node pressure head are also listed in Table A3.4.

Source Code and Its Development

The source code for the analysis of a single-input source water distribution pipe network system is listed in Table A3.5. The line by line explanation of the source code is provided in the following text.

Line 100
Comment line for the name of the program, "Single-input source water distribution network analysis program."

Line 101
Comment line indicating that the next lines are for dimensions listing parameters requiring memory storages. (The $*$ is used for continuity of code lines.)

Line 102:106
The dimensions (memory storages) are provided for a 200-pipe network. The users can modify the memory size as per their requirements. The explanation for notations for which dimensions are provided is given below:

AK(I) = Multiplier for pipe head-loss computation
AL(I) = Length of pipe I
D(I) = Pipe diameter
DQ(K) = Discharge correction in loop K
F(I) = Friction factor for pipe I
FK(I) = Form-loss coefficient (k_f) due to pipe fittings and valve
H(J) = Terminal nodal pressure at node J
HA(M) = Input point head
IK(K,L) = Pipes in loop K, where L = 1, NLP(K)
IKL(I,1&2) = Loops 1 & 2 of pipe I
INP(M) = Input node of Mth input point, in case of single input source total input point ML = 1
IP(J,L) = Pipes connected to node J, where L = 1, NIP(J)
JK(K,L) = Nodes in loop K, where L = 1, NLP(K)
JLP(I,1&2) = Nodes 1 & 2 of pipe I; suffix 1 for lower-magnitude node and 2 for higher, however, this limitation can be eliminated by simple modification to code as described in an earlier section

TABLE A3.5. Single-Input Source Water Distribution System Source Code

Line		Single-Input Source Water Distribution Network Analysis Program
100	C	Single input source looped and branched network analysis program
101	C	Memory storage parameters (* line continuity)
102		DIMENSION JLP(200,2),IKL(200,2), AL(200),FK(200),D(200),
103	*	PP(200),Z(200),INP(1),HA(1), IP(200,10),NIP(200),
104	*	JN(200,10),S(200,10),QQ(200), Q(200),IK(200,10),
105	*	JK(200,10),NLP(200),SN(200,10), F(200),KD(200),
106	*	AK(200),DQ(200),H(200)
107	C	Input and output files
108		OPEN(UNIT=1,FILE='APPENDIX.DAT') ! data file
109		OPEN(UNIT=2,FILE='APPENDIX.OUT') ! output file
110	C	Read data for total pipes, nodes, loops, and input source ML=1
111		READ(1,*)IL,JL,KL,ML
112		WRITE (2,916)
113		WRITE (2,901)
114		WRITE (2,201) IL,JL,KL,ML
115		WRITE (2,250)
116		PRINT 916
117		PRINT 901
118		PRINT 201, IL,JL,KL,ML
119		PRINT 250
120	C	Read data for pipes- pipe number, pipe nodes 1&2, pipe loop 1&2, pipe length,
121	C	formloss coefficient due to valves& fitting, population load on pipe and pipe

(Continued)

TABLE A3.5 *Continued*

122	C	diameter. Note: Pipe node 1
		is lower number of the two
		nodes of a pipe.
123		WRITE (2, 917)
124		WRITE(2,902)
125		PRINT 917
126		PRINT 902
127		DO 1 I=1,IL
128		READ(1,*) IA,(JLP(IA,J),J=1,2),
		(IKL(IA,K),K=1,2),AL(IA),
129	*	FK(IA),PP(IA),D(IA)
130		WRITE(2,202) IA,(JLP(IA,J),J=1,2),
		(IKL(IA,K),K=1,2),AL(IA),
131	*	FK(IA),PP(IA),D(IA)
132		PRINT 202, IA,(JLP(IA,J),J=1,2),
		(IKL(IA,K),K=1,2),AL(IA),
133	*	FK(IA),PP(IA),D(IA)
134	1	CONTINUE
135		WRITE (2,250)
136		PRINT 250
137	C	Read data for nodal elevations
138		WRITE (2,918)
139		WRITE(2,903)
140		PRINT 918
141		PRINT 903
142		DO 2 J=1,JL
143		READ(1,*)JA, Z(JA)
144		WRITE(2,203) JA, Z(JA)
145		PRINT 203, JA, Z(JA)
146	2	CONTINUE
147		WRITE (2,250)
148		PRINT 250
149	C	Read data for input source node
		number and source input head
150		WRITE (2,919)
151		WRITE(2,904)
152		PRINT 919
153		PRINT 904
154		READ(1,*) M, INP(M), HA(M)

(Continued)

TABLE A3.5 *Continued*

```
155          WRITE (2,204) M, INP(M), HA(M)
156          PRINT 204, M, INP(M), HA(M)
157          WRITE (2,250)
158          PRINT 250

159    C        input parameters - rate of water
                supply and peak factor
160    C     - - - - - - - - - - - - - - - - - -
161          RTW=150.0    !Rate of water supply
                             (liters/person/day)
162          QPF=2.5      !Peak factor for
                             design flows
163    C     - - - - - - - - - - - - - - - - - -
164          CRTW=86400000.0 !Discharge conversion factor -
                                 Liters/day to m³/s
165          G=9.78          !Gravitational constant
166          PI=3.1415926    !Value of Pi
167          GAM=9780.00     !Weight density
168    C     - - - - - - - - - - - - - - - - - -

169    C     Initialize pipe flows by assigning
             zero flow rate

170          DO 4 I=1, IL
171          QQ(I)=0.0
172    4     CONTINUE

173    C        Identify all the pipes connected to a node J

174          DO 5 J=1,JL
175          IA=0
176          DO 6 I=1,IL
177          IF(.NOT.(J.EQ.JLP(I,1).OR
             .J.EQ.JLP(I,2)))GO TO 6
178          IA=IA+1
179          IP(J,IA)=I
180          NIP(J)=IA
181    6     CONTINUE
182    5     CONTINUE

183    C     Write and print pipes connected to a node J

184          Write (2,920)
185          WRITE(2,905)
```

TABLE A3.5 *Continued*

186		PRINT 920
187		PRINT 905
188		DO 7 J=1,JL
189		WRITE (2,205)J,NIP(J),(IP(J,L), L=1,NIP(J))
190		PRINT 205,J,NIP(J),(IP(J,L), L=1,NIP(J))
191	7	CONTINUE
192		WRITE (2,250)
193		PRINT 250
194	C	Identify all the nodes connected a node J through connected pipes
195		DO 8 J=1,JL
196		DO 9 L=1,NIP(J)
197		IPE=IP(J,L)
198		DO 10 LA=1,2
199		IF(JLP(IPE,LA).NE.J) JN(J,L)=JLP(IPE,LA)
200	10	CONTINUE
201	9	CONTINUE
202	8	CONTINUE
203	C	Write and print all the nodes connected to a node J
204		WRITE (2,921)
205		PRINT 921
206		WRITE(2,906)
207		PRINT 906
208		DO 60 J =1, JL
209		WRITE (2,206) J,NIP(J),(JN(J,L),L=1,NIP(J))
210		PRINT 206,J,NIP(J),(JN(J,L),L=1,NIP(J))
211	60	CONTINUE
212		WRITE (2,250)
213		PRINT 250
214	C	Identify loop pipes and loop nodes
215		DO 28 K=1,KL
216		DO 29 I=1,IL
217		IF(.NOT.((K.EQ.IKL(I,1)).OR. (K.EQ.IKL(I,2)))) GO TO 29
218		JK(K,1)=JLP(I,1)

(Continued)

TABLE A3.5 *Continued*

```
219        JB=JLP(I,1)
220        IK(K,1)=I
221        JK(K,2)=JLP(I,2)
222        GO TO 54
223     29 CONTINUE
224     54 NA=1
225        JJ=JK(K,NA+1)
226        II=IK(K,NA)
227     56 DO 30 L=1,NIP(JJ)
228        II=IK(K,NA)
229        IKL1=IKL(IP(JJ,L),1)
230        IKL2=IKL(IP(JJ,L),2)
231        IF(.NOT.((IKL1.EQ.K).OR.(IKL2.EQ.K)))
           GO TO 30
232        IF(IP(JJ,L).EQ.II) GO TO 30
233        NA=NA+1
234        NLP(K)=NA
235        IK(K,NA)=IP(JJ,L)
236        IF(JLP(IP(JJ,L),1).NE.JJ)
           JK(K,NA+1)=JLP(IP(JJ,L),1)
237        IF(JLP(IP(JJ,L),2).NE.JJ)
           JK(K,NA+1)=JLP(IP(JJ,L),2)
238        II=IK(K,NA)
239        JJ=JK(K,NA+1)
240        GO TO 57
241     30 CONTINUE
242     57 IF(JJ.NE.JB) GO TO 56
243     28 CONTINUE

245      C Write and print loop forming pipes

246        WRITE (2,250)
247        PRINT 250
248        WRITE(2,922)
249        WRITE (2,909)
250        PRINT 922
251        PRINT 909
252        DO 51 K=1,KL
253        WRITE (2, 213)K,NLP(K),(IK(K,NC),
           NC=1,NLP(K))
254        PRINT 213,K,NLP(K),(IK(K,NC),
           NC=1,NLP(K))
255     51 CONTINUE
```

(*Continued*)

TABLE A3.5 *Continued*

```
256   C      Write and print loop forming nodes
257          WRITE (2,250)
258          PRINT 250
259          WRITE(2,923)
260          WRITE (2,910)
261          PRINT 923
262          PRINT 910
263          DO 70 K=1, KL
264          WRITE (2, 213)K,NLP(K),(JK(K,NC),
             NC=1,NLP(K))
265          PRINT 213,K,NLP(K),(JK(K,NC),
             NC=1,NLP(K))
266   70 CONTINUE

267   C        Assign sign convention to
               pipes to apply continuity equations

268          DO 20 J=1,JL
269          DO 20 L=1,NIP(J)
270          IF(JN(J,L).LT.J) S(J,L)=1.0
271          IF(JN(J,L).GT.J) S(J,L)=-1.0
272   20 CONTINUE

273   C        Estimate nodal water demands-Transfer
               pipe loads to nodes

274          DO 73 J=1,JL
275          Q(J)=0.0
276          DO 74 L=1,NIP(J)
277          II=IP(J,L)
278          JJ=JN(J,L)
279          IF(J.EQ.INP(1)) GO TO 73
280          IF(JJ.EQ.INP(1)) GO TO 550
281          Q(J)=Q(J)+PP(II)*RTW*QPF/(CRTW*2.0)
282          GO TO 74
283   550 Q(J)=Q(J)+PP(II)*RTW*QPF/CRTW
284   74   CONTINUE
285   73   CONTINUE

286   C        Calculate input source point
               discharge (inflow)

287          SUM=0.0
288          DO 50 J=1,JL
```

(*Continued*)

TABLE A3.5 *Continued*

```
289        IF(J.EQ.INP(1)) GO TO 50
290        SUM=SUM+Q(J)
291     50 CONTINUE
292        QT=SUM
293        Q(INP(1))=-QT

294  C       Print and write nodal discharges

295        WRITE(2,907)
296        PRINT 907
297        WRITE (2,233)(J, Q(J),J=1,JL)
298        PRINT 233,(J,Q(J),J=1,JL)
299        WRITE (2,250)
300        PRINT 250

301  C       Initialize nodal terminal
             pressures by assigning zero head

302  69   DO 44 J=1,JL
303        H(J)=0.0
304  44   CONTINUE

305  C       Initialize pipe flow discharges
             by assigning zero flow rates

306        DO 45 I=1,IL
307        QQ(I)=0.0
308  45   CONTINUE

309  C   Assign arbitrary flow rate of
         0.01 m3/s to one of the loop pipes in
310  C   all the loops to apply
         continuity equation. [Change to
         0.1 m3/s to see impact].

311        DO 17 KA=1,KL
312        KC=0
313        DO 18 I=1,IL
314        IF(.NOT.(IKL(I,1).EQ.KA).OR.
             (IKL(I,2).EQ.KA))
315      * GO TO 18
316        IF(QQ(I).NE.0.0) GO TO 18
317        IF(KC.EQ.1) GO TO 17
318        QQ(I)=0.01
319        KC=1
```

(*Continued*)

TABLE A3.5 *Continued*

```
320   18   CONTINUE
321   17   CONTINUE

322   C        Apply continuity equation first
               at nodes having single connected pipe

323        DO 11 J=1,JL
324        IF(NIP(J).EQ.1) QQ(IP(J,1))=S(J,1)*Q(J)
325   11   CONTINUE

326   C        Now apply continuity equation at nodes
               having only one of its pipes with
327   C        unknown (zero) discharge till all the
               branch pipes have known discharges

328        NE=1
329        DO 12 J=1,JL
330        IF(J.EQ.INP(1)) GO TO 12
331        NC=0
332        DO 13 L=1,NIP(J)
333        IF(.NOT.((IKL(IP(J,L),1).EQ.0).AND.
           (IKL(IP(J,L),2).EQ.0)))
334    *   GO TO 13
335        NC=NC+1
336   13   CONTINUE
337        IF(NC.NE.NIP(J)) GO TO 12
338        DO 16 L=1,NIP(J)
339        IF(QQ(IP(J,L)).EQ.0.0) NE=0
340   16   CONTINUE
341        ND=0
342        DO 14 L=1,NIP(J)
343        IF(QQ(IP(J,L)).NE.0.0) GO TO 14
344        ND=ND+1
345        LD=L
346   14   CONTINUE
347        IF(ND.NE.1) GO TO 12
348        QQ(IP(J,LD))=S(J,LD)*Q(J)
349        DO 15 L= 1,NIP(J)
350        IF(IP(J,LD).EQ.IP(J,L)) GO TO 15
351        QQ(IP(J,LD))=QQ(IP(J,LD))
           -S(J,L)*QQ(IP(J,L))
352   15   CONTINUE
353   12   CONTINUE
354        IF(NE.EQ.0) GO TO 11
```

(Continued)

TABLE A3.5 *Continued*

```
355  C     Identify nodes that have one
           pipe with zero discharge

356  55    DO 21 J=1,JL
357        IF(J.EQ.INP(1)) GO TO 21
358        KD(J)=0
359        DO 22 L=1,NIP(J)
360        IF(QQ(IP(J,L)).NE.0.0) GO TO 22
361        KD(J)=KD(J)+1
362        LA=L
363  22    CONTINUE
364        IF(KD(J).NE.1) GO TO 21
365        SUM=0.0
366        DO 24 L=1,NIP(J)
367        SUM=SUM+S(J,L)*QQ(IP(J,L))
368  24    CONTINUE
369        QQ(IP(J,LA))=S(J,LA)*(Q(J)-SUM)
370  21    CONTINUE

371        DO 25 J=1,JL
372        IF(KD(J).NE.0) GO TO 55
373  25    CONTINUE

374  C     Write and print pipe discharges
           based on only continuity equation

375        WRITE (2,250)
376        PRINT   250
377        WRITE (2, 908)
378        PRINT 908
379        WRITE (2,210)(II,QQ(II),II=1,IL)
380        PRINT 210,(II,QQ(II),II=1,IL)

381  C     Allocate sign convention to loop
           pipes to apply loop discharge
382  C     corrections using Hardy-Cross method

383        DO 32 K=1,KL
384        DO 33 L=1,NLP(K)
385        IF(JK(K,L+1).GT.JK(K,L)) SN(K,L)=1.0
386        IF(JK(K,L+1).LT.JK(K,L)) SN(K,L)=-1.0
387  33    CONTINUE
```

(Continued)

TABLE A3.5 *Continued*

```
388   32   CONTINUE
389   C    Calculate friction factor using Eq. 2.6c

390   58     DO 34 I=1,IL
391          FAB=4.618*(D(I)/(ABS(QQ(I))*10.0**6))**0.9
392          FAC=0.00026/(3.7*D(I))
393          FAD=ALOG(FAB+FAC)
394          FAE=FAD**2
395          F(I)=1.325/FAE
396          EP=8.0/PI**2
397          AK(I)=(EP/(G*D(I)**4))*(F(I)*
             AL(I)/D(I)+FK(I))
398   34   CONTINUE

399   C      Loop discharge correction using
             Hardy-Cross method

400          DO 35 K=1,KL
401          SNU=0.0
402          SDE=0.0
403          DO 36 L=1,NLP(K)
404          IA=IK(K,L)
405          BB=AK(IA)*ABS(QQ(IA))
406          AA=SN(K,L)*AK(IA)*QQ(IA)*ABS(QQ(IA))
407          SNU=SNU+AA
408          SDE=SDE+BB
409   36   CONTINUE
410          DQ(K)=-0.5*SNU/SDE
411          DO 37 L=1,NLP(K)
412          IA=IK(K,L)
413          QQ(IA)=QQ(IA)+SN(K,L)*DQ(K)
414   37   CONTINUE
415   35   CONTINUE

416   C    Check for DQ(K) value for all the loops

417          DO 40 K=1,KL
418          IF(ABS(DQ(K)).GT.0.0001) GO TO 58
419   40     CONTINUE

420   C    Write and print input source node
           peak discharge
```

(Continued)

TABLE A3.5 *Continued*

```
421        WRITE (2,250)
422        PRINT 250
423        WRITE(2,913)
424        PRINT 913
425        WRITE (2,914) INP(1), Q(INP(1))
426        PRINT 914, INP(1), Q(INP(1))

427  C     Calculations for terminal pressure heads,
           starting from input source node

428        H(INP(1))=HA(1)
429    59 DO 39 J=1,JL
430        IF(H(J).EQ.0.0) GO TO 39
431        DO 41 L=1,NIP(J)
432        JJ=JN(J,L)
433        II=IP(J,L)
434        IF(JJ.GT.J) SI=1.0
435        IF(JJ.LT.J) SI=-1.0
436        IF(H(JJ).NE.0.0) GO TO 41
437        AC=SI*AK(II)*QQ(II)*ABS(QQ(II))
438        H(JJ)=H(J)-AC+Z(J)-Z(JJ)
439    41 CONTINUE
440    39 CONTINUE
441        DO 42 J=1,JL
442        IF(H(J).EQ.0.0) GO TO 59
443    42    CONTINUE

444  C            Write and print final pipe discharges

445        WRITE (2,250)
446        PRINT 250
447        WRITE (2,912)
448        PRINT 912
449        WRITE (2,210)(I,QQ(I),I=1,IL)
450        PRINT 210,(I,QQ(I),I=1,IL)

451  C     Write and print nodal terminal pressure heads

452        WRITE (2,250)
453        PRINT 250
454        WRITE (2,915)
455        PRINT 915
456        WRITE (2,229)(J,H(J),J=1,JL)
457        PRINT 229,(J,H(J),J=1,JL)
```

(Continued)

TABLE A3.5 *Continued*

```
458   201 FORMAT(5I5)
459   202 FORMAT(5I6,2F9.1,F8.0,2F9.3)
460   203 FORMAT(I5,2X,F8.2)
461   204 FORMAT(I5,I10,F10.2)
462   205 FORMAT(I5,1X,I5,3X,10I5)
463   206 FORMAT(I5,1X,I5,3X,10I5)
464   210 FORMAT(4(2X,'QQ('I3')='F6.4))
465   213 FORMAT(1X,2I4,10I7)
466   229 FORMAT(4(2X,'H('I3')='F6.2))
467   230 FORMAT(3(3X,'q(jim('I2'))='F9.4))
468   233 FORMAT(4(2X,'Q('I3')='F6.4))
469   250 FORMAT(/)
470   901 FORMAT(3X,'IL',3X,'JL',3X,'KL',3X,'ML')
471   902 FORMAT(4X,'i'3X,'J1(i)'2X'J2(i)'1X,'
          K1(i)'1X,'K2(i)'3X'L(i)'
472       * 6X'k_f(i)'2X'P(i)',5X'D(i)')
473   903 FORMAT(4X,'j',5X,'Z(j)')
474   904 FORMAT(4X,'m',6X,'INP(m)',3X,'HA(m)')
475   905 FORMAT(3X,'j',3X,'NIP(j)'5X'(IP(j,L),
          L=1,NIP(j)-Pipes to node)')
476   906 FORMAT(3X,'j',3X,'NIP(j)'5X'(JN(j,L),
          L=1,NIP(j)-Nodes to node)')
477   907 FORMAT(3X,'Nodal discharges - Input
          source node -tive discharge')
478   908 FORMAT(3x,'Pipe discharges
          based on continuity equation only')
479   909 FORMAT( 4X,'k',1X,'NLP(k)'2X'
          (IK(k,L),L=1,NLP(k)-Loop pipes)')
480   910 FORMAT( 4X,'k',1X,'NLP(k)'2X'
          (JK(k,L),L=1,NLP(k)-Loop nodes)')
481   911 FORMAT(2X,'Pipe friction factors
          using Swamee (1993) eq.')
482   912 FORMAT (2X, 'Final pipe discharges (m3/s)')
483   913 FORMAT (2X, 'Input source
          node and its discharge (m3/s)')
484   914 FORMAT (3X,'Input source node=['I3']',
          2X,'Input discharge='F8.4)
485   915 FORMAT (2X,'Nodal terminal
          pressure heads (m)')
486   916 FORMAT (2X, 'Total network size info')
487   917 FORMAT (2X, 'Pipe links data')
488   918 FORMAT (2X, 'Nodal elevation data')
489   919 FORMAT (2X, 'Input source nodal data')
490   920 FORMAT (2X, 'Information on pipes
          connected to a node j')
```

(Continued)

TABLE A3.5 *Continued*

```
491   921   FORMAT (2X, 'Information on nodes
                   connected to a node j')
492   922   FORMAT (2X, 'Loop forming pipes')
493   923   FORMAT (2X, 'Loop forming nodes')
494         CLOSE(UNIT=1)
495         STOP
496         END
```

JN(J,L) = Nodes connected to a node (through pipes), where L = 1, NIP(J)
KD(J) = A counter to count pipes with unknown discharges at node J
NIP(J) = Number of pipes connected to node J
NLP(K) = Total pipes or nodes in the loop
PP(I) = Population load on pipe I
Q(J) = Nodal water demand or withdrawal at node J
QQ(I) = Discharge in pipe I
S(J,L) = Sign convention for pipes at node J to apply continuity
equation (+1 or −1)
SN(K,L) = Sign convention for loop pipe discharges (+1 or −1)
Z(J) = Nodal elevation
Line 107
Comment line indicating next lines are for input and output files.
Line 108
Input data file "APPENDIX.DAT" that contains Tables A3.1, A3.2, A3.3, and A3.4.
Line 109
Output file "APPENDIX.OUT" that contains output specified by WRITE commands.
Line 110
Comment for READ command for network size – pipes, nodes, loops, and input point.
Line 111
READ statement – read data from unit 1 (data file "APPENDIX.DAT") for total pipes, total nodes, total loops, and input point source. (1,*) explains 1 is for unit 1 "APPENDIX.DAT" and * indicates free format used in unit 1.
Line 112
WRITE in unit 2 (output file "APPENDIX.OUT") FORMAT 916; that is, "Total pipe size info." See output file and FORMAT 916 in the code.
Line 113
WRITE command, write in unit 2 (output file) FORMAT 901; that is, "IL JL KL ML" to clearly read output file. See output file.
Line 114
WRITE command, write in output file–total pipes, nodes, loops and input source point (example for given data: 55 33 23 1).
Line 115
WRITE command, write in output file FORMAT 250; that is, provide next 2 lines blank. This is to separate two sets of WRITE statements. See output file.

Line 116:119
These lines are similar to Lines 112:115 but provide output on screen.
Line 120:122
Comment lines for next set of data in input file and instructions to write in output file and also to print on screen.
Line 123:136
READ, WRITE, and PRINT the data for pipes, both their nodes, both loops, pipe length (m), total pipe form-loss coefficient due to fittings and valves, population load (numbers), and pipe diameter (m). DO statement is used here (Lines 127 & 134) to read pipe by pipe data. See input data Table A3.6 and output file Table A3.7.

TABLE A3.6. Input Data File APPENDIX.DAT

55	33	23	1					
1	1	2	2	0	380	0	500	0.150
2	2	3	4	0	310	0	385	0.125
3	3	4	5	0	430	0.2	540	0.125
4	4	5	6	0	270	0	240	0.080
5	1	6	1	0	150	0	190	0.050
6	6	7	0	0	200	0	500	0.065
7	6	9	1	0	150	0	190	0.065
8	1	10	1	2	150	0	190	0.200
9	2	11	2	3	390	0	490	0.150
10	2	12	3	4	320	0	400	0.050
11	3	13	4	5	320	0	400	0.065
12	4	14	5	6	330	0	415	0.080
13	5	14	6	7	420	0	525	0.080
14	5	15	7	0	320	0	400	0.050
15	9	10	1	0	160	0	200	0.080
16	10	11	2	0	120	0	150	0.200
17	11	12	3	8	280	0	350	0.200
18	12	13	4	9	330	0	415	0.200
19	13	14	5	11	450	0.2	560	0.080
20	14	15	7	14	360	0.2	450	0.065
21	11	16	8	0	230	0	280	0.125
22	12	19	8	9	350	0	440	0.100
23	13	20	9	10	360	0	450	0.100
24	13	22	10	11	260	0	325	0.250
25	14	22	11	13	320	0	400	0.250
26	21	22	10	12	160	0	200	0.250
27	22	23	12	13	290	0	365	0.250
28	14	23	13	14	320	0	400	0.065
29	15	23	14	15	500	0	625	0.100
30	15	24	15	0	330	0	410	0.050

(*Continued*)

TABLE A3.6 *Continued*

31	16	17	0	0	230	0	290	0.050
32	16	18	8	0	220	0	275	0.125
33	18	19	8	16	350	0	440	0.065
34	19	20	9	17	330	0	410	0.050
35	20	21	10	19	220	0	475	0.100
36	21	23	12	19	250	0	310	0.100
37	23	24	15	20	370	0	460	0.100
38	18	25	16	0	470	0	590	0.065
39	19	25	16	17	320	0	400	0.080
40	20	25	17	18	460	0	575	0.065
41	20	26	18	19	310	0	390	0.065
42	23	27	19	20	330	0	410	0.200
43	24	27	20	21	510	0	640	0.050
44	24	28	21	0	470	0	590	0.100
45	25	26	18	0	300	0	375	0.065
46	26	27	19	0	490	0	610	0.080
47	27	29	22	0	230	0	290	0.200
48	27	28	21	22	290	0	350	0.200
49	28	29	22	23	190	0	240	0.150
50	29	30	23	0	200	0	250	0.050
51	28	31	23	0	160	0	200	0.100
52	30	31	23	0	140	0	175	0.050
53	31	32	0	0	250	0	310	0.065
54	32	33	0	0	200	0	250	0.050
55	7	8	0	0	200	0	250	0.065
1	101.85							
2	101.90							
3	101.95							
4	101.60							
5	101.75							
6	101.80							
7	101.80							
8	101.40							
9	101.85							
10	101.90							
11	102.00							
12	101.80							
13	101.80							
14	101.90							
15	100.50							
16	100.80							
17	100.70							
18	101.40							
19	101.60							

(Continued)

TABLE A3.6 *Continued*

20	101.80
21	101.85
22	101.95
23	101.80
24	101.10
25	101.40
26	101.20
27	101.70
28	101.90
29	101.70
30	101.80
31	101.80
32	101.80
33	100.40

1   22   20.00

TABLE A3.7. Output File APPENDIX.OUT

Total network size info

IL	JL	KL	ML
55	33	23	1

Pipe links data

i	J1(i)	J2(i)	K1(i)	K2(i)	L(i)	$k_f(i)$	P(i)	D(i)
1	1	2	2	0	380.0	.0	500.	.150
2	2	3	4	0	310.0	.0	385.	.125
3	3	4	5	0	430.0	.2	540.	.125
4	4	5	6	0	270.0	.0	240.	.080
5	1	6	1	0	150.0	.0	190.	.050
6	6	7	0	0	200.0	.0	500.	.065
7	6	9	1	0	150.0	.0	190.	.065
8	1	10	1	2	150.0	.0	190.	.200
9	2	11	2	3	390.0	.0	490.	.150
10	2	12	3	4	320.0	.0	400.	.050
11	3	13	4	5	320.0	.0	400.	.065
12	4	14	5	6	330.0	.0	415.	.080
13	5	14	6	7	420.0	.0	525.	.080
14	5	15	7	0	320.0	.0	400.	.050
15	9	10	1	0	160.0	.0	200.	.080
16	10	11	2	0	120.0	.0	150.	.200
17	11	12	3	8	280.0	.0	350.	.200
18	12	13	4	9	330.0	.0	415.	.200
19	13	14	5	11	450.0	.2	560.	.080

(Continued)

TABLE A3.7   *Continued*

20	14	15	7	14	360.0	.2	450.	.065
21	11	16	8	0	230.0	.0	280.	.125
22	12	19	8	9	350.0	.0	440.	.100
23	13	20	9	10	360.0	.0	450.	.100
24	13	22	10	11	260.0	.0	325.	.250
25	14	22	11	13	320.0	.0	400.	.250
26	21	22	10	12	160.0	.0	200.	.250
27	22	23	12	13	290.0	.0	365.	.250
28	14	23	13	14	320.0	.0	400.	.065
29	15	23	14	15	500.0	.0	625.	.100
30	15	24	15	0	330.0	.0	410.	.050
31	16	17	0	0	230.0	.0	290.	.050
32	16	18	8	0	220.0	.0	275.	.125
33	18	19	8	16	350.0	.0	440.	.065
34	19	20	9	17	330.0	.0	410.	.050
35	20	21	10	19	220.0	.0	475.	.100
36	21	23	12	19	250.0	.0	310.	.100
37	23	24	15	20	370.0	.0	460.	.100
38	18	25	16	0	470.0	.0	590.	.065
39	19	25	16	17	320.0	.0	400.	.080
40	20	25	17	18	460.0	.0	575.	.065
41	20	26	18	19	310.0	.0	390.	.065
42	23	27	19	20	330.0	.0	410.	.200
43	24	27	20	21	510.0	.0	640.	.050
44	24	28	21	0	470.0	.0	590.	.100
45	25	26	18	0	300.0	.0	375.	.065
46	26	27	19	0	490.0	.0	610.	.080
47	27	29	22	0	230.0	.0	290.	.200
48	27	28	21	22	290.0	.0	350.	.200
49	28	29	22	23	190.0	.0	240.	.150
50	29	30	23	0	200.0	.0	250.	.050
51	28	31	23	0	160.0	.0	200.	.100
52	30	31	23	0	140.0	.0	175.	.050
53	31	32	0	0	250.0	.0	310.	.065
54	32	33	0	0	200.0	.0	250.	.050
55	7	8	0	0	200.0	.0	250.	.065

Nodal elevation data
   j   Z(j)
   1   101.85
   2   101.90
   3   101.95
   4   101.60

(*Continued*)

TABLE A3.7 *Continued*

5	101.75
6	101.80
7	101.80
8	101.40
9	101.85
10	101.90
11	102.00
12	101.80
13	101.80
14	101.90
15	100.50
16	100.80
17	100.70
18	101.40
19	101.60
20	101.80
21	101.85
22	101.95
23	101.80
24	101.10
25	101.40
26	101.20
27	101.70
28	101.90
29	101.70
30	101.80
31	101.80
32	101.80
33	100.40

Input source nodal data

m	INP(m)	HA(m)
1	22	20.00

Information on pipes connected to a node j

j	NIP(j)	(IP(j,L),L=1,NIP(j)-Pipes to node)		
1	3	1	5	8
2	4	1	2	9	10
3	3	2	3	11
4	3	3	4	12
5	3	4	13	14
6	3	5	6	7
7	2	6	55
8	1	55

(Continued)

TABLE A3.7 *Continued*

9	2	7	15				
10	3	8	15	16			
11	4	9	16	17	21		
12	4	10	17	18	22		
13	5	11	18	19	23	24	
14	6	12	13	19	20	25	28
15	4	14	20	29	30		
16	3	21	31	32			
17	1	31					
18	3	32	33	38			
19	4	22	33	34	39		
20	5	23	34	35	40	41	
21	3	26	35	36			
22	4	24	25	26	27		
23	6	27	28	29	36	37	42
24	4	30	37	43	44		
25	4	38	39	40	45		
26	3	41	45	46			
27	5	42	43	46	47	48	
28	4	44	48	49	51		
29	3	47	49	50			
30	2	50	52				
31	3	51	52	53			
32	2	53	54				
33	1	54					

```
Information on nodes connected to a node j
    j    NIP(j)       (JN(j,L),L=1,NIP(j)-Nodes to node)
    1       3         2    6   10
    2       4         1    3   11   12
    3       3         2    4   13
    4       3         3    5   14
    5       3         4   14   15
    6       3         1    7    9
    7       2         6    8
    8       1         7
    9       2         6   10
   10       3         1    9   11
   11       4         2   10   12   16
   12       4         2   11   13   19
   13       5         3   12   14   20   22
   14       6         4    5   13   15   22   23
   15       4         5   14   23   24
   16       3        11   17   18
```

(*Continued*)

TABLE A3.7   *Continued*

17	1	16				
18	3	16	19	25		
19	4	12	18	20	25	
20	5	13	19	21	25	26
21	3	22	20	23		
22	4	13	14	21	23	
23	6	22	14	15	21	24	27
24	4	15	23	27	28	
25	4	18	19	20	26	
26	3	20	25	27		
27	5	23	24	26	29	28
28	4	24	27	29	31	
29	3	27	28	30		
30	2	29	31			
31	3	28	30	32		
32	2	31	33			
33	1	32				

Loop forming pipes

k	NLP(k)	(IK(k,L),L=1,NLP(k)-Loop pipes)				
1	4	5	7	15	8	
2	4	1	9	16	8	
3	3	9	17	10		
4	4	2	11	18	10	
5	4	3	12	19	11	
6	3	4	13	12		
7	3	13	20	14		
8	5	17	22	33	32	21
9	4	18	23	34	22	
10	4	23	35	26	24	
11	3	19	25	24		
12	3	26	27	36		
13	3	25	27	28		
14	3	20	29	28		
15	3	29	37	30		
16	3	33	39	38		
17	3	34	40	39		
18	3	40	45	41		
19	5	35	36	42	46	41
20	3	37	43	42		
21	3	43	48	44		
22	3	47	49	48		
23	4	49	50	52	51	

(Continued)

TABLE A3.7 *Continued*

Loop forming nodes

k	NLP(k)	(JK(k,L),L=1,NLP(k)-Loop nodes)				
1	4	1	6	9	10	
2	4	1	2	11	10	
3	3	2	11	12		
4	4	2	3	13	12	
5	4	3	4	14	13	
6	3	4	5	14		
7	3	5	14	15		
8	5	11	12	19	18	16
9	4	12	13	20	19	
10	4	13	20	21	22	
11	3	13	14	22		
12	3	21	22	23		
13	3	14	22	23		
14	3	14	15	23		
15	3	15	23	24		
16	3	18	19	25		
17	3	19	20	25		
18	3	20	25	26		
19	5	20	21	23	27	26
20	3	23	24	27		
21	3	24	27	28		
22	3	27	29	28		
23	4	28	29	30	31	

Nodal discharges - Input source node -tive discharge
Q(1)= .0019 Q(2)= .0039 Q(3)= .0029 Q(4)= .0026
Q(5)= .0025 Q(6)= .0019 Q(7)= .0016 Q(8)= .0005
Q(9)= .0008 Q(10)= .0012 Q(11)= .0028 Q(12)= .0035
Q(13)= .0054 Q(14)= .0068 Q(15)= .0041 Q(16)= .0018
Q(17)= .0006 Q(18)= .0028 Q(19)= .0037 Q(20)= .0050
Q(21)= .0026 Q(22)=-.0909 Q(23)= .0064 Q(24)= .0046
Q(25)= .0042 Q(26)= .0030 Q(27)= .0050 Q(28)= .0030
Q(29)= .0017 Q(30)= .0009 Q(31)= .0015 Q(32)= .0012
Q(33)= .0005

Pipe discharges based on continuity equation only
QQ(1)= .0100 QQ(2)= .0100 QQ(3)= .0100 QQ(4)= .0100
QQ(5)= .0100 QQ(6)= .0022 QQ(7)= .0059 QQ(8)= -.0219

(Continued)

TABLE A3.7 *Continued*

```
QQ(  9)=-.0139 QQ( 10)= .0100 QQ( 11)=-.0029 QQ( 12)=-.0026
QQ( 13)=-.0025 QQ( 14)= .0100 QQ( 15)= .0051 QQ( 16)=-.0180
QQ( 17)=-.0446 QQ( 18)= .0743 QQ( 19)= .0460 QQ( 20)= .0141
QQ( 21)= .0100 QQ( 22)=-.1124 QQ( 23)= .0100 QQ( 24)= .0100
QQ( 25)= .0100 QQ( 26)=-.1009 QQ( 27)= .0100 QQ( 28)= .0100
QQ( 29)= .0100 QQ( 30)= .0100 QQ( 31)= .0006 QQ( 32)= .0075
QQ( 33)=-.0053 QQ( 34)=-.0824 QQ( 35)=-.0974 QQ( 36)= .0009
QQ( 37)= .0146 QQ( 38)= .0100 QQ( 39)=-.0390 QQ( 40)= .0100
QQ( 41)= .0100 QQ( 42)= .0100 QQ( 43)= .0100 QQ( 44)= .0100
QQ( 45)=-.0232 QQ( 46)=-.0162 QQ( 47)= .0100 QQ( 48)=-.0111
QQ( 49)= .0017 QQ( 50)= .0100 QQ( 51)=-.0058 QQ( 52)= .0091
QQ( 53)= .0018 QQ( 54)= .0005 QQ( 55)= .0005

Input source node and its discharge (m3/s)
  Input  source node=[ 22] Input discharge= -.0909
Final pipe discharges (m3/s)
QQ(  1)= .0011 QQ(  2)= .0004 QQ(  3)=-.0009 QQ(  4)=-.0005
QQ(  5)= .0016 QQ(  6)= .0022 QQ(  7)=-.0025 QQ(  8)=-.0046
QQ(  9)=-.0028 QQ( 10)=-.0004 QQ( 11)=-.0016 QQ( 12)=-.0030
QQ( 13)=-.0026 QQ( 14)=-.0005 QQ( 15)=-.0033 QQ( 16)=-.0091
QQ( 17)=-.0202 QQ( 18)=-.0282 QQ( 19)=-.0012 QQ( 20)= .0013
QQ( 21)= .0056 QQ( 22)= .0041 QQ( 23)= .0033 QQ( 24)=-.0373
QQ( 25)=-.0154 QQ( 26)=-.0106 QQ( 27)= .0275 QQ( 28)= .0006
QQ( 29)=-.0031 QQ( 30)=-.0002 QQ( 31)= .0006 QQ( 32)= .0032
QQ( 33)=-.0002 QQ( 34)=-.0008 QQ( 35)=-.0056 QQ( 36)= .0024
QQ( 37)= .0033 QQ( 38)= .0005 QQ( 39)= .0011 QQ( 40)= .0015
QQ( 41)= .0016 QQ( 42)= .0178 QQ( 43)=-.0002 QQ( 44)=-.0013
QQ( 45)=-.0011 QQ( 46)=-.0025 QQ( 47)= .0045 QQ( 48)= .0057
QQ( 49)=-.0022 QQ( 50)= .0006 QQ( 51)= .0036 QQ( 52)=-.0003
QQ( 53)= .0018 QQ( 54)= .0005 QQ( 55)= .0005

Nodal terminal pressure heads (m)
H(  1)= 17.60 H(  2)= 17.54 H(  3)= 17.48 H(  4)= 18.00
H(  5)= 17.94 H(  6)= 14.31 H(  7)= 12.21 H(  8)= 12.46
H(  9)= 16.31 H( 10)= 17.13 H( 11)= 17.09 H( 12)= 17.97
H( 13)= 19.49 H( 14)= 19.90 H( 15)= 19.93 H( 16)= 17.51
H( 17)= 16.73 H( 18)= 16.74 H( 19)= 16.64 H( 20)= 18.55
H( 21)= 20.06 H( 22)= 20.00 H( 23)= 19.74 H( 24)= 19.51
H( 25)= 16.45 H( 26)= 17.41 H( 27)= 19.21 H( 28)= 18.92
H( 29)= 19.18 H( 30)= 18.41 H( 31)= 18.54 H( 32)= 16.80
H( 33)= 17.62
```

Line 137

Comment line for nodal elevations.

Line 138:148

READ, WRITE, and PRINT nodal elevations.

Line 149

Comment for input source point and input source head data.

Line 150:158

READ, WRITE, and PRINT input source node and input head.

Line 159

Comment for input data for rate of water supply and peak flow factor.

Line 160

Comment line – separation of a block.

Line 161:162

Input data for rate of water supply RTW (liters/person/day) and QPF peak flow factor.

Line 163

Comment line – separation of a block.

Line 164

CRTW is a conversion factor from liters/day to m^3/s.

Line 165:167

Input value for gravitational constant (G), (PI), and weight density of water (GAM).

Line 168

Comment – line for block separation.

Line 169

Comment line for initializing pipe flows by assigning zero discharges.

Line 170:172

Initialize pipe discharges $QQ(I) = 0.0$ for all the pipes in the network using DO statement.

Line 173

Comment line for identifying pipes connected to a node.

Line 174:182

The algorithm coded in these lines is described below:

Check at each node J for pipes that have either of their nodes JLP(I,1) or JLP(I,2) equal to node J. First such pipe is IP(J,1) and the second IP(J,2) and so on. The total pipes connected to node J are NIP(J).

See Fig. A3.2; for node J = 1 scanning for pipe nodes JLP(I,1) and JLP(I,2) starting with pipe I = 1, one will find that JLP(1,1) = 1. Thus, the first pipe connected to node 1 is pipe number 1. Further scanning for pipes, one will find that pipes 2, 3, and 4 do not have any of their nodes equal to 1. Further investigation will indicate that pipes 5 and 8 have one of their nodes equal to 1. No other pipe in the whole network has one of its nodes equal to 1. Thus, only three pipes 1, 5, and 8 have one of their nodes as 1 and are connected to node 1. The total number of connected pipes NIP(J = 1) = 3.

The first DO loop (Line 174) is for node by node investigation. The second DO loop (Line 176) is for pipe by pipe scanning. Line 177 checks if pipe node JLP(I,1) or JLP(I,2) is equal to node J. If the answer is negative, then go to the next pipe. If the answer is positive, increase the value of the counter IA by 1 and record the connected

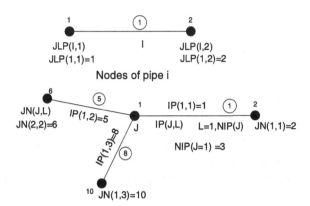

Figure A3.2. Pipes and nodes connected to a node J.

pipe IP(I, IA) = I. Check all the pipes in the network. At the end, record total pipes NIP(J) = IA connected to node J (Line 180). Repeat the process at all the other nodes.
Line 183
Comment line for write and print pipes connected to various nodes.
Line 184:193
DO statement (Lines 188 & 191) has been used to WRITE and PRINT node by node total pipes NIP(J) connected to a node J and connected pipes IP(J,L), where L = 1, NIP(J). Other statements (Lines 192 & 193) are added to separate the different sections of output file to improve readability.
Line 194
Comment statement to indicate that the next program lines are to identify nodes connected to a node J through connected pipes.
Line 195:202
First DO statement (Line 195) is for nodes, 1 to JL. Second DO statement is for total pipes meeting at node J, where index L = 1 to NIP(J). Third DO statement is for two nodes of a pipe I, thus index LA = 1, 2. Then check if JLP(I,LA) ≠ J, which means other node JN(J,L) of node J is JLP(I,LA). Thus repeating the process for all the nodes and connected pipes at each node, all the connected nodes JN(J,L) to node J are identified.
Line 203
Comment line that the next code lines are for write and print nodes connected to a node J.
Line 204:213
DO statement (Line 208) is used to WRITE and PRINT node by node the other nodes JN(J,L) connected to node J, where L = 1, NIP(J). Total pipes connected at node J are NIP(J).
Line 214
Comment line that the next code lines are to identify loop pipes and loop nodes.
Line 215:243
Line 215 is for DO statement to move loop by loop using index K.
Line 216 is for DO statement to move pipe by pipe using index I.

Line 217 is for IF statement checking if any loop of pipe I is equal to index K, if not go to next pipe otherwise go to next line.

Line 218 first node JK(K,1) of the loop K = JLP(I,1).

Line 219 is for renaming JLP(I,1) as JB (Starting node of loop).

Line 220 is for first pipe IK(K,1) of Kth loop = I.

Line 221 is for second node of loop JK(K,2) = JLP(I,2).

Line 222 is GO statement (go to Line 224).

Line 224 initiate the counter NA for loop pipes = 1.

Line 225 redefines JK(K,NA+1) as JJ, which is the other node of pipe IK(K,NA) and

Line 226 redefines IK(K,NA) as II.

Line 227 is a DO statement to check at node JJ for next pipe and next node of loop K.

Line 228 redefines IK(K,NA) as II.

Line 229 defines first loop IKL(IP(JJ,L),1) of pipe IP(JJ,L) as IKL1.

Line 230 defines second loop IKL(IP(JJ,L),2) of pipe IP(JJ,L) as IKL2.

Line 231 checks if any of the pipe IP(JJ,L)'s loops equal to loop index K (Line 215). If not, go to next pipe of node JJ otherwise go to next Line 232.

Line 232 checks if pipe IP(JJ,L) is the same pipe II as in Line 228, which has been already identified as Kth loop pipe, then go to next pipe at node JJ otherwise go to next Line 233.

Line 233 Here index NA is increased by 1, that is, NA = NA + 1.

Line 234 for total number of pipes in Kth loop (NLP(K) = NA).

Line 235 for next loop pipe IK(K,NA) = IP(JJ,L).

Line 236 and 237 will check for node of pipe IK(K,NA), which is not equal to node JJ, that node of pipe IK(K,NA) will be JN(K,NA + 1).

Line 238 and 239 redefine II and JJ with new values of loop pipe and loop node.

Line 240 is a GO statement to transfer execution to line 242.

Line 242 is for checking if node JJ is equal to node JB (starting loop forming node), if not repeat the process from Line 227 with new JJ and II values and repeat the process until node JJ = node JB. A this stage, all the loop-forming pipes are identified.

See Fig. A3.1, starting from loop index K = 1 (Line 215) check for pipes (Line 216) if any of pipe I's loops equal to K. The pipe 5 has its first loop IKL(5,1) = 1, thus JK(1,1) = 1 (Line 218) and the first pipe of loop 1 is IK(1,1) = 5 (Line 220). The second node of 1st loop (K = 1) is JK(1,2) = JLP(5,2) = 6 (second node of pipe 5 is node 6). Now check at node JJ = 6. At this node, NIP(JJ) = 3 thus 3 pipes (5, 6, and pipe 7) are connected at node 6. Now again check for pipe having one of its loop equal to 1 (Line 231), the pipe 5 is picked up first. Now check if this pipe has been picked up already in previous step (Line 232). If yes, skip this pipe and check for the next pipe connected at node 6. Next pipe is pipe 6, which has none of the loops equal to loop 1. Skip this pipe. Again moving to next pipe 7, it has one of its loops equal to 1. The next loop pipe IK(1,2) = 7 and next node JN(1,3) = 9. Repeating the process at node 9, IK(1,3) = 15 and JK(1,4) = 10 are identified. Until this point, the node JJ = 10 is not equal to starting node JB = 1, thus the process is repeated again, which identifies IK(1,4) = 8. At this stage, the algorithm for identifying loop-forming pipes for loop 1 stops as now JJ = JB. This will result

IK(1,1) = 5, IK(1,2) = 7, IK(1,3) = 15 and IK(1,4) = 8. Total NLP(1) = 4

JK(1,1) = 1, JK(1,2) = 6, JK(1,3) = 9 and JK(1,4) = 10.
Line 245
Comment line that the next lines are for write and print loop-forming pipes.
Line 246:255
WRITE and PRINT command for loop pipes.
Line 256
Comment line for write and print loop-forming nodes.
Line 257:266
WRITE and PRINT loop-wise total nodes in a loop and loop nodes.
Line 267
Comment line for assigning sign convention to pipes for applying continuity equation.
Line 268:272
Assign sign convention to pipes meeting at node J based on the magnitude of the other node of the pipe. The sign S(J,L) is positive (1.0) if the magnitude of the other node JN(J,L) is less than node J or otherwise negative (-1.0).
Line 273
Comment line that the next lines are for calculating nodal water demand by transferring pipe population loads to nodes.
Line 274:285
In these lines, the pipe population load is transferred equally to its both nodes (Line 281). In case of a pipe having one of its nodes as input point node, the whole population load is transferred to the other node (Line 283). Finally, nodal demands are calculated for all the nodes by summing the loads transferred from connected pipes. Lines 279 and 280 check the input point node. The population load is converted to peak demand by multiplying by peak factor (Line 162) and rate of water supply per person per day (Line 161). The product is divided by a conversation factor CRTW (Line 164) for converting daily demand rate to m^3/s.
Line 286
Comment line indicating that the next code lines are to estimate input source node discharge.
Line 287:293
The discharge of the input point source is the sum of all the nodal point demands except input source, which is a supply node. The source node has negative discharge (inflow) whereas demand nodes have positive discharge (outflow).
Line 294
Comment line that the next code lines are for write and print nodal discharges (demand).
Line 295:300
WRITE and PRINT nodal water demands
Check the FORMAT 233 and the output file, the way the nodal discharges Q are written.
Line 301
Comment line that the next code lines are about initializing terminal nodal pressures.
Line 302:304
Initialize all the nodal terminal heads at zero meter head.
Line 305
Comment line for next code lines.

Line 306:308
Initialize all the pipe discharges $QQ(I) = 0.0\,m^3/s$.
Line 309:310
Comment line indicating that the next code lines are for assigning an arbitrary discharge $(0.01\,m^3/s)$ in one of the pipes of a loop. Such pipes are equal to the number of total loops KL. Change the arbitrary flow value between 0.01 and $0.1\,m^3/s$ to see the impact on final pipe flows if any.
Line 311:321
First DO statement is for moving loop by loop. Here KA is used as loop index number instead of K. Second DO statement is for checking pipe by pipe if the pipe's first loop IKL(I,1) or the second loop IKL(I,2) is equal to the loop index KA (Line 314). If not, go to the next pipe and repeat the process again. If any of the pipe's loop IKL(I,1) or IKL(I,2) is equal to KA, then go to next line. Line 316 checks if a pipe has been assigned arbitrary discharge previously, then go to the next pipe of that loop. Line 317 checks if a pipe of the loop has been assigned a discharge, if so go to next loop. Line 318 assigns pipe discharge = 0.01. Lines 312, 317, and 319 check that only single pipe in a loop is assigned with this arbitrary discharge to apply continuity equation.
Line 322
Comment line stating that continuity equation is applied first at nodes with $NIP(J) = 1$.
Line 323:325
Check for nodes having only one connected pipe; the pipe discharge at such nodes is $QQ(IP(J,1) = S(J,1) \times Q(J)$.
See Fig. A3.3 for node 33. Node 33 has only one pipe $NIP(33) = 1$ and the connected pipe $IP(33,1) = 54$. The sign convention at node 33 is $S(33,1) = 1.0$. It is positive as the other node $[JN(33,1) = 32]$ of node 33 has lower magnitude. As the nodal withdrawals are positive, so $Q(33)$ will be positive. Thus $QQ(54) = Q(33)$. Meaning thereby, the discharge in pipe 54 is positive and flows from lower-magnitude node to higher-magnitude node.
Line 326:327
Comment line for applying continuity equation at nodes having one of its pipes with unknown discharge. Repeat the process until all the branched pipes have nonzero discharges.
Line 328:354
Here at Line 328, NE is a counter initialized equal to 1, which will check if any pipe at node J has zero discharge. See Line 339. If any pipe at node J has zero discharge, the NE

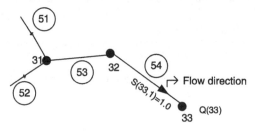

Figure A3.3. Nodal discharge computation.

value will change from 1 to 0. Now see Line 354. If NE = 0, the process is repeated until all the branch pipes have nonzero discharges.

Line 329 is for DO statement to move node by node. Next Line 330 is to check if the node J under consideration is an input node; if so leave this node and go to next node.

Line 331 is for a counter NC initialized equal to 0 for checking if any of the node pipes is a member of any loop (Lines 332:336). If so (Line 337) leave this node and go to next node.

Next Lines 338:340 are to check if any of the pipe discharge connected to node J is equal to zero, the counter is redefined NE = 0.

Line 341 is again for a counter ND, initialized equal to 0 is to check and count number of pipes having zero discharges. This counting process takes place in Lines 342:346.

Line 347 is to check if only one pipe has unknown discharge. If not go to next node, otherwise execute the next step.

Line 348 is for calculating pipe discharge QQ(IP(J,LD)) where LD stands for pipe number at node J with zero discharge. The discharge component from nodal demand Q(J) is transferred to pipe discharge QQ(IP(J,LD)) = S(J,LD) × Q(J).

Lines 349:352 add algebraic discharges of other pipes connected at node J to QQ(J,LD), that is,

$$QQ(IP(J,LD)) = QQ(IP(J,LD)) - \sum_{L=1,L\neq LD}^{NIP(j)-1} S(J,L) \times QQ(IP(J,L))$$

Line 354 checks if any of the branched pipes has zero discharge, then repeat the process from Line 328. IF statement of Line 354 will take the execution back to Line 325, which is continue command. The repeat execution will start from Line 328. Repeating the process, the discharges in all the branch pipes of the network can be estimated.

Line 355
Comment line indicating that the next lines are for identifying nodes that have only one pipe with unknown discharge. This will cover looped network section.

Line 356:373
Line 356 is for a DO loop statement for node by node command execution.

Line 357 checks if the node J under consideration is the input source point, if so go to next node.

Line 358 is for counter KD(J), which is initialized at 0. It counts the number of pipes with zero discharge at node J.

Lines 359:363 are for counting pipes with zero discharges.

Line 364 is an IF statement to check KD(J) value, if not equal to 1 then go to next node. KD(J) = 1 indicates that at node J, only one pipe has zero discharge and this pipe is IP(J,LA).

Lines 365:368 are for algebraic sum of pipe discharges at node J. At this stage, the discharge in pipe IP(J,LA) is zero and will not impact SUM estimation.

Line 369 is for estimating discharge in pipe IP(J,LA) by applying continuity equation

$$QQ(IP(J,LA)) = S(J,LA) \times (Q(J) - SUM), \text{ where}$$

$$SUM = \sum_{L=1}^{NIP(J)} S(J,L) \times QQ(IP(J,LA))$$

Lines 371:373 are for checking if any of the nodes has any pipe with zero discharge, if so repeat the process from Line 356 again.
Line 374
Comment line for write and print commands.
Line 375:380
WRITE and PRINT pipe discharges after applying continuity equation.
Line 381:382
Comment lines for allocation of sign convention for loop pipes for loop discharge correction. Hardy Cross method has been applied here.
Line 383:388
Loop-wise sign conventions are allocated as described below:
If loop node JK(K,L + 1) is greater than JK(K,L), then allocate SN(K,L) = 1.0 or otherwise if JK(K,L + 1) is less than JK(K,L), then allocate SN(K,L) = −1.0
Line 389
Comment line for calculating friction factor in pipes using Eq. (2.6c).
Line 390:398
Using Eq. (2.6c), the friction factor in pipes F(I) is

$$f_i = \frac{1.325}{\left[\ln\left(\frac{\varepsilon_i}{3.7D_i}\right) + 4.618\left(\frac{vD_i}{Q_i}\right)^{0.9}\right]^2}$$

See list of notations for notations used in the above equation.
Line 395 calculates finally the friction factor F(I) in pipe I.
Line 397 is for calculating head-loss multiplier AK(I) in pipe I, which is K_i Eq. (3.15) and head-loss multiplier due to pipe fittings and valves derived from Eq. (2.7b)

$$AK_i = \frac{8f_i L_i}{\pi^2 g D_i^5} + k_{fi} \frac{8}{\pi^2 g D_i^4}, \text{ where } f_i = F(I) \text{ and } k_{fi} \text{ is FK(I)}$$

Line 399
Comment line for loop discharge corrections using Hardy Cross method. Readers can modify this program using other methods described in Chapter 3 (Section 3.7).
Line 400:415
This section of code calculates discharge correction in loop pipes using Eq. (3.17).

Line 412 calculates loop discharge correction

$$DQ(K) = -0.5\frac{SNU}{SDE} = -0.5\frac{\displaystyle\sum_{\text{loop K}} K_i Q_i |Q_i|}{\displaystyle\sum_{\text{loop K}} K_i |Q_i|}$$

where Line 407 calculates SNU and Line 408 calculates SDE.
Lines 411:414 apply loop discharge corrections to loop pipes.

Line 416
Comment line for checking the magnitude of discharge correction DK(K).

Line 417:419
Check loop-wise discharge correction if the magnitude of any of the discharge correction is greater than $0.0001\,\text{m}^3/\text{s}$, then repeat the process for loop discharge correction. The user can modify this limit; however, the smaller the value, the higher the accuracy but more computer time.

Line 420
Comment line for write and print peak source node discharge.

Line 421:426
WRITE and PRINT input point discharge (peak flow).

Line 427
Comment line for next block of code, which is for nodal terminal pressures heads calculation.

Line 428:443
Line 428 equates the terminal pressure H (INP(1)) of input point node equal to given source node pressure head (HA(1)). This section calculates nodal terminal pressure heads starting from a node that has known terminal head. The algorithm will start from input point node as the terminal pressure head of this node is known then calculates the terminal pressure heads of connected nodes JN(J,L) through pipes IP(J,L). The sign convention SI is allotted (Lines 434:435) to calculate pressure head based on the magnitude of JN(J,L). Line 438 calculates the terminal pressure at node JJ, that is JN(J,L).
The code can be modified by deleting Lines 441 and 442 and modifying Line 444 to AC = -S(J,L)*AK(II)*QQ(II)*ABS(QQ(II)). Try and see why this will also work?
Line 442 will check if any of the terminal head is zero, if so repeat the process.

Line 444
Comment line for write and print final pipe discharges.

Line 445:450
WRITE and PRINT final pipe discharges.

Line 451
Comment line that the next section of code is for write and print terminal pressure heads.

Line 452:457
WRITE and PRINT terminal pressure heads of all the nodes.

Line 458:493
The various FORMAT commands used in the code development are listed in this section. See the output file for information on these formats.

Line 495:496

STOP and END the program.

The input and output files obtained using this software are attached as Table A3.6 and Table A3.7.

MULTI-INPUT WATER DISTRIBUTION NETWORK ANALYSIS PROGRAM

The multi-input water distribution network analysis program is described in this section. The city water distribution systems are generally multi-input source networks. A water distribution network as shown in Fig. A3.1 is modified to introduce two additional input source points at nodes 11 and 28. The modified network is shown in Fig. A3.4. The source code is provided in Table A3.10.

Data Set

The water distribution network has 55 pipes (i_L), 33 nodes (j_L), 23 loops (k_L), and 3 input sources (m_L). The revised data set is shown in Table A3.8.

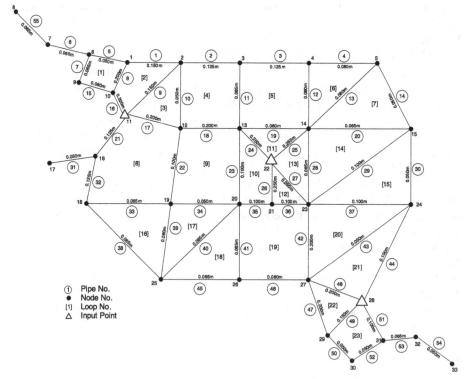

Figure A3.4. Multi-input source water distribution system.

TABLE A3.8. Pipe Network Size

i_L	j_L	k_L	m_L
IL	JL	KL	ML
55	33	23	3

The network pipe data in Table A3.2 and nodal elevation data in Table A3.3 are also applicable for multi-input water distribution network. The final set of data for input source nodes $S(m)$ and input heads $h_0(m)$ is provided in Table A3.9 for this network.

Source Code and Its Development

The source code for the analysis of a multi-input source water distribution pipe network system is listed in Table A3.10. The line by line explanation of the source code is provided in the following text.

Line 100
Comment line for the name of the program, "Multi-input source water distribution network analysis program."

Line 101:107
Same as explained for single-input source network.
The dimensions for input point source INP(10) and input point head (HA(10) are modified to include up to 10 input sources.

Line 108
Input data file "APPENDIXMIS.DAT" contains Tables A3.8, A3.2, A3.3, and A3.9.

Line 109
Output file "APPENDIXMIS.OUT" that contains output specified by WRITE commands. User can modify the names of input and output files as per their choice.

Line 110:153
Same as explained for single input source code.

Line 154:160
A DO loop has been introduced in Lines 154 and 158 to cover multi-input source data. For remaining lines, the explanation is the same as provided for single-input source program Lines 154:158.

TABLE A3.9. Input Source Data

m	$S(m)$	$h_0(m)$
M	INP(M)	HA(M)
1	11	20
2	22	20
3	28	20

TABLE A3.10. Multi-input Source Water Distribution System Analysis: Source Code

```
Line       Multi-input Source Water Distribution
           Network Analysis Program
100  C     Multi-input source looped and branched
           network analysis program
101  C     Memory storage parameters
           (* line continuity)

102        DIMENSION JLP(200,2),IKL(200,2),AL(200),
           FK(200),D(200),
103      *  PP(200),Z(200),INP(10),HA(10),
           IP(200,10),NIP(200),
104      *  JN(200,10),S(200,10),QQ(200),Q(200),
           IK(200,10),
105      *  JK(200,10),NLP(200),SN(200,10),
           F(200),KD(200),
106      *  AK(200),DQ(200),H(200)

107  C     Input and output files

108        OPEN(UNIT=1,FILE='APPENDIXMIS.DAT')
           !data fille
109        OPEN(UNIT=2,FILE='APPENDIXMIS.OUT')
           !output file

110  C     Read data for total pipes, nodes, loops,
           and input source

111        READ(1,*)IL,JL,KL,ML
112        WRITE (2, 916)
113        WRITE (2,901)
114        WRITE (2,201) IL,JL,KL,ML
115        WRITE (2,250)
116        PRINT 916
117        PRINT 901
118        PRINT 201, IL,JL,KL,ML
119        PRINT 250

120  C Read data for pipes- pipe number, pipe nodes
         1&2, pipe loop 1&2, pipe length,
121  C formloss coefficient due to fitting, pipe
         population load and pipe diameter
```

(Continued)

TABLE A3.10 *Continued*

```
122  C    Note: Pipe node 1 is lower magnitude number
          of the two nodes of a pipe.

123       WRITE (2,917)
124       WRITE(2,902)
125       PRINT 917
126       PRINT 902
127       DO 1 I=1,IL
128       READ(1,*)IA,(JLP(IA,J),J=1,2),
          IKL(IA,K),K=1,2),AL(IA),
129  *    FK(IA),PP(IA),D(IA)
130       WRITE(2,202) IA,(JLP(IA,J),J=1,2),
          (IKL(IA,K),K=1,2),AL(IA),
131  *    FK(IA),PP(IA),D(IA)
132       PRINT 202,IA,(JLP(IA,J),J=1,2),
          IKL(IA,K),K=1,2),AL(IA),
133  *    FK(IA),PP(IA),D(IA)
134  1    CONTINUE
135       WRITE (2,250)
136       PRINT 250

137  C        Read data for nodal elevations
138           WRITE (2, 918)
139           WRITE(2,903)
140           PRINT 918
141           PRINT 903
142           DO 2 J=1,JL
143           READ(1,*)JA, Z(JA)
144           WRITE(2,203) JA, Z(JA)
145           PRINT 203, JA, Z(JA)
146  2        CONTINUE
147           WRITE (2,250)
148           PRINT 250

149  C        Read data for input source node number
             and source input head

150           WRITE (2,919)
151           WRITE(2,904)
152           PRINT 919
153           PRINT 904

154       DO 3 M=1,ML
```

TABLE A3.10 *Continued*

```
155          READ(1,*)MA,INP(MA),HA(MA)
156          WRITE (2,204) MA,INP(MA),HA(MA)
157          PRINT 204,MA,INP(MA),HA(MA)
158     3    CONTINUE
159          WRITE (2,250)
160          PRINT 250

161  C    Input parameters rate of water supply
          and peak factor
162  C    - - - - - - - - - - - - - - - - - - - - -
163          RTW=150.0           ! Rate of water supply
                                   (liters/person/day)
164          QPF=2.5             ! Peak factor for design
                                   flows

165  C    - - - - - - - - - - - - - - - - - - - - -
166          CRTW=86400000.0  ! Discharge conversion
                                  factor -Liters/day to m3/s
167          G=9.78              ! Gravitational constant
168          PI=3.1415926        ! Value of Pi
169          GAM=9780.00         ! Weight density
170  C    - - - - - - - - - - - - - - - - - - - - -
171          FF=60.0             ! Initial error in input
                                   head and computed head
172          NFF=1               ! Counter for discharge
                                   correction

173  C    Initialize pipe flows by assigning zero
          flow rate

174          DO 4 I=1,IL
175          QQ(I)=0.0
176     4    CONTINUE

177  C    Identify all the pipes connected to a node J

178          DO 5 J=1,JL
179          IA=0
180          DO 6 I=1,IL
181          IF(.NOT.(J.EQ.JLP(I,1).OR.J.EQ.JLP(I,2)))
             GO TO 6
182          IA=IA+1
183          IP(J,IA)=I
184          NIP(J)=IA
```

(Continued)

TABLE A3.10 *Continued*

```
185   6      CONTINUE
186   5      CONTINUE

187   C      Write and print pipes connected to a node J

188          Write (2,920)
189          WRITE(2,905)
190          PRINT 920
191          PRINT 905
192          DO 7 J=1,JL
193          WRITE (2,205)J,NIP(J),(IP(J,L),L=1,NIP(J))
194          PRINT 205,J,NIP(J),(IP(J,L),L=1,NIP(J))
195   7      CONTINUE
196          WRITE (2,250)
197          PRINT 250

198   C      Identify all the nodes connected a node J
             through connected pipes IP(J,L)

199          DO 8 J=1,JL
200          DO 9 L=1,NIP(J)
201          IPE=IP(J,L)
202          DO 10 LA=1,2
203          IF(JLP(IPE,LA).NE.J) JN(J,L)=JLP(IPE,LA)
204   10     CONTINUE
205   9      CONTINUE
206   8      CONTINUE

207   C      Write and print all the nodes connected
             to a node J

208          WRITE (2,921)
209          PRINT 921
210          WRITE(2,906)
211          PRINT 906
212          DO 60 J =1, JL
213          WRITE (2,206) J,NIP(J),(JN(J,L),L=1,NIP(J))
214          PRINT 206,J,NIP(J),(JN(J,L),L=1,NIP(J))
215   60     CONTINUE
216          WRITE (2,250)
217          PRINT 250

218   C      Identify loop pipes and loop nodes
```

(Continued)

TABLE A3.10   *Continued*

```
219        DO 28 K=1,KL
220        DO 29 I=1,IL
221        IF(.NOT.((K.EQ.IKL(I,1)).OR.
           (K.EQ.IKL(I,2)))) GO TO 29
222        JK(K,1)=JLP(I,1)
223        JB=JLP(I,1)
224        IK(K,1)=I
225        JK(K,2)=JLP(I,2)
226        GO TO 54
227     29 CONTINUE
228     54 NA=1
229        JJ=JK(K,NA+1)
230        II=IK(K,NA)
231     56 DO 30 L=1,NIP(JJ)
232        II=IK(K,NA)
233        IKL1=IKL(IP(JJ,L),1)
234        IKL2=IKL(IP(JJ,L),2)
235        IF(.NOT.((IKL1.EQ.K).OR.(IKL2.EQ.K)))
           GO TO 30
236        IF(IP(JJ,L).EQ.II) GO TO 30
237        NA=NA+1
238        NLP(K)=NA
239        IK(K,NA)=IP(JJ,L)
240        IF(JLP(IP(JJ,L),1).NE.JJ) JK
           (K,NA+1)=JLP(IP(JJ,L),1)
241        IF(JLP(IP(JJ,L),2).NE.JJ)JK
           (K,NA+1)=JLP(IP(JJ,L),2)
242        II=IK(K,NA)
243        JJ=JK(K,NA+1)
245        GO TO 57
246     30 CONTINUE
247     57 IF(JJ.NE.JB) GO TO 56
248     28 CONTINUE

249 C   Write and print loop forming pipes

250        WRITE(2,922)
251        WRITE (2,909)
252        PRINT 922
253        PRINT 909
254        DO 51 K=1,KL
255        WRITE (2, 213)K,NLP(K),(IK(K,NC),NC=1,NLP(K))
256        PRINT 213,K,NLP(K),(IK(K,NC),NC=1,NLP(K))
257     51 CONTINUE
258        WRITE (2,250)
```

(Continued)

TABLE A3.10   *Continued*

```
259        PRINT 250

260   C    Write and print loop forming nodes

261        WRITE(2,923)
262        WRITE (2,910)
263        PRINT 923
264        PRINT    910
265        DO 70 K=1, KL
266        WRITE (2,  213)K,NLP(K),(JK(K,NC),NC=1,NLP(K))
267        PRINT 213,K,NLP(K),(JK(K,NC),NC=1,NLP(K))
268     70 CONTINUE
269        WRITE (2,250)
270        PRINT 250

271   C    Assign sign convention to pipes
               to apply continuity equations

272        DO 20 J=1,JL
273        DO 20 L=1,NIP(J)
274        IF(JN(J,L).LT.J) S(J,L)=1.0
275        IF(JN(J,L).GT.J) S(J,L)=-1.0
276     20 CONTINUE

277   C Estimate nodal water demands -Transfer
          pipe population loads to nodes

278        DO 73 J=1,JL
279        Q(J)=0.0
280        DO 74 L=1,NIP(J)
281        II=IP(J,L)
282        JJ=JN(J,L)
283        DO 75 M=1,ML
284        IF(J.EQ.INP(M)) GO TO 73
285        IF(JJ.EQ.INP(M)) GO TO 550
286     75 CONTINUE
287        Q(J)=Q(J)+PP(II)*RTW*QPF/(CRTW*2.0)
288        GO TO 74
289    550 Q(J)=Q(J)+PP(II)*RTW*QPF/CRTW
290     74 CONTINUE
291     73 CONTINUE

292   C    Calculate input source point
               discharge (inflow)
```

(*Continued*)

TABLE A3.10 *Continued*

```
293          SUM=0.0
295          DO 50 J=1,JL
296          DO 61 M=1,ML
297          IF(J.EQ.INP(M)) GO TO 50
298    61    CONTINUE
299          SUM=SUM+Q(J)
300    50    CONTINUE
301          QT=SUM
302          DO 67 M=1,ML
303          AML=ML
304          Q(INP(M))=-QT/AML
305    67    CONTINUE

306    C     Initial input point discharge correction AQ

307          AQ=QT/(3.0*AML)

308    C     Print and write nodal discharges

309          WRITE(2,907)
310          PRINT 907
311          WRITE (2,233)(J, Q(J),J=1,JL)
312          PRINT 233,(J,Q(J),J=1,JL)
313          WRITE (2,250)
314          PRINT 250

315    C     Allocate sign convention to loop pipes
             to apply loop discharge
316    C     corrections using Hardy-Cross method

317          DO 32 K=1,KL
318          DO 33 L=1,NLP(K)
319          IF(JK(K,L+1).GT.JK(K,L)) SN(K,L)=1.0
320          IF(JK(K,L+1).LT.JK(K,L)) SN(K,L)=-1.0
321    33    CONTINUE
322    32    CONTINUE

323    C     Initialize nodal terminal pressures by
             assigning zero head

324    69    DO 44 J=1,JL
325          H(J)=0.0
326    44    CONTINUE
```

(Continued)

TABLE A3.10 *Continued*

327	C	Initialize pipe flow discharges by assigning zero flow rates
328		DO 45 I=1,IL
329		QQ(I)=0.0
330	45	CONTINUE
331	C	Assign arbitrary flow rate of 0.01 m3/s to one of the loop pipes in
332	C	all the loops to apply continuity equation. Change to 0.1 m3/sto see impact.
333		DO 17 KA=1,KL
334		KC=0
335		DO 18 I=1,IL
336		IF(.NOT.(IKL(I,1).EQ.KA).OR. (IKL(I,2)).EQ.KA)
337		* GO TO 18
338		F(QQ(I).NE.0.0) GO TO 18
339		IF(KC.EQ.1) GO TO 17
340		QQ(I)=0.01
341		KC=1
342	18	CONTINUE
343	17	CONTINUE
344	C	Apply continuity equation first at nodes having single pipe connected and
345	C	then at nodes having only one of its pipes with unknown (zero) discharge
346	C	till all the branch pipes have known (non-zero) discharges
347		DO 11 J=1,JL
348		IF(NIP(J).EQ.1) QQ(IP(J,1))=S(J,1)*Q(J)
349	11	CONTINUE
350		NE=1
351		DO 12 J=1,JL
352		IF(J.EQ.INP(1)) GO TO 12
353		NC=0
354		DO 13 L=1,NIP(J)
355		IF(.NOT.((IKL(IP(J,L),1).EQ.0).AND. (IKL(IP(J,L),2).EQ.0)))

(Continued)

TABLE A3.10 *Continued*

```
356          * GO TO 13
357          NC=NC+1
358    13    CONTINUE
359          IF(NC.NE.NIP(J)) GO TO 12
360          DO 16 L=1,NIP(J)
361          IF(QQ(IP(J,L)).EQ.0.0) NE=0
362    16    CONTINUE
363          ND=0
364          DO 14 L=1,NIP(J)
365          IF(QQ(IP(J,L)).NE.0.0) GO TO 14
366          ND=ND+1
367          LD=L
368    14    CONTINUE
369          IF(ND.NE.1) GO TO 12
370          QQ(IP(J,LD))=S(J,LD)*Q(J)
371          DO 15 L= 1,NIP(J)
372          IF(IP(J,LD).EQ.IP(J,L)) GO TO 15
373          QQ(IP(J,LD))=QQ(IP(J,LD))-S(J,L)
             *QQ(IP(J,L))
374    15    CONTINUE
375    12    CONTINUE
376          IF(NE.EQ.0) GO TO 11

377    C     Identify nodes that have one pipe with
             unknown (zero) discharge

378    55    DO 21 J=1,JL
379          IF(J.EQ.INP(1)) GO TO 21
380          KD(J)=0
381          DO 22 L=1,NIP(J)
382          IF(QQ(IP(J,L)).NE.0.0) GO TO 22
383          KD(J)=KD(J)+1
384          LA=L
385    22    CONTINUE
386          IF(KD(J).NE.1) GO TO 21
387          SUM=0.0
388          DO 24 L=1,NIP(J)
389          SUM=SUM+S(J,L)*QQ(IP(J,L))
390    24    CONTINUE
391          QQ(IP(J,LA))=S(J,LA)*(Q(J)-SUM)
392    21    CONTINUE
393          DO 25 J=1,JL
394          IF(KD(J).NE.0) GO TO 55
395    25    CONTINUE
```

TABLE A3.10 *Continued*

396	C	Write and print pipe discharges based on only continuity equation

397	C	WRITE (2, 908)
398	C	PRINT 908
399	C	WRITE (2,210)(II,QQ(II),II=1,IL)
400	C	PRINT 210,(II,QQ(II),II=1,IL)
401	C	WRITE (2,250)
402	C	PRINT 250

403	C	Calculate friction factor using Eq.2.6c

```
404   58   DO 34 I=1,IL
405        FAB=4.618*(D(I)/(ABS(QQ(I))*10.0**6))**0.9
406        FAC=0.00026/(3.7*D(I))
407        FAD=ALOG(FAB+FAC)
408        FAE=FAD**2
409        F(I)=1.325/FAE
410        EP=8.0/PI**2
411        AK(I)=(EP/(G*D(I)**4))*(F(I)*AL(I)/
           D(I)+FK(I))
412   34   CONTINUE
```

413	C	Loop discharge correction using Hardy-Cross method

```
414        DO 35 K=1,KL
415        SNU=0.0
416        SDE=0.0
417        DO 36 L=1,NLP(K)
418        IA=IK(K,L)
419        BB=AK(IA)*ABS(QQ(IA))
420        AA=SN(K,L)*AK(IA)*QQ(IA)*ABS(QQ(IA))
421        SNU=SNU+AA
422        SDE=SDE+BB
423   36   CONTINUE
424        DQ(K)=-0.5*SNU/SDE
425        DO 37 L=1,NLP(K)
426        IA=IK(K,L)
427        QQ(IA)=QQ(IA)+SN(K,L)*DQ(K)
428   37   CONTINUE
429   35   CONTINUE
430        DO 40 K=1,KL
```

(Continued)

TABLE A3.10 *Continued*

```
431        IF(ABS(DQ(K)).GT.0.0001) GO TO 58
432   40   CONTINUE

433   C    Calculations for terminal pressure heads,
           starting from input source node
434   C    with maximum piezometric head

435            HAM=0.0
436            DO 68 M=1,ML
437            HZ=HA(M)+Z(INP(M))
438            IF(HZ.LT.HAM) GO TO 68
439            MM=M
440            HAM=HZ
441   68       CONTINUE

442        H(INP(MM))=HA(MM)
443   59   DO 39 J=1,JL
444        IF(H(J).EQ.0.0) GO TO 39
445        DO 41 L=1,NIP(J)
446        JJ=JN(J,L)
447        II=IP(J,L)
448        IF(JJ.GT.J) SI=1.0
449        IF(JJ.LT.J) SI=-1.0
450        IF(H(JJ).NE.0.0) GO TO 41
451        AC=SI*AK(II)*QQ(II)*ABS(QQ(II))
452        H(JJ)=H(J)-AC+Z(J)-Z(JJ)
453   41   CONTINUE
454   39   CONTINUE
455        DO 42 J=1,JL
456        IF(H(J).EQ.0.0) GO TO 59
457   42   CONTINUE

458   C    Write and print final pipe discharges

459   C    WRITE (2,912)
460   C    PRINT912
461   C    WRITE (2,210)(I,QQ(I),I=1,IL)
462   C    PRINT 210,(I,QQ(I),I=1,IL)
463   C    WRITE (2,250)
464   C    PRINT 250

465   C    Write and print nodal terminal
           pressure heads

466   C    WRITE (2,915)
```

(*Continued*)

TABLE A3.10 *Continued*

```
467  C     PRINT 915
468  C     WRITE (2,229)(J,H(J),J=1,JL)
469  C     PRINT 229,(J,H(J),J=1,JL)
470  C     WRITE (2,250)
471  C     PRINT 250

472  C     Write and print input point discharges and
           estimated input point heads

473        WRITE (2,924)
474        PRINT 924
475        WRITE (2,230)(M,Q(INP(M)),M=1,ML)
476        PRINT 230,(M,Q(INP(M)),M=1,ML)
477        WRITE (2,234)(M,H(INP(M)),M=1,ML)
478        PRINT 234,(M,H(INP(M)),M=1,ML)
479        WRITE (2,250)
480        PRINT 250

481        IF (ML.EQ.1) GO TO 501

482  C     Check error between input point calculated
           heads and input heads

483        AEFF=0.0
484        DO 64 M=1,ML
485        AFF=100.0*ABS((HA(M)-H(INP(M)))/HA(M))
486        IF(AEFF.LT.AFF) AEFF=AFF
487  64    CONTINUE

488  C     Input discharge correction based on input
           point head

489        DO 71 M=1,ML
490        IF(HA(M).GT.H(INP(M)))
           Q(INP(M))=Q(INP(M))-AQ
491        IF(HA(M).LT.H(INP(M)))
           Q(INP(M))=Q(INP(M))+AQ
492  71    CONTINUE

493  C     Estimate input discharge for input source
           node with maximum piezometric head

494        SUM=0.0
495        DO 72 M=1,ML
496        IF(M.EQ.MM) Go TO 72
497        SUM=SUM+Q(INP(M))
```

(*Continued*)

TABLE A3.10 *Continued*

```
498   72    CONTINUE
499         Q(INP(MM))=-(QT+SUM)

500         NFF=NFF+1
501         IF (NFF.GE.5) AQ=0.75*AQ
502         IF (NFF.GE.5) NFF=1
503         IF(AEFF.LE.0.5) GO TO 501
504         IF( AEFF.GT.FF) GO To 69

505         AQ=0.75*AQ
506         FF=FF/2.0
507         IF( FF.GT.0.5) Go To 69
508   501   CONTINUE

509   C     Write and print final pipe discharges

510         WRITE (2,912)
511         PRINT   912
512         WRITE (2,210)(I,QQ(I),I=1,IL)
513         PRINT 210,(I,QQ(I),I=1,IL)
514         WRITE (2,250)
515         PRINT 250

516   C     Write and print nodal terminal
            pressure heads

517         WRITE (2,915)
518         PRINT 915
519         WRITE (2,229)(J,H(J),J=1,JL)
520         PRINT 229,(J,H(J),J=1,JL)
521         WRITE (2,250)
522         PRINT 250

523   201   FORMAT(5I5)
524   202   FORMAT(5I6,2F9.1,F8.0,2F9.3)
525   203    FORMAT(I5,2X,F8.2)
526   204    FORMAT(I5,I10,F10.2)
527   205    FORMAT(I5,1X,I5,3X,10I5)
528   206    FORMAT(I5,1X,I5,3X,10I5)
529   210    FORMAT(4(2X,'QQ('I3')='F6.4))
530   213    FORMAT(1X,2I4,10I7)
531   229    FORMAT(4(2X,'H('I3')='F6.2))
532   230    FORMAT(3(3X,'Q(INP('I2'))='F9.4))
533   233    FORMAT(4(2X,'Q('I3')='F6.4))
534   234    FORMAT(3(3X,'H(INP('I2'))='F9.2))
```

(Continued)

TABLE A3.10 *Continued*

```
535    250    FORMAT(/)
536    901    FORMAT(3X,'IL',3X,'JL',3X,'KL',3X,'ML')
537    902    FORMAT(4X,'i'3X,'J1(i)'2X'J2(i)'1X,
              'K1(i)'1X,'K2(i)'3X'L(i)'
538      *    6X'kv(i)'2X'P(i)',5X'D(i)')
539    903    FORMAT(4X,'j',5X,'Z(j)')
540    904    FORMAT(4X,'m',6X,'INP(M)',3X,'HA(M)')
541    905    FORMAT(3X,'j',3X,'NIP(j)'5X'(IP(J,L),
              L=1,NIP(j)-Pipes to node)')
542    906    FORMAT(3X,'j',3X,'NIP(j)'5X'(JN(J,L),
              L=1,NIP(j)-Nodes to node)')
543    907    FORMAT(3X,'Nodal discharges - Input source
              node -tive discharge')
544    908    FORMAT(3x,'Pipe discharges based on
              continuity equation only')
545    909    FORMAT( 4X,'k',1X,'NLP(k)'2X'(IK(K,L),
              L=1,NLP(k)-Loop pipes)')
546    910    FORMAT( 4X,'k',1X,'NLP(k)'2X'(JK(K,L),
              L=1,NLP(k)-Loop nodes)')
547    911    FORMAT(2X,'Pipe friction factors using
              Swamee and Jain eq.')
548    912    FORMAT (2X, 'Final pipe discharges
              (m3/s)')
549    913    FORMAT (2X, 'Input source node and its
              discharge (m3/s)')
550    914    FORMAT (3X,'Input   source node=['I3']',
              2X,'Input discharge='F8.4)
551    915    FORMAT (2X,'Nodal terminal pressure
              heads (m)')
552    916    FORMAT (2X, 'Total network size info')
553    917    FORMAT (2X, 'Pipe links data')
554    918    FORMAT (2X, 'Nodal elevation data')
555    919    FORMAT (2X, 'Input source nodal data')
556    920    FORMAT (2X, 'Information on pipes
              connected to a node j')
557    921    FORMAT (2X, 'Information on nodes
              connected to a node j')
558    922    FORMAT (2X, 'Loop forming pipes')
559    923    FORMAT (2X, 'Loop forming nodes')
560    924    FORMAT (2X, 'Input source point
              discharges & calculated heads')
561           CLOSE(UNIT=1)
562           STOP
563           END
```

TABLE A3.11. Output File APPENDIXMIS.OUT

Total network size info

IL	JL	KL	ML
55	33	23	3

Pipe links data

i	J1(i)	J2(i)	K1(i)	K2(i)	L(i)	kf(i)	P(i)	D(i)
1	1	2	2	0	380.0	.0	500.	.150
2	2	3	4	0	310.0	.0	385.	.125
3	3	4	5	0	430.0	.2	540.	.125
4	4	5	6	0	270.0	.0	240.	.080
5	1	6	1	0	150.0	.0	190.	.050
6	6	7	0	0	200.0	.0	500.	.065
7	6	9	1	0	150.0	.0	190.	.065
8	1	10	1	2	150.0	.0	190.	.200
9	2	11	2	3	390.0	.0	490.	.150
10	2	12	3	4	320.0	.0	400.	.050
11	3	13	4	5	320.0	.0	400.	.065
12	4	14	5	6	330.0	.0	415.	.080
13	5	14	6	7	420.0	.0	525.	.080
14	5	15	7	0	320.0	.0	400.	.050
15	9	10	1	0	160.0	.0	200.	.080
16	10	11	2	0	120.0	.0	150.	.200
17	11	12	3	8	280.0	.0	350.	.200
18	12	13	4	9	330.0	.0	415.	.200
19	13	14	5	11	450.0	.2	560.	.080
20	14	15	7	14	360.0	.2	450.	.065
21	11	16	8	0	230.0	.0	280.	.125
22	12	19	8	9	350.0	.0	440.	.100
23	13	20	9	10	360.0	.0	450.	.100
24	13	22	10	11	260.0	.0	325.	.250
25	14	22	11	13	320.0	.0	400.	.250
26	21	22	10	12	160.0	.0	200.	.250
27	22	23	12	13	290.0	.0	365.	.250
28	14	23	13	14	320.0	.0	400.	.065
29	15	23	14	15	500.0	.0	625.	.100
30	15	24	15	0	330.0	.0	410.	.050
31	16	17	0	0	230.0	.0	290.	.050
32	16	18	8	0	220.0	.0	275.	.125
33	18	19	8	16	350.0	.0	440.	.065
34	19	20	9	17	330.0	.0	410.	.050
35	20	21	10	19	220.0	.0	475.	.100
36	21	23	12	19	250.0	.0	310.	.100
37	23	24	15	20	370.0	.0	460.	.100

(Continued)

TABLE A3.11 *Continued*

38	18	25	16	0	470.0	.0	590.	.065
39	19	25	16	17	320.0	.0	400.	.080
40	20	25	17	18	460.0	.0	575.	.065
41	20	26	18	19	310.0	.0	390.	.065
42	23	27	19	20	330.0	.0	410.	.200
43	24	27	20	21	510.0	.0	640.	.050
44	24	28	21	0	470.0	.0	590.	.100
45	25	26	18	0	300.0	.0	375.	.065
46	26	27	19	0	490.0	.0	610.	.080
47	27	29	22	0	230.0	.0	290.	.200
48	27	28	21	22	290.0	.0	350.	.200
49	28	29	22	23	190.0	.0	240.	.150
50	29	30	23	0	200.0	.0	250.	.050
51	28	31	23	0	160.0	.0	200.	.100
52	30	31	23	0	140.0	.0	175.	.050
53	31	32	0	0	250.0	.0	310.	.065
54	32	33	0	0	200.0	.0	250.	.050
55	7	8	0	0	200.0	.0	250.	.065

Nodal elevation data

j	Z(j)
1	101.85
2	101.90
3	101.95
4	101.60
5	101.75
6	101.80
7	101.80
8	101.40
9	101.85
10	101.90
11	102.00
12	101.80
13	101.80
14	101.90
15	100.50
16	100.80
17	100.70
18	101.40
19	101.60
20	101.80
21	101.85
22	101.95

(*Continued*)

TABLE A3.11 *Continued*

23	101.80
24	101.10
25	101.40
26	101.20
27	101.70
28	101.90
29	101.70
30	101.80
31	101.80
32	101.80
33	100.40

Input source nodal data

m	INP(M)	HA(M)
1	11	20.00
2	22	20.00
3	28	20.00

Information on pipes connected to a node j

j	NIP(j)	(IP(J,L),L = 1,NIP(j)-Pipes to node)					
1	3	1	5	8			
2	4	1	2	9	10		
3	3	2	3	11			
4	3	3	4	12			
5	3	4	13	14			
6	3	5	6	7			
7	2	6	55				
8	1	55					
9	2	7	15				
10	3	8	15	16			
11	4	9	16	17	21		
12	4	10	17	18	22		
13	5	11	18	19	23	24	
14	6	12	13	19	20	25	28
15	4	14	20	29	30		
16	3	21	31	32			
17	1	31					
18	3	32	33	38			
19	4	22	33	34	39		
20	5	23	34	35	40	41	
21	3	26	35	36			
22	4	24	25	26	27		
23	6	27	28	29	36	37	42
24	4	30	37	43	44		

(Continued)

TABLE A3.11 *Continued*

25	4	38	39	40	45
26	3	41	45	46	
27	5	42	43	46	47	48
28	4	44	48	49	51
29	3	47	49	50	
30	2	50	52		
31	3	51	52	53	
32	2	53	54		
33	1	54			

Information on nodes connected to a node j

j   NIP(j)   (JN(J,L),L = 1,NIP(j)-Nodes to node)

j	NIP(j)						
1	3	2	6	10			
2	4	1	3	11	12		
3	3	2	4	13			
4	3	3	5	14			
5	3	4	14	15			
6	3	1	7	9			
7	2	6	8				
8	1	7					
9	2	6	10				
10	3	1	9	11			
11	4	2	10	12	16		
12	4	2	11	13	19		
13	5	3	12	14	20	22	
14	6	4	5	13	15	22	23
15	4	5	14	23	24		
16	3	11	17	18			
17	1	16					
18	3	16	19	25			
19	4	12	18	20	25		
20	5	13	19	21	25	26	
21	3	22	20	23			
22	4	13	14	21	23		
23	6	22	14	15	21	24	27
24	4	15	23	27	28		
25	4	18	19	20	26		
26	3	20	25	27			
27	5	23	24	26	29	28	
28	4	24	27	29	31		
29	3	27	28	30			
30	2	29	31				
31	3	28	30	32			
32	2	31	33				
33	1	32					

(Continued)

TABLE A3.11 *Continued*

```
Loop forming pipes
 k NLP(k)   (IK(K,L),L = 1,NLP(k)-Loop pipes)
 1    4      5       7      15       8
 2    4      1       9      16       8
 3    3      9      17      10
 4    4      2      11      18      10
 5    4      3      12      19      11
 6    3      4      13      12
 7    3     13      20      14
 8    5     17      22      33      32      21
 9    4     18      23      34      22
10    4     23      35      26      24
11    3     19      25      24
12    3     26      27      36
13    3     25      27      28
14    3     20      29      28
15    3     29      37      30
16    3     33      39      38
17    3     34      40      39
18    3     40      45      41
19    5     35      36      42      46      41
20    3     37      43      42
21    3     43      48      44
22    3     47      49      48
23    4     49      50      52      51
```

```
Loop forming nodes
 k NLP(k)   (JK(K,L),L = 1,NLP(k)-Loop nodes)
 1    4      1       6       9      10
 2    4      1       2      11      10
 3    3      2      11      12
 4    4      2       3      13      12
 5    4      3       4      14      13
 6    3      4       5      14
 7    3      5      14      15
 8    5     11      12      19      18      16
 9    4     12      13      20      19
10    4     13      20      21      22
11    3     13      14      22
12    3     21      22      23
13    3     14      22      23
14    3     14      15      23
```

(*Continued*)

TABLE A3.11 *Continued*

15	3	15	23	24		
16	3	18	19	25		
17	3	19	20	25		
18	3	20	25	26		
19	5	20	21	23	27	26
20	3	23	24	27		
21	3	24	27	28		
22	3	27	29	28		
23	4	28	29	30	31	

```
Nodal discharges - Input souce node -tive discharge
  Q(  1)= .0019 Q(  2)= .0049 Q(  3)= .0029 Q(  4)= .0026
  Q(  5)= .0025 Q(  6)= .0019 Q(  7)= .0016 Q(  8)= .0005
  Q(  9)= .0008 Q( 10)= .0015 Q( 11)=-.0303 Q( 12)= .0042
  Q( 13)= .0054 Q( 14)= .0068 Q( 15)= .0041 Q( 16)= .0024
  Q( 17)= .0006 Q( 18)= .0028 Q( 19)= .0037 Q( 20)= .0050
  Q( 21)= .0026 Q( 22)=-.0303 Q( 23)= .0064 Q( 24)= .0058
  Q( 25)= .0042 Q( 26)= .0030 Q( 27)= .0058 Q( 28)=-.0303
  Q( 29)= .0022 Q( 30)= .0009 Q( 31)= .0019 Q( 32)= .0012
  Q( 33)= .0005 Q(

Input source point discharges & calculated heads
  Q(INP( 1))=  -.0303  Q(INP( 2))=  -.0303  Q(INP( 3))=  -.0303
  H(INP( 1))=   20.00  H(INP( 2))=   20.09  H(INP( 3))=   20.74

Input source point discharges & calculated heads
  Q(INP( 1))=  -.0505  Q(INP( 2))=  -.0202  Q(INP( 3))=  -.0202
  H(INP( 1))=   20.00  H(INP( 2))=   18.79  H(INP( 3))=   18.90

Input source point discharges & calculated heads
  Q(INP( 1))=  -.0353  Q(INP( 2))=  -.0278  Q(INP( 3))=  -.0278
  H(INP( 1))=   20.00  H(INP( 2))=   19.89  H(INP( 3))=   20.38

Input source point discharges & calculated heads
  Q(INP( 1))=  -.0353  Q(INP( 2))=  -.0335  Q(INP( 3))=  -.0221
  H(INP( 1))=   20.00  H(INP( 2))=   19.89  H(INP( 3))=   20.06

Input source point discharges & calculated heads
  Q(INP( 1))=  -.0353  Q(INP( 2))=  -.0377  Q(INP( 3))=  -.0178
  H(INP( 1))=   20.00  H(INP( 2))=   19.89  H(INP( 3))=   19.88

Input source point discharges & calculated heads
  Q(INP( 1))=  -.0306  Q(INP( 2))=  -.0401  Q(INP( 3))=  -.0202
  H(INP( 1))=   20.00  H(INP( 2))=   20.08  H(INP( 3))=   20.16
```

(Continued)

TABLE A3.11 *Continued*

```
Input source point discharges & calculated heads
 Q(INP( 1))=  -.0341  Q(INP( 2))=  -.0383  Q(INP( 3))=  -.0184
 H(INP( 1))=   20.00  H(INP( 2))=   19.94  H(INP( 3))=   19.94

Final pipe discharges (m3/s)
QQ(  1)= .0038 QQ(  2)= .0044 QQ(  3)= .0022 QQ(  4)= .0011
QQ(  5)= .0016 QQ(  6)= .0022 QQ(  7)=-.0025 QQ(  8)=-.0073
QQ(  9)=-.0053 QQ( 10)=-.0002 QQ( 11)=-.0008 QQ( 12)=-.0015
QQ( 13)=-.0016 QQ( 14)= .0002 QQ( 15)=-.0033 QQ( 16)=-.0121
QQ( 17)= .0093 QQ( 18)= .0006 QQ( 19)= .0002 QQ( 20)= .0010
QQ( 21)= .0074 QQ( 22)= .0042 QQ( 23)= .0033 QQ( 24)=-.0091
QQ( 25)=-.0106 QQ( 26)=-.0078 QQ( 27)= .0108 QQ( 28)= .0000
QQ( 29)=-.0027 QQ( 30)=-.0002 QQ( 31)= .0006 QQ( 32)= .0044
QQ( 33)= .0006 QQ( 34)=-.0004 QQ( 35)=-.0044 QQ( 36)= .0008
QQ( 37)= .0030 QQ( 38)= .0009 QQ( 39)= .0016 QQ( 40)= .0011
QQ( 41)= .0012 QQ( 42)=-.0005 QQ( 43)=-.0004 QQ( 44)=-.0027
QQ( 45)=-.0006 QQ( 46)=-.0023 QQ( 47)=-.0015 QQ( 48)=-.0075
QQ( 49)= .0043 QQ( 50)= .0006 QQ( 51)= .0040 QQ( 52)=-.0003
QQ( 53)= .0018 QQ( 54)= .0005 QQ( 55)= .0005

Nodal terminal pressure heads (m)
 H(  1)= 20.00 H(  2)= 19.78 H(  3)= 19.44 H(  4)= 19.65
 H(  5)= 19.24 H(  6)= 16.72 H(  7)= 14.62 H(  8)= 14.87
 H(  9)= 18.72 H( 10)= 19.99 H( 11)= 20.00 H( 12)= 20.05
 H( 13)= 20.05 H( 14)= 19.92 H( 15)= 20.41 H( 16)= 20.30
 H( 17)= 19.53 H( 18)= 19.39 H( 19)= 18.80 H( 20)= 19.09
 H( 21)= 20.03 H( 22)= 19.94 H( 23)= 20.02 H( 24)= 19.93
 H( 25)= 18.42 H( 26)= 18.60 H( 27)= 20.12 H( 28)= 19.94
 H( 29)= 20.12 H( 30)= 19.36 H( 31)= 19.44 H( 32)= 17.69
 H( 33)= 18.52
```

Line 161:170
Same explanation as provided for single-input source analysis program Lines 159:168.
Line 171
FF is an initially assumed percent error in calculated terminal head and input head at input points. The maximum piezometric head input source point is considered as a reference point in calculating terminal heads at other source nodes.
Line 172
NFF is a counter to check the number of iterations before the discharge correction AQ is modified.

Line 173:276
Same explanation as provided for single-input source program Lines 169:272.
Line 277:291
Same explanation as provided for single-input source network Lines 277:285; however, a Do loop has been introduced in Lines 283 and 286 to cover all the input points in a multi-input source network.
Line 292:305
Same explanation as provided for single-input source program for Lines 286:293; however, a DO loop has been introduced to divide the total water demand equally at all the input source points initially. This DO loop is in Lines 302 and 305.
Line 306
Comment line for input point discharge correction.
Line 307
AQ is a discharge correction applied at input points except at input point with maximum piezometric head. See explanation for Lines 489:492 below. User may change the denominator multiplier 3 to change AQ.
Line 308:432
Same explanation as provided for single-source network program Lines 294:419.
Line 433:434
Comment line indicating that the next code lines are for terminal pressure head computations starting with source node having maximum piezometric head.
Line 435:441
These lines identify the source node MM with maximum piezometric head.
Line 442:457
Same explanation as provided for single-source network program Lines 428:443.
In Line 442, the known terminal head is the input head of source point having maximum piezometric head. On the other hand in the single-input source network, the terminal pressure computations started from input point node.
Line 458:471
Same explanation as provided for single-source network program Lines 444:457.
The lines are blocked here to reduce output file size. User can unblock the code by removing comment C to check intermediate pipe flows and terminal pressure.
Line 472
Comment line for next code lines about write and print input point discharges and estimated input point heads.
Line 473:480
WRITE and PRINT input point discharges at each iteration.
WRITE and PRINT input point pressure heads at each iteration.
The process repeats until error FF is greater than 0.5 (Line 507).
Line 481
The multi-input, looped network program also works for single-input source network. In case of single-input source network, Lines 482:507 are inoperative.
Line 482
Comment line for next code of lines that checks error between computed heads and input heads at input points.

Line 483:487

Calculate error AFF between HA(M) and H(INP(M)). The maximum error AEFF is also identified here.

Line 488

Comment for next code lines are about discharge correction at input points.

Line 489:492

DO loop is introduced to apply discharge correction at input points. If input head at a source point is greater than the calculated terminal head, the discharge at this input point is reduced by an amount AQ. On the other hand if input head at a source point is less than the calculated terminal head, the discharge at this input point is increased by an amount AQ.

Line 493

Comment line that the next code lines are for estimating input discharge for input point with maximum piezometric head.

Line 494:499

Q(INP(MM)) is estimated here, which is the discharge of the input point with maximum piezometric head.

Line 500

NFF is a counter to count the number of iterations for input point discharge correction.

Line 501

If counter NFF is greater than or equal to 5, the discharge correction is reduced to 75%.

Line 502

If counter NFF is greater than or equal to 5, redefine NFF = 1.

Line 503

If AEFF (maximum error, see Line 486) is less than or equal to 0.5, go to Line 508, which will stop the program after final pipe discharges and nodal heads write and print commands.

Line 504

If AEFF (maximum error) is greater than assumed error (FF), start the computations again from Line 324, otherwise go to next line.

Line 505

Redefine the discharge correction.

Line 506

Redefine the initially assumed error.

Line 507

If FF is greater than 0.5, start the computations from Line 324 again or otherwise continue to next line.

Line 509:522

Same explanation as provided for single-input source Lines 444:457.

Line 523:560

Various FORMAT commands used in the code development are listed in this section. See the input and output file for the information on these formats.

Line 562:563

STOP and END commands of the program.

The software and the output files are as listed in Table A3.6 and Table A3.7.

INDEX